Condensed MATTER THEORIES

VOLUME 1

A Continuation Order Plan is available for this series. A continuation order will bring delivery of each new volume immediately upon publication. Volumes are billed only upon actual shipment. For further information please contact the publisher.

Condensed MATTER THEORIES

VOLUME 1

Edited by

F. B. Malik
Southern Illinois University at Carbondale
Carbondale, Illinois

SPRINGER SCIENCE+BUSINESS MEDIA, LLC

Library of Congress Cataloging in Publication Data

International Workshop on Condensed Matter Physics (9th: 1985: San Francisco, Calif.)
Condensed matter theories.

"Proceedings of the Ninth International Workshop on Condensed Matter Physics, held August 5-10, 1985, in San Francisco, California, sponsored by the U.S. Army Research Office at Durham by a grant, and by the San Francisco State University and the Southern Illinois University at Carbondale"—T.p. verso.
Includes bibliographies and index.
1. Condensed matter—Congresses. 2. Quantum liquids—Congresses. I. Malik, F. B. II. San Francisco State University. III. Southern Illinois University at Carbondale. IV. Title.
QC173.4.C65I47 1985 530.4'1 86-9400
ISBN 978-1-4615-6709-7

Conference Organizing Committee

J. W. Clark (U.S.A.)
M. de Llano (U.S.A.)
F. B. Malik (U.S.A.), Chairperson
C. W. Woo (U.S.A.)

Proceedings of the Ninth International Workshop on Condensed Matter Physics, held August 5-10, 1985, in San Francisco, California, sponsored by the U.S. Army Research Office at Durham by a grant, and by the San Francisco State University and the Southern Illinois University at Carbondale

Originally published by Plenum Press, New York in 1986

ISBN 978-1-4615-6709-7 ISBN 978-1-4615-6707-3 (eBook)
DOI 10.1007/978-1-4615-6707-3

To my parents

PREFACE

This the first volume of the proceedings of the International Workshop on Condensed Matter Theories published by a commercial publisher and of a series which is planned to appear anually. It is a tribute to the group of scientists who started this workshop as the Pan American Workshop on Condensed Matter Theories in 1977 and helped to develop it to a significant annual international workshop. Many scientists' efforts have contributed to this important development and it is impossible to name all of them. But at least three persons are to be singled out: Professors Manuel de Llano and Angel Plastino who conceived the idea of the annual workshop and carried it forward, and Professor John W. Clark who has been the prime driving force behind it in recent years.

The Workshop started in 1977 in São Paolo, Brazil, as the first Pan American Workshop on Condensed Matter Theories with the idea of bringing together scientists from the Western Hemisphere working in many different areas of Condensed Matter Theories for the purpose of cross-fertilization of ideas used in different areas and fostering collaborations among them. The next five Workshops were held at Trieste, Italy, in 1978; in Buenos Aires, Argentina, in 1979; in Caracas, Venezuela, in 1980; in Mexico City, Mexico, in 1981; and in St. Louis, Missouri, U.S.A., in 1982. At that point it became clear that the Workshop had reached an international dimension - a large number of scientists from countries other than those in the Western Hemisphere began participating actively and hence, in 1983, it evolved into an international workshop and met in Altenburg in the Federal Republic of Germany. In 1984 it took place in Granada, Spain, and the ninth one was in San Francisco in 1985.

This edited book basically includes the invited talks presented at the last Workshop. It contains, however, one contribution which the authors could not personally present at the conference and omits one presentation at the speaker's request. The paper not presented at the conference but included here was sent to referees and subsequently approved by the organizing committee for inclusion.

The articles are not arranged in the same sequence as the talks. Articles on a similar topic have been grouped together. However, the grouping is done only in a very broad sense. I would like to thank Dr. John W. Clark for his assistance in this matter.

This approximate classification reveals, however, one of the main spirits of this workshop. Given a proper forum, scientists working in such diverse arease as band structure calculations an neutron stars can still come together, understand each other's research and borrow ideas from one area to another. In these days of specialization, this is an uncommon thing but important for the cross-fertilization of different fields and the overall understanding of the physical world surrounding us. The success of this

Workshop demonstrates that a common language exists among scientists doing research in many highly specialized diversed areas. Another goal of the Workshop was to foster collaborative investigations and many such collaborations have since occurred.

At this stage , I would like to express my special thanks to those who made the ninth Workshop and these proceedings possible. Certainly without the active collaboration of the other members of the organizing and international advisory committees, this workshop would not have taken place. I thank all of them for their assistance. The financial support of the U.S. Army Research Office at Durham, North Carolina, U.S.A., and the Office of Research and Development Administration of Southern Illinois University at Carbondale were of paramount importance for the success of the Workshop and are thankfully acknowledged. The administrative support of the host institution, San Francisco State University, was also an important factor. The active support of the office of the President of San Francisco State University, particularly President Chia-Wei Woo and Mrs. Elogeane Grossman and of the office of the President of Southern Illinois University, particularly President Albert Somit are much appreciated. The editorial staff of Plenum Publishing Corporation have been most helpful. The secretarial assistance of the physics department of Southern Illinois University and of the Institute for Theoretical Physics of the Univesity of Tübingen are much appreciated. I am also grateful to Professor A. Faessler for the invitation to spend the fall of 1985 at his institute where the final manuscripts have been prepared.

Carbondale, Illinois, U.S.A. F.B. Malik

CONTENTS

(Asterisk next to name identifies the speaker)

I. FORMAL METHODS

II. MONTE CARLO METHODS

III. QUANTUM FLUIDS AT T=0

IV. QUANTUM FLUIDS AT T>O

V. ELECTRONIC SYSTEMS AND SOLIDS

VI. NEW FORMS OF CONDENSED MATTER

VII. NUCLEAR FORCES AND MATTER

A COMPARISON BETWEEN CORRELATED BASIS FUNCTIONS METHOD AND THE DENSITY FUNCTIONAL THEORY[a]

Ruibao Tao[*,b] and Chia-Wei Woo[†]

[*]Department of Physics
Fudan University
Shanghai, China

[†]Department of Physics
San Francisco State University
San Francisco, California 94132, U.S.A.

INTRODUCTION

The method of correlated basis functions (CBF) has been applied to treating a wide variety of homogeneous many body systems,[1-4] including liquid and solid helium, nuclear matter, and Coulomb systems. That the method could be successful for Bose and Fermi systems alike has to do with its apparent ability to sum both ring and ladder diagrams, as demonstrated in an analysis by Sim and Woo[5] in 1970. In that work, the pair correlation function and ground state energy for a weakly interacting Bose gas were first calculated exactly in the perturbation theory using the formalism of Hugenholtz and Pines. The same quantities were then obtained using the CBF approach. Results from the two methods were compared order by order in powers of the density and the interaction strength. This helped determine what perturbative diagrams were summed by the CBF. Indeed, from this analysis a systematic scheme was identified that enabled us[6,7] to use the Hugenholtz-Pines theory to suggest optimum three-particle and higher-order factors for correlated wave functions. Thus, diagrammatic perturbation theory and CBF became intermingled.

We now have another opportunity to bring together two distinct many body formalisms. This time for inhomogeneous systems. On the one hand, we have the conventional density functional theory (DFT), invented and popularized by Kohn and co-workers. On the other, we have again the CBF: a late contender in the field of metal surfaces[8] but an early leader for inhomogeneous helium systems.[9] An interesting development worth some notice is that the inventor of the density functional approach is now employing the CBF for treating metal surfaces.[10]

Once again we return to a weakly interacting Bose gas. The reason is that it is one of the few systems which offer us exact solutions in the perturbation theory, against which approximate theories can be compared. We recognize that the conclusions drawn from considering the weakly interacting Bose gas cannot be generalized to liquid helium or metal surfaces. However, it constitutes a first step, and provides us with useful indications and directions.

The system under consideration, being weakly interacting, cannot be inhomogeneous in the ground state. The inhomogeneity must be generated and maintained by an external field. We introduce such a weak external field, with a strength parameter ε, to cause a weak inhomogeneity in the system of bosons whose interactions are characterized by a Fourier-transformable potential with a strength λ. As in Ref. 5, we now carry out calculations in powers of ε and λ. The external field is, of course, an added complication. The density is now a function, and cannot serve as an expansion parameter. The calculation has to be done with three methods: perturbation, DFT, and CBF, rather than two. In the DFT, we use first the local density functional (LDF), which is supposed to work well in the region of mild inhomogeneities; we then take into account the second-order nonlocal density correction.

On account of space limitations, we present here only a general sketch of our work and the results. The details, which contain (unfortunately for all such formal analyses) pages of equations, will be published elsewhere.

PERTURBATION THEORY

Our system consists of N bosons interacting via a pairwise potential $\lambda v(|\vec{r}_i - \vec{r}_j|)$, and placed under an external field $\varepsilon U_{ext}(\vec{r}_i)$ which has no effect on the normalizing volume Ω. The Hamiltonian of the system is:

$$H = \sum_{i=1}^{N} \frac{-1}{2}\nabla_i^2 + \left[\frac{1}{2}\sum_{\substack{i\neq j\\=1}}^{N} \lambda v(|\vec{r}_i - \vec{r}_j|) + \sum_{i=1}^{N} \varepsilon U_{ext}(\vec{r}_i)\right] \equiv H_o + V_I, \quad (1)$$

where $\hbar^2/m$ has been set to unity.

The ground state solution of the corresponding Schrödinger equation is given by

$$|\psi\rangle = |\phi_o\rangle + \sum_{m=1}^{\infty}\left(\left[\frac{\hat{P}}{-\hat{H}_o}\hat{V}_I\right]^m |\phi_o\rangle\right)_n , \quad (2)$$

where $\hat{H}_o$ and $\hat{V}_I$ are second-quantized, n in the subscript denotes graphs which do not contain self-energy, $|\phi_o\rangle$ denotes the ground state of $\hat{H}_o$ (with energy eigenvalue $E_o = 0$), and $\hat{P}$ stands for a projection operator. The ground state energy is then

$$E = \langle \phi_o|\hat{H}|\psi\rangle \equiv E_o + \Delta E = \Delta E$$

$$\equiv \sum_{\alpha=1}^{\infty} \varepsilon^{\alpha} \Delta E_{\alpha} \equiv \sum_{\alpha=1}^{\infty} \sum_{\beta=0}^{\infty} \varepsilon^{\alpha} \lambda^{\beta} \Delta E_{\alpha}{}^{(\beta)} = \sum_{m=0}^{\infty} (-1)^m \langle\phi_o|\hat{V}_I\left(\frac{\hat{P}}{-\hat{H}_o}\hat{V}_I\right)^m|\phi_o\rangle_L , \quad (3)$$

the subscript L provides that only connected clusters are included in the sum.

All terms of order ε^1 with U_{ext} appearing just once vanish:

$$\Delta E_1{}^{(\beta)} = 0, \quad \beta = 0,1,2. \quad (4)$$

In order ε^2, we obtain after much algebra,

$$\Delta E_2{}^{(0)} = -2n_o \int \frac{U_{ext}(\vec{k})U_{ext}(-\vec{k})}{k^2} d\vec{k} , \quad (5)$$

$$\Delta E_2{}^{(1)} = 8n_o{}^2 \int \frac{v(\vec{k})}{k^4} U_{ext}(\vec{k})U_{ext}(-\vec{k})\, d\vec{k} , \quad (6)$$

$$\Delta E_2{}^{(2)} = -32n_o{}^3 \int \frac{v^2(\vec{k})}{k^6} U_{ext}(\vec{k})U_{ext}(-\vec{k})\, d\vec{k}$$

$$-4n_o{}^2 \int \frac{v(\vec{k}_1+\vec{k}_2)v(\vec{k}_2)}{k_1{}^4 k_2{}^2} U_{ext}(\vec{k}_1)U_{ext}(-\vec{k}_1)\, d\vec{k}_1 d\vec{k}_2$$

$$-4n_o{}^2 \int \frac{v(\vec{k}_2)[v(\vec{k}_2) + v(\vec{k}_1+\vec{k}_2)]}{k_1{}^2 k_2{}^2 [k_2{}^2 + (\vec{k}_1+\vec{k}_2)^2]} U_{ext}(\vec{k}_1)U_{ext}(-\vec{k}_1)\, d\vec{k}_1 d\vec{k}_2$$

$$-2n_o{}^2 \int \frac{v^2(\vec{k}_1)}{k_1{}^4 (k_1{}^2 + k_2{}^2)} U_{ext}(\vec{k}_1-\vec{k}_2)U_{ext}(\vec{k}_2-\vec{k}_1)\, d\vec{k}_1 d\vec{k}_2$$

$$-2n_o{}^2 \int \frac{v(\vec{k}_1)v(\vec{k}_2)}{k_1{}^2 k_2{}^2 (k_1{}^2 + k_2{}^2)} U_{ext}(\vec{k}_1-\vec{k}_2)U_{ext}(\vec{k}_2-\vec{k}_1)\, d\vec{k}_1 d\vec{k}_2 \quad (7)$$

where $n_o = N/\Omega$. For the purpose of this paper we have not gone beyond order $\varepsilon^2\lambda^2$.

DENSITY FUNCTIONAL THEORY

Ebner and Saam[11] generalized the local density functional method (LDF) to treating Bose liquids. The ground state energy functional

$$E[n] \approx \int e_H\left(n(\vec{r})\right) d\vec{r} + \varepsilon \int n(\vec{r}) U_{ext}(\vec{r}) dr + \frac{1}{2}\int \left(\nabla\sqrt{n(\vec{r})}\right)^2 d\vec{r} \qquad (8)$$

contains an energy density e_H of a homogeneous system at the local density $n(\vec{r})$. Apart from a constant,

$$e_H(n) = -\frac{1}{2} n^2 A \equiv -\frac{1}{2} n^2 \lambda^2 \int \frac{v^2(\vec{k})}{k^2} d\vec{k}. \qquad (9)$$

Let us now expand $n(\vec{r})$ in powers of ε:

$$n(\vec{r}) = n_o + \sum_{\alpha=1}^{\infty} \varepsilon^{\alpha} n_{\alpha}(\vec{r}). \qquad (10)$$

Recall that the nature of our external field is such that

$$\int n(\vec{r}) d\vec{r} = \int n_o \, d\vec{r} .$$

Thus

$$\int n_{\alpha}(\vec{r}) d\vec{r} = 0, \ \alpha \neq 0 . \qquad (11)$$

Casting E[n] in momentum representation, and varying E with respect to $n_1(\vec{k})$, we find order by order in the expansion

$$n_{\alpha}(\vec{k}) = \sum_{\beta=0}^{\infty} \lambda^{\beta} n_{\alpha}^{(\beta)}(\vec{k}) , \qquad (12)$$

$$n_1^{(0)}(\vec{k}) = \frac{-4n_o U_{ext}(\vec{k})}{k^2} , \qquad (13)$$

$$n_1^{(1)}(\vec{k}) = 0 , \qquad (14)$$

$$n_1^{(2)}(\vec{k}) = \frac{4n_o n_1^{(0)}(\vec{k})}{k^2} \int \frac{v^2(\vec{p})}{p^2} d\vec{p} . \qquad (15)$$

And $\Delta E_1^{(\beta)} = 0$,

$$\Delta E_2^{(0)} = -2n_o \int \frac{U_{ext}(\vec{k}) U_{ext}(-\vec{k})}{k^2} d\vec{k}, \qquad (16)$$

$$\Delta E_2^{(1)} = 0 , \qquad (17)$$

$$\Delta E_2^{(2)} = -8n_o^2 \int \frac{v^2(\vec{k}_1) U_{ext}(\vec{k}_2) U_{ext}(-\vec{k}_2)}{k_1^2 k_2^4} d\vec{k}_1 d\vec{k}_2 . \qquad (18)$$

Compare to Eqs. (5)-(7). In order $\varepsilon^2\lambda^o$, the result here is identical to that of the perturbation theory. In higher orders there is no similarity at all.

Going beyond local density approximation, one adds a nonlocal correction term to the energy functional $\delta E[n]$:

$$\delta E[n] \approx -\frac{1}{4}\int d\vec{r}d\vec{r}'[n(\vec{r}) - n(\vec{r}')]^2 \int d\vec{k}e^{i\vec{k}\cdot(\vec{r}-\vec{r}')} [\chi^{-1}_{\vec{k}}(n_o)-\chi^{(o)^{-1}}_{\vec{k}}(n_o)].$$

Or, $$\delta E[n_\alpha(\vec{k})] \approx \varepsilon^2 \frac{1}{2}\int [\chi^{-1}_{\vec{k}}(n_o) - \chi^{(o)^{-1}}_{\vec{k}}(n_o)]n_1(\vec{k})n_1(-\vec{k})d\vec{k}, \tag{19}$$

where

$$\begin{aligned}\chi^{-1}_{\vec{k}}(n_o) - \chi^{(o)^{-1}}_{\vec{k}}(n_o) &= \frac{k^2}{4n_o}[S_H^{-2}(\vec{k})-1]\\ &= \lambda\frac{-k^2}{2}g_H^{(1)}(\vec{k}) + \lambda^2\frac{k^2}{4}\left\{-2g_H^{(2)}(\vec{k})+3n_o[g_H^{(1)}(\vec{k})]^2\right\}\\ &\quad + O(\lambda^3).\end{aligned} \tag{20}$$

$S_H(\vec{k})$ denotes the liquid structure factor of the homogeneous Bose system, and $g_H^{(\beta)}(\vec{k})$ represents the βth term in the expansion of $\frac{1}{n_o}[S_H(\vec{k})-1]$ in powers of λ. $\left\{g_H^{(\beta)}(\vec{k})\right\}$ are taken then from the exact expressions given in Ref. 5.

Variation of $E + \delta E$ with respect to $n_1(\vec{k})$ and separating into various powers of λ as before give us a new set of $\left\{n_1^{(\beta)}(\vec{k})\right\}$ and new energy expressions. While $\Delta E_2^{(o)}$ is unchanged, we find

$$\Delta E_2^{(1)} = 8n_o^2\int\frac{v(\vec{k})}{k^4}U_{ext}(\vec{k})\,U_{ext}(-\vec{k})\,d\vec{k}, \tag{21}$$

$$\begin{aligned}\Delta E_2^{(2)} = &-8n_o^2\int\frac{v^2(\vec{k}_1)}{k_1^2}d\vec{k}_1\int\frac{U_{ext}(\vec{k}_2)U_{ext}(-\vec{k}_2)}{k_2^4}d\vec{k}_2\\ &+ 4n_o^2\int\frac{v^2(\vec{k}_1)}{k_1^4}d\vec{k}_1\int\frac{U_{ext}(\vec{k}_2)U_{ext}(-\vec{k}_2)}{k_2^2}d\vec{k}_2\\ &- 8n_o^2\int\frac{v(\vec{k}_2)v(\vec{k}_1+\vec{k}_2)U_{ext}(\vec{k}_1)U_{ext}(-\vec{k}_1)}{k_1^4\,k_2^2}\,d\vec{k}_1d\vec{k}_2\\ &- 8n_o^2\int\frac{v(\vec{k}_2)v(\vec{k}_1+\vec{k}_2)U_{ext}(\vec{k}_1)U_{ext}(-\vec{k}_1)}{k_1^2\,(\vec{k}_1+\vec{k}_2)^2\,k_2^2}\,d\vec{k}_1d\vec{k}_2\\ &- 32n_o^3\int\frac{v^2(\vec{k})}{k^6}U_{ext}(\vec{k})U_{ext}(-\vec{k})\,d\vec{k}.\end{aligned} \tag{22}$$

$\Delta E_2^{(1)}$ becomes now identical to the result of perturbation theory, Equation (6). $\Delta E_2^{(2)}$ contains one correct term, some terms which have the correct structure but wrong coefficients, and other unlinked integrals.

CORRELATED BASIS FUNCTION METHOD

As in Ref. 8, a properly symmetrized correlated wave function is used to describe the inhomogeneous system. In this case, one takes

$$\Psi(1,2,\ldots,N) = \prod_{i=1}^{N} \phi(\vec{r}_i) \prod_{\substack{j>k \\ =1}}^{N} \exp \frac{1}{2} u(\vec{r}_j,\vec{r}_k). \tag{23}$$

As usual in all CBF calculations, one defines the ℓ-particle distribution function as:

$$\begin{aligned}\rho(1,2,\ldots,\ell) &= n(\vec{r}_1)n(\vec{r}_2)\ldots n(\vec{r}_\ell)\; g(\vec{r}_1,\vec{r}_2,\ldots,\vec{r}_\ell) \\ &= \frac{N!}{(N-\ell)!} \; \frac{\int \Psi^2 d\vec{r}_{\ell+1}\ldots d\vec{r}_N}{\int \Psi^2 d\vec{r}_1 d\vec{r}_2\ldots d\vec{r}_N} \; .\end{aligned} \tag{24}$$

The ground state energy can then be expressed in terms of $n(\vec{r})$, $u(\vec{r}_1,\vec{r}_2)$, and the distribution functions $\left\{g(\vec{r}_1,\vec{r}_2,\ldots,\vec{r}_\ell)\right\}$, as follows:

$$\begin{aligned} E = &\frac{1}{2} \int [\nabla\sqrt{n(\vec{r})}]^2 \, d\vec{r} + \varepsilon \int U_{ext}(\vec{r}) n(\vec{r}) d\vec{r} \\ &+ \frac{\lambda}{2} \int n(\vec{r}_1)n(\vec{r}_2) \; v(|\vec{r}_1-\vec{r}_2|) \; g(\vec{r}_1,\vec{r}_2) \; d\vec{r}_1 d\vec{r}_2 \\ &+ \frac{1}{8} \int n(\vec{r}_1)n(\vec{r}_2) \; [\nabla_1 u(\vec{r}_1,\vec{r}_2)] \; g(\vec{r}_1,\vec{r}_2) \; d\vec{r}_1 d\vec{r}_2 \\ &- \frac{1}{8} \int n(\vec{r}_1) \left\{ \int n(\vec{r}_2) [\nabla_1 u(\vec{r}_1,\vec{r}_2)] \; g(\vec{r}_1,\vec{r}_2) \; d\vec{r}_2 \right\}^2 d\vec{r}_1 \\ &+ \frac{1}{8} \int n(\vec{r}_1)n(\vec{r}_2)n(\vec{r}_3) \; [\nabla_1 u(\vec{r}_1,\vec{r}_2) \cdot \nabla_1 u(\vec{r}_1,\vec{r}_3)] \; g(\vec{r}_1,\vec{r}_2,\vec{r}_3) d\vec{r}_1 d\vec{r}_2 d\vec{r}_3 . \end{aligned} \tag{25}$$

The task at hand is to express $g(\vec{r}_1,\vec{r}_2)$ and $g(\vec{r}_1,\vec{r}_2,\vec{r}_3)$ in terms of $n(\vec{r})$ and $u(\vec{r}_1,\vec{r}_2)$, and then minimize E with respect to the latter.

First we introduce a closure approximation for $g(\vec{r}_1,\vec{r}_2,\vec{r}_3)$: Kirkwood superposition approximation

$$g(\vec{r}_1,\vec{r}_2,\vec{r}_3) \approx g(\vec{r}_1,\vec{r}_2) \; g(\vec{r}_2,\vec{r}_3) \; g(\vec{r}_3,\vec{r}_1) \; ; \tag{26}$$

the approximation (26) is exact to order λ^2, which is all we seek at this time.

Next we expand all expressions in powers of ε. For $n(\vec{r})$ we have Eq. (10). For $u(\vec{r}_1,\vec{r}_2)$ and $g(\vec{r}_1,\vec{r}_2)$ we have

$$u(\vec{r}_1,\vec{r}_2) = u_H(r_{12}) + \sum_{\alpha=1}^{\infty} \varepsilon^\alpha u_\alpha(\vec{r}_1,\vec{r}_2) \ . \tag{27}$$

$$g(\vec{r}_1,\vec{r}_2) = g_H(r_{12}) + \sum_{\alpha=1}^{\infty} \varepsilon^\alpha g_\alpha(\vec{r}_1,\vec{r}_2) = 1 + O(\lambda) \ . \tag{28}$$

Substituting these expressions in E, we obtain, to order ε^2, the energy as a functional of $u_H(r_{12})$, $g_H(r_{12})$, $n_1(\vec{r})$, $u_1(\vec{r}_1,\vec{r}_2)$, $u_2(\vec{r}_1,\vec{r}_2)$, $g_1(\vec{r}_1,\vec{r}_2)$, and $g_2(\vec{r}_1,\vec{r}_2)$. The optimum forms of $u_H(r_{12})$ and $g_H(r_{12})$ are known from Ref. 5. $g_1(\vec{r}_1,\vec{r}_2)$ and $g_2(\vec{r}_1,\vec{r}_2)$ are to be related to $n_1(\vec{r})$, $u_1(\vec{r}_1,\vec{r}_2)$, and $u_2(\vec{r}_1,\vec{r}_2)$ through the BBGKY equation:

$$\begin{aligned}\nabla_1 g(\vec{r}_1,\vec{r}_2,\ldots,\vec{r}_\ell) &= g(\vec{r}_1,\vec{r}_2,\ldots,\vec{r}_\ell) \sum_{i=2}^{\ell} \nabla_1 u(\vec{r}_1,\vec{r}_\ell) \\ &+ \int n(\vec{r}_{\ell+1}) [g(\vec{r}_1,\vec{r}_2,\ldots,\vec{r}_{\ell+1}) \\ &- g(\vec{r}_1,\vec{r}_{\ell+1}) g(\vec{r}_1,\vec{r}_2,\ldots,\vec{r}_\ell)]\ \nabla_1 u(\vec{r}_1,\vec{r}_{\ell+1}) d\vec{r}_{\ell+1} \ .\end{aligned} \tag{29}$$

After expanding all in powers of λ and in momentum representation, we find, up to order $\varepsilon^2\lambda^2$, that the energy can be arranged neatly to permit the variation with respect to $n_\alpha^{(\beta)}(\vec{k})$ and $u_\alpha^{(\beta)}(\vec{k}_1,\vec{k}_2)$ to proceed in a convenient stepwise fashion. The final results show

$$\Delta E_1^{(\beta)} = 0, \ \beta = 0,1,2. \tag{30}$$

$$\Delta E_2^{(o)} = -2n_o \int \frac{U_{ext}(\vec{k}) U_{ext}(-\vec{k})}{k^2} d\vec{k} \ , \tag{31}$$

$$\Delta E_2^{(1)} = 8n_o \int \frac{v(\vec{k})}{k^4} U_{ext}(\vec{k}) U_{ext}(-\vec{k}) d\vec{k} \ , \tag{32}$$

$$\begin{aligned}\Delta E_2^{(2)} &= -\ 32n_o^3 \int \frac{v^2(\vec{k})}{k^6} U_{ext}(\vec{k}) U_{ext}(-\vec{k}) d\vec{k} \\ &- 16n_o^2 \int \frac{v(\vec{k}_1+\vec{k}_2) v(\vec{k}_2)}{k_1^4 k_2^2} U_{ext}(\vec{k}_1) U_{ext}(-\vec{k}_1) d\vec{k}_1 d\vec{k}_2 \\ &+ 8n_o^2 \int \frac{(\vec{k}_1+\vec{k}_2)\cdot\vec{k}_2}{k_1^4 k_2^2 (\vec{k}_1+\vec{k}_2)^2} v(\vec{k}_1+\vec{k}_2) v(\vec{k}_2) U_{ext}(\vec{k}_1) U_{ext}(-\vec{k}_1) d\vec{k}_1 d\vec{k}_2 .\end{aligned} \tag{33}$$

Equations (30)-(32) are identical to the results of perturbation theory, equations (4)-(6). $\Delta E_2^{(2)}$ picks up parts, but not all, of Eq. (7). It contains no unphysical unlinked structure, but the last term is one that cannot be expressed in a perturbative diagram.

Our major conclusion is that, even in its simplest form the CBF method is already giving resutls better than the DFT. In particular, it is superior to DFT in the local density approximation, even though the latter is designed for systems with slow and small density variations like the one under consideration.

This work will be extended to higher orders.

REFERENCES AND FOOTNOTES

a. Work supported in part by a grant from Exxon Research and Engineering Company.
b. Visiting Scholar, San Francisco State University.

1. E. Feenberg, "Theory of Quantum Fluids," Academic, New York (1969).
2. C.-W. Woo, Chapter 5, in: "Physics of Liquid and Solid Helium, I," K.H. Bennemann and J.B. Ketterson, eds., Wiley, New York (1976).
3. C.E. Campbell, in: "Progress in Liquid Physics," C.A. Croxton, ed., Wiley, New York (1978).
4. H.L. Kummel and M.L. Ristig, ed. "Recent Progress in Many-Body Theories," and J.G. Zabolitzky, ed., Springer, Berlin (1981-84)
5. H.K. Sim and C.-W. Woo, Phys. Rev. A2:2032 (1970); Phys. Rev. A2:2024 (1970).
6. C.-W. Woo, Phys. Rev. Lett. 28:1442 (1972); Phys. Rev. A6:2312 (1972).
7. S. Chakravarty and C.-W. Woo, Phys. Rev. B13:105 (1976).
8. X. Sun, T. Li, and C.-W. Woo, in: "Recent Progress in Many Body Theories," J.G. Zabolitzky, ed., Springer, Berlin (1981); Acta Phys. Sin. 31:1466,1474 (1982); X. Sun, T. Li, M. Farjam, and C.-W. Woo, Phys. Rev. B27:3913 (1983); X. Sun, M. Farjam, and C.-W. Woo, Phys. Rev. B28:5599 (1973)
9. Y.M. Shih and C.-W. Woo, Phys. Rev. Lett. 30:478 (1973); C.C. Chang and M. Cohen, Phys. Rev. A8:1930,3131 (1973).
10. E. Krotscheck, G.-X. Qian, and W. Kohn, Phys. Rev. B31:4245 (1985); also new preprint.
11. C. Ebner and W.F. Saam, Phys. Rev. B12:923 (1975).

PLANAR THEORY MADE PLAINER

Roger Alan Smith

Physics Department
Texas A&M University
College Station, TX 77843

INTRODUCTION

In recent years, A. D. Jackson, A. Lande, and I have been approaching the many-body problem from the point of view of summing an important set of Feynman diagrams, the planar or parquet diagrams. It had long been recognized that the hypernetted-chain (HNC) variational theory approach and the summation of rings and ladders in perturbation theory were both dealing with diagrams of very similar topological structure. Early work by Sim, Buchler, and Woo[1] showed that the optimized hypernetted-chain (HNC) variational theory for bosons summed all ring and ladder diagrams exactly and in addition generated terms whose structure was that of other legitimate diagrams. Neither the numerical factors nor the diagrams generated this way were identified. Ref. [2] showed that the diagrams generated by the optimized HNC theory were a subset of the parquet diagrams and in general were not generated with the right numerical factors. Roughly speaking, the boson parquet theory sums ring diagrams in which there can be ladders between bubbles and ladder diagrams in which the rungs of the ladders can be chains of bubbles. These two types of diagrams are embedded in each other in a self-consistent way. A first effort was made in this work to generate an approximate sum of parquet diagrams. For the energy of liquid ^{4}He, the results were comparable to those obtained from optimized Jastrow theory; the corresponding liquid-structure function was reasonable except for the behavior at small k. A number of corrections to the approximation were included in ref. [3]. These corrections brought the energy down to close to the ''experimental'' Monte-Carlo results, but failed to make any significant changes to the liquid structure function. A more careful analysis of the diagrammatic structure led to a different approximation scheme in [4]. With this new approximation scheme, which more carefully analyzed the contribution of ladder diagrams to the liquid structure function, we were able to derive all results of the optimized boson HNC theory from the summation of Feynman diagrams without ever thinking of a variational function. A simple, elegant, and important factor in this derivation was the recognition[5] that for a large class of theories including HNC, the computation of the energy by coupling constant integration or by direct evaluation will give the same result.

The references cited above all pertain to boson systems. For the parquet method to be used for nuclear matter or liquid ^{3}He, it needs to be expanded in two ways which are essentially orthogonal: it must include exchange effects and spin-dependence of the bare and effective interactions. Even in ^{3}He, where the underlying interaction is spin-independent (ignoring nuclear spins), the Pauli principle will generate effective interactions which are very different in the density and spin channels. Reference [6] took the first step in indicating how to carry out the exact summation of crossing-symmetric diagrams starting with some set of irreducible diagrams. These give a very clear prescription of how the summation can be carried out formally, but the results are not in a form which even suggests approximation schemes comparable to those developed for boson systems. In the next section of this paper, the previous results will be discussed and used to derive a new set of equations which makes the structure of the diagrams much more clear and suggests how one could make useful approximations.

The inclusion of spin-dependence is straightforward in the parquet scheme. Since the theory builds up a two-body vertex, one only has to know how to do the spin algebra in the s, t, and u channels. A manuscript describing this process in detail and giving applications to bosons is being prepared[7]; highlights of this work will be discussed in the third section of this contribution.

The final section of this contribution will present an overall perspective of the parquet theory: what it has accomplished, what is presently being worked on, and what needs to be done to apply the parquet methods to a number of very interesting physical systems.

PARQUET DIAGRAMS FOR FERMION SYSTEMS

In this section, perturbation theory is done using Feynman propagators. All diagrams and operations are written somewhat schematically, and the precise meaning may be determined from the definitions in standard references[8]. I want to emphasize the structure of the diagrams summed. In general, the propagators may be the full single-particle Green's function (dressed with the full self-energy).

The basic object in the parquet theory is the two-body vertex shown in figs. 1a-c. It is shown as a box to which two incoming and two outgoing legs can be attached. Three more boxes could be drawn with the arrows going in the opposite directions. The first-order diagrams which are possible are shown in figs. 1d-i. As was pointed out in [6], exchanges of the external legs will change diagrams from one category to another. In these figures, the light lines indicate propagators and the heavy lines the interaction.

The parquet diagrams are constructed by taking any two boxes and joining them together by connecting two legs from one box to the other box. Half of the possible ways of doing this are illustrated in fig. 1j-l. The remaining ways involve connecting the incoming and outgoing legs in the only other way possible. Cutting the two added links will separate the diagram from top to bottom in the s channel, from left to right in the t channel, and one diagonal from the other in the u channel. It is relatively easy to see that any diagram is either reducible in one of these channels or it is irreducible.

There are many symmetries which may be observed by rearranging the legs of the propagators. For example, the s-exchange of a diagram of type T gives a diagram of type U. The phases associated with these interchanges are given in [6].

The parquet approach is to use this construction process to construct a completely crossing-symmetric two-body vertex. This may then be used to construct the dynamic liquid structure function $S(k,\omega)$ as in fig. 1m, the proper self-energy $\Sigma^*(k,\omega)$ as in fig. 1n, or the energy E as in fig. 1o.

Ref. [6] derived an equation summing (self-consistently) all diagrams reducible in the s channel:

$$S = (I + T)s(I + S + T + U), \tag{1}$$

where I represents the input set of irreducible diagrams (including exchange terms). The corresponding equations for the T and U diagrams are straightforwardly obtained by permuting S, T, and U. These equations are still quite complicated, because the diagrams appearing on the right-hand side could have any of the orientations of the diagrams of fig. 1a-c, but the s operation can only join outgoing to ingoing lines. In the remainder of this section, we will transform these results into an appealing and useful form.

The important step is to divide diagrams of type 1a into those of type 1 in which the ingoing line on the left goes out on the left (eventually) and those of type 2 in which it goes out on the right. The symmetries then make it possible to construct any diagram as a rearrangement of diagrams of type S_1, T_1, and T_2 or, equivalently S_1, T_1, and U_1 given an input set of irreducible diagrams I_1. We adopt the latter set of diagrams as preserving maxi-

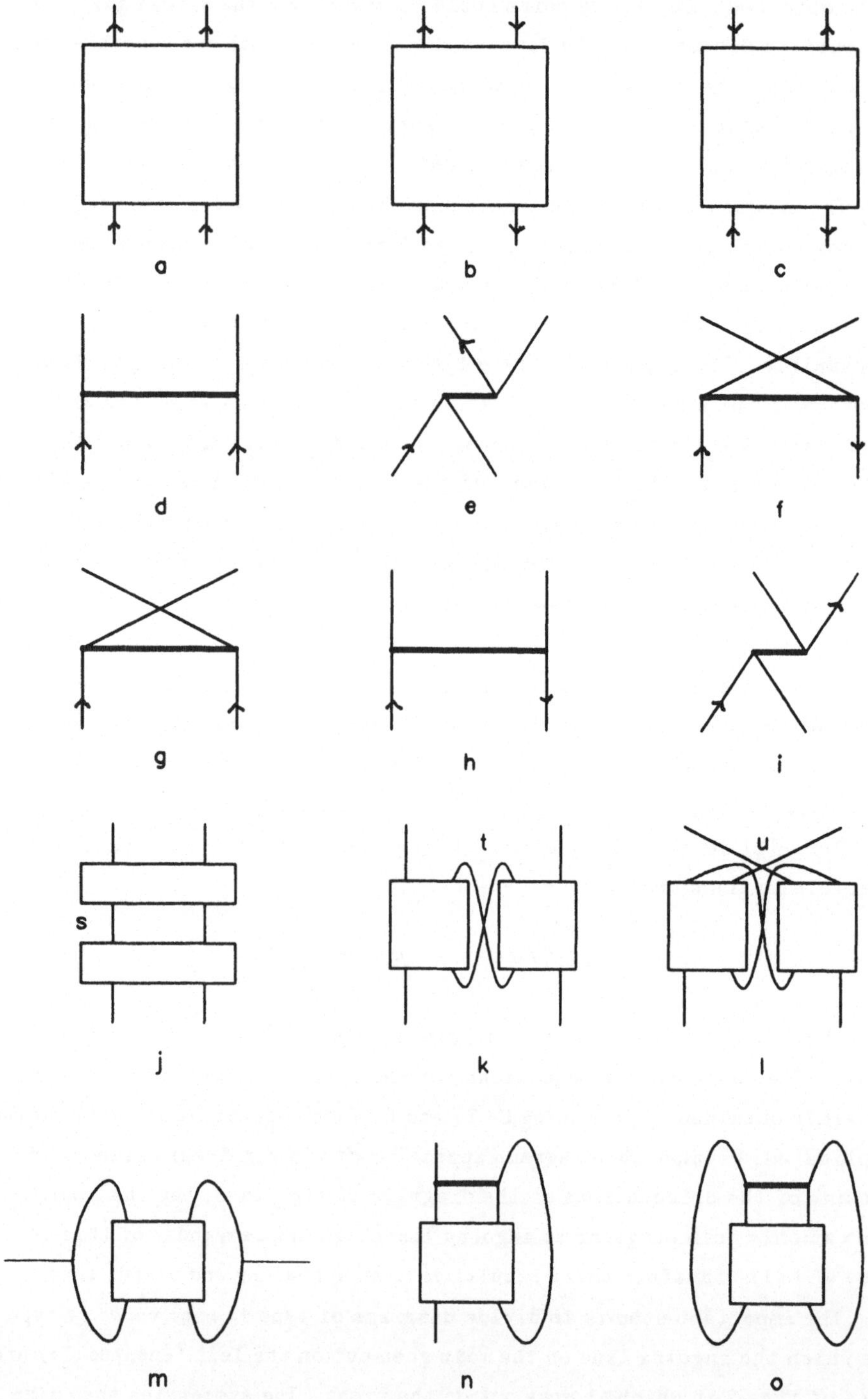

Fig. 1. Diagrams illustrating the text.

mum symmetry and then drop the 1 subscript in what follows. By looking carefully at eq. (1) for the s channel, we obtain the equation

$$S = (I + T + U)s(I + S + T + U). \tag{2}$$

All terms in this equation are of type 1. Typical diagrams have just the form of particle-particle ladders as shown in fig. 2a. To clarify the structure, only the bare interaction appears in these links; the general ladder would have rungs of type T and U as well.

The equation corresponding to eq. (1) for the u channel similarly gives

$$U = (I + S + T)u(I + S + T + U). \tag{3}$$

These diagrams are particle-hole ladders such as that in fig. 2b.

The equation resulting from the t-channel analog of eq. (1) is more complicated. One can identify three separate types of diagrams which can be constructed this way. We write

$$T = C + L + R, \tag{4}$$

where C denotes chain diagrams in which the operation creates a new fermion loop, L and R denote vertex corrections applied to the left or right of a diagram. The corresponding equations which follow from a careful analysis of the structure are

$$\begin{aligned} C &= \pm(I + S + U)t(I + S + T + U) \\ L &= (I + S + T)l(I + S + T + U) \\ R &= (I + S + U)r(I + S + T + U). \end{aligned} \tag{5}$$

Typical diagrams are illustrated in figs. 2c-e. Only the C diagrams have the $\pm$ sign due to statistics; these are the only diagrams in which a closed loop of propagators is formed.

If I is the set of all irreducible diagrams, then the set of all diagrams is given by $I + S + T + U$. Furthermore, the diagrammatic structure of the diagrams in fig. 2 suggests ways of making clean approximations to the exact parquet sum. I anticipate that such approximations can give results which are superior to the usual Fermi HNC methods.

As an illustration of the diagrams which can be summed, I have generated the solution of eqs. (2-5) graphically through third order in fig. 3. In this figure, the heavy line indicates an interaction and the lighter line indicates the propagator.

SPIN AND TENSOR FORCES

The formal inclusion of spin and tensor forces for problems involving bosons has been carried out in ref. [7]. I wish to mention the main results here, as they are directly applicable to the fermion calculations as well.

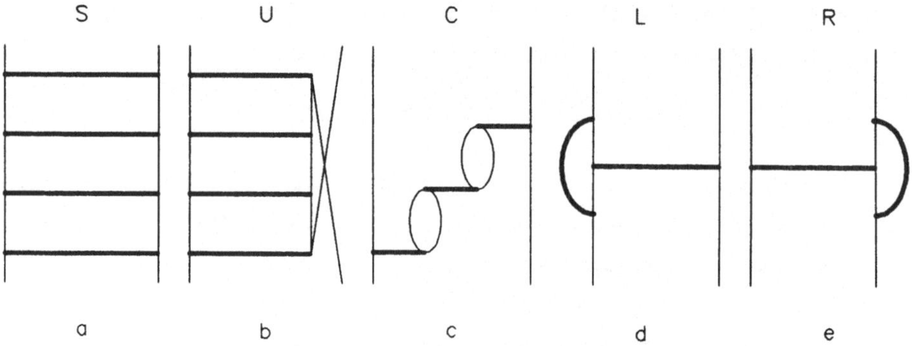

Fig. 2. Representative Diagrams

The approximate parquet summation has been carried out for a v_6 potential, that is one which has spin, isospin, and tensor components. This results in six ladder channels and six ring channels. The recoupling to go from one set of six channels to the other is straightforward and causes no problems formally or numerically. In the ring channels, there are possibilities for instabilities to appear at zero momentum or finite momentum in any of the six channels. We have previously seen that these instabilities are associated with physical instabilites of the system for central forces; the same can be said in the presence of these spin-dependent forces. For example, one would expect a pion-condensate instability to show up in the tensor-isospin channel at finite momentum. The spin-dependent parquet thus treats the ground state but is aware of nearby instabilities.

There was no problem in constructing the energy functional with for a v_6 potential. This made it possible to construct the energy without having to perform a coupling constant integration. The paired-phonon algorithm for solving the coupled channels is also applicable in the spin-dependent case.

SUMMARY

I have discussed in this contribution two important steps which have been taken to extend the parquet method to fermion systems and for systems with spin- and isospin-dependent forces. Other extensions are interesting as well and deserve future attention. The work done by E. Krotscheck and collaborators[9-11] on inhomogeneous systems can rather clearly be cast in parquet form. Some of the very elegant but technically intricate manipulations which he performed can be completely bypassed in the parquet approach. I would like to close with fig. 4 which starts with the exact theory and shows the parquet perspective of what has been done and what remains to be done to deal with interesting systems. The heavy connecting lines show the present state of the theory; the light connecting lines show what re-

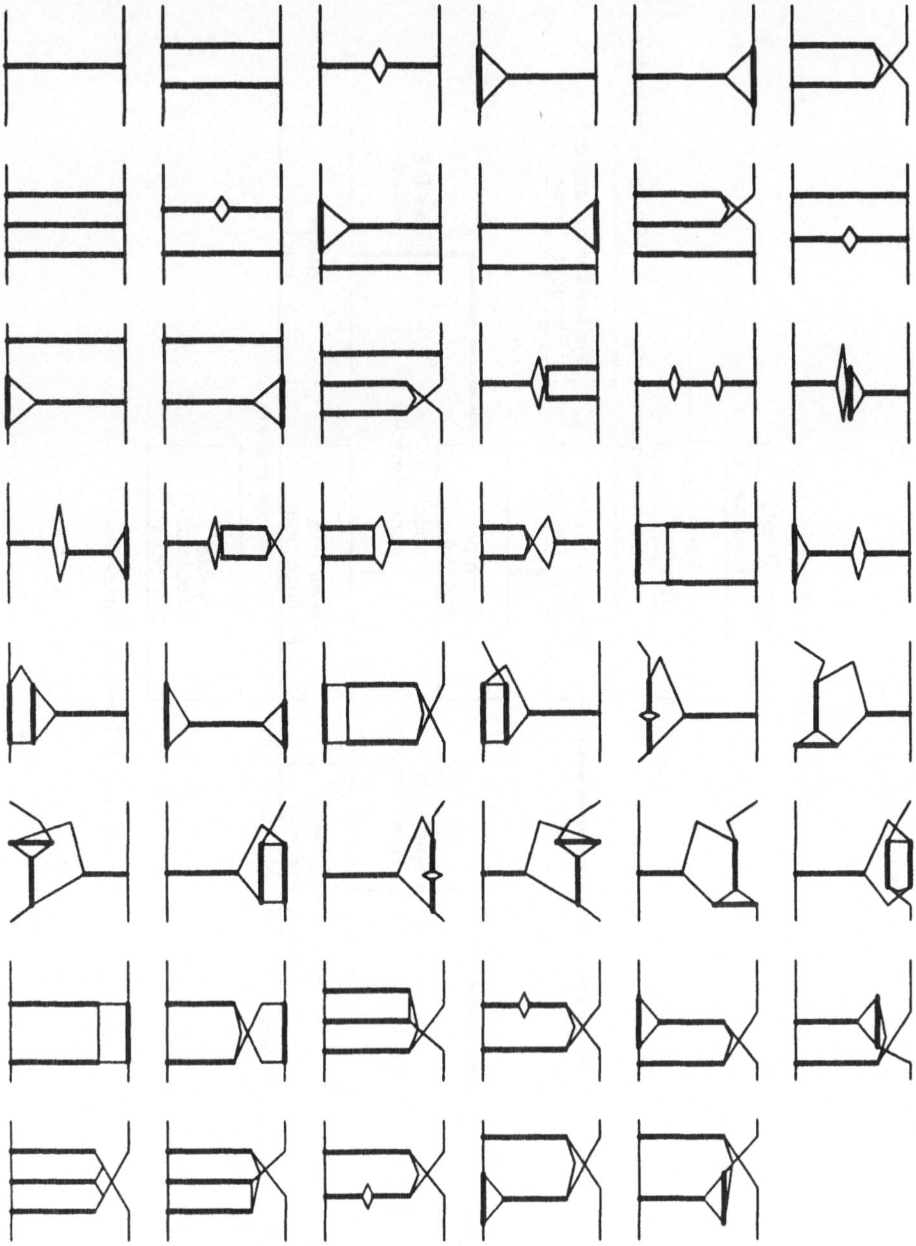

Fig. 3. Parquet diagrams to third order

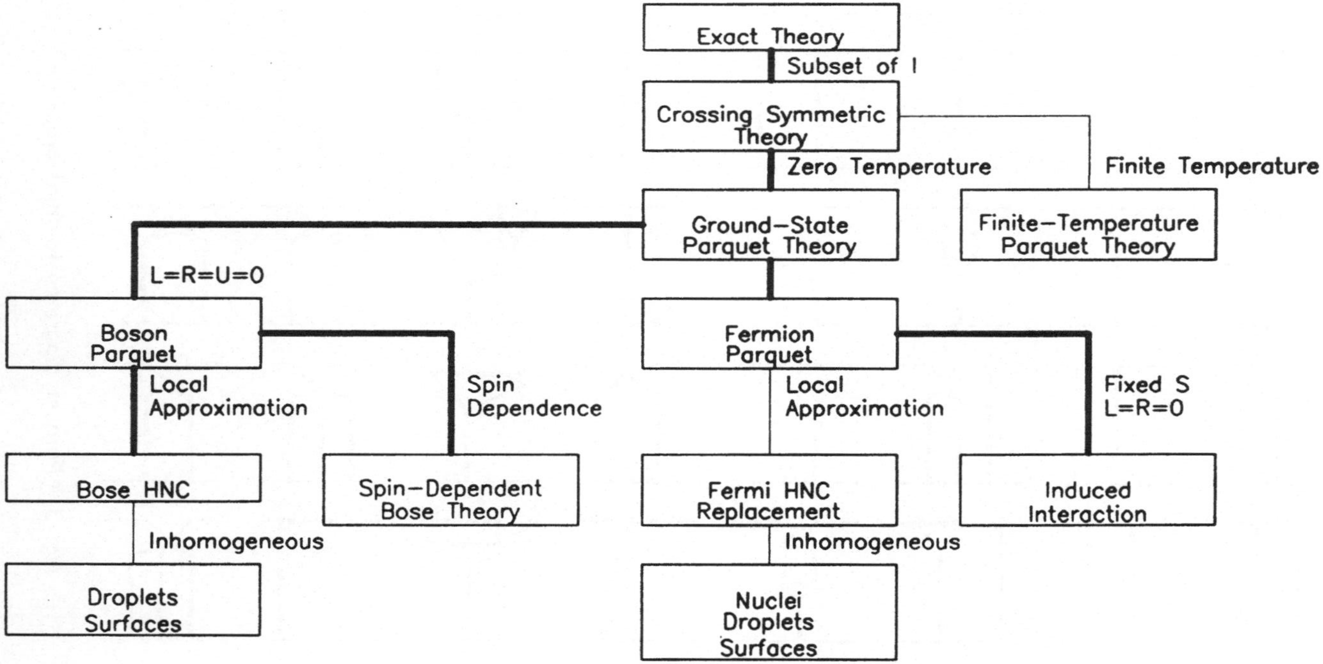

Fig. 4. A Parquet Perspective

mains to be done. Of course, additional structure could be added to this diagram at many points.

I have collaborated with A. D. Jackson, A. Lande, and R. Guitink on various aspects of the parquet theory and am particularly grateful to E. Krotscheck for many enlightening discussions. The work done here was supported in part by the NSF under grant 8206325.

REFERENCES

1. H. K. Sim, C.-W. Woo and J. R. Buchler, Phys. Rev. **A2**, 2024 (1970).
2. A. D. Jackson, A. Lande and R. A. Smith, Phys. Rep. **86**, 55 (1982).
3. A. D. Jackson, R. Guitink, A. Lande and R. A. Smith, Phys. Rev. **B31**, 403 (1985).
4. A. D. Jackson, A. Lande and R. A. Smith, Phys. Rev. Lett. **54**, 1469 (1985).
5. E. Krotscheck, private communication.
6. A. Lande and R. A. Smith, Phys. Lett. **131B**, 253(1983).
7. R. A. Smith and A. D. Jackson, to be published.
8. A. L. Fetter and J. D. Walecka, Quantum Theory of Many-Particle Systems (McGraw Hill, NY, 1971).
9. E. Krotscheck, G.-X. Qian and W. Kohn, Phys. Rev. **B31**, 4245(1985).
10. E. Krotscheck, G.-X. Qian and W. Kohn, Phys. Rev. **B31**, 4258(1985).
11. E. Krotscheck, G.-X. Qian and W. Kohn, Phys. Rev. **B31**, 4267(1985).

PAIRING CORRELATIONS, COHERENT STATES, AND BLACK HOLES

R.F. Bishop and A. Vourdas

Department of Mathematics
University of Manchester Institute
of Science and Technology
P.O. Box 88
Manchester M60 1QD, England

1. INTRODUCTION

Coherent states were first introduced into quantum mechanics by Schrödinger[1] in 1926, when he explicitly constructed a set of states $|A\rangle$ for the simple harmonic oscillator, which obeyed the classical equations of motion: $\langle A|\hat{x}(t)|A\rangle = x_{c\ell} \equiv x_o \sin(\omega t+\phi)$, $\langle A|\hat{p}(t)|A\rangle = p_{c\ell} \equiv m\dot{x}_{c\ell}$; and which did *not* display the otherwise rather universal quantum phenomenon of "spreading of the wavepacket" or dispersion. The concept of coherent state as it is understood today was first introduced by Glauber,[2] who showed how the states constructed by Schrödinger could be used to provide a good quantum-mechanical description of a coherent beam of light as obtained from a laser. Since then, these states have been very profitably further exploited in the field of quantum optics by Glauber and Sudarshan,[2,3] and many others that followed them.

In order to set the stage for the later developments, some of the properties of these "ordinary" or "atomic" or "Glauber" coherent states are discussed in Section 2. Various of these properties have been used more recently to try to generalize the concept of coherent states.[4-8]. Of particular importance here, we mention that various group-theoretical generalizations of coherent states have been given.[4-7] Such coherent states are associated with irreducible representations of various Lie groups. (We should also mention parenthetically, although it is of little consequence for present purposes, that the group-theoretical generalizations are rather divorced from the original motivation of Schrödinger[1] -- namely to discover quantal states that follow the classical motion. Nieto and his co-workers[8] have sought generalized coherent states which retain this property.)

We adopt the viewpoint that a quantum-mechanical coherent state has the physical property of being non-dispersive and periodic, and the mathematical property of being an element in an orbit in some irreducible space of a Lie group.[6,7] In this respect, we shall see that the Glauber states are the coherent states of the Heisenberg-Weyl group whose generators, in one dimension, are the boson creation and destruction operators ($a^\dagger$ and a respectively) and the identity operator I. These three operators together span the Heisenberg-Weyl algebra $W(1)$: $[a,a^\dagger] = I$.

The physical motivation that now leads us to the particular generalized coherent states that we discuss in Section 3, is the desire to find a rather general framework in which to discuss *pairing correlations* in the first instance, and more generally in a later extension, clustering phenomena of more than two particles. Restricting ourselves to identical bosons, we are led to consider the pairing operators: $a^{\dagger 2}$, a^2, and $a^\dagger a$. We show in Section 3 that these operators provide a simple realization of the Lie algebra SU(1,1), and we therefore introduce a set of generalized coherent states associated with the corresponding Lie group. We should mention that other coherent states of the group SU(1,1) have been introduced and studied in previous works,[5,6,9] which have provided hints to their relationship with the Bogoliubov transformation. In the present work we demonstrate this relationship very clearly. In particular we show that simply starting with the idea of pairing operators leads inexorably via the concepts of coherent states to the Bogoliubov transformation itself. Just as the Glauber coherent states may be viewed (and see Section 2) as eigenstates of the original single-boson destruction operator a, we demonstrate that our generalized SU(1,1) coherent states are also eigenstates of some new single-boson destruction operator b, which is itself generated from the operators a and $a^\dagger$ by a canonical Bogoliubov transformation. In other words, we show how the generalized pairing coherent states may be viewed as ordinary (or Glauber) coherent states of the Bogoliubov quasiparticles. Although they are ordinary coherent states with respect to the operators b and $b^\dagger$, we also show that they have interesting properties in relation to the original operators a and $a^\dagger$, which have not previously received much attention. We believe that these generalized coherent states, and their properties that we discuss here, will be very useful for practical calculations in the many fundamental problems where the Bogoliubov transformation plays a key role -- see, for example, Refs. [10-12].

From the motivation already discussed it is apparent that the physical systems to which our results may be applied, will be described by hamiltonians which are at least approximately bilinear in the underlying boson fields. Fairly obvious examples include superfluidity, the parametric excitation of a quantum oscillator,[13] and the pair production of spin-zero particles in a uniform but time-varying electric field. As a rather less obvious example we discuss in Section 4 the application of our results to a uniformly accelerated (Rindler) observer moving in flat (Minkowski) four-space. In particular, if the operators a and $a^\dagger$ are now associated with the quanta appropriate, say, to the solutions of the massless Klein-Gordon equation in the Minkowski metric of an inertial observer, then we show that the operators b and $b^\dagger$ discussed above may be associated with the corresponding solutions in the so-called Rindler metric appropriate to an observer undergoing uniform proper acceleration. Some previously known results for this important example in relativity can then be rather simply demonstrated using our generalized coherent state results. In particular we may easily show how the Minkowski (inertial) vacuum appears to the Rindler (accelerated) observer as a black-body radiator of quanta with a Planckian (Bose-Einstein) distribution corresponding to a temperature proportional to the proper acceleration. The relationship of this result to the phenomenon of Hawking radiation from black holes[14] (and hence to the ultimate form of condensed matter!) is pointed out.

2. THE ORDINARY COHERENT STATES

The ordinary coherent states are defined with respect to the simple harmonic oscillator with hamiltonian

$$H = p^2/2m + \tfrac{1}{2}m\omega^2 x^2 \tag{1}$$

The energy eigenstates $|n\rangle$ have eigenergies

$$E_n = (n+\tfrac{1}{2})\hbar\omega \quad , \tag{2}$$

and are readily given in terms of Hermite polynomials as,

$$\langle x|n\rangle = (\alpha/\pi^{\frac{1}{2}}2^n n!)^{-\frac{1}{2}} \exp(-\tfrac{1}{2}\alpha^2x^2)H_n(\alpha x) \quad , \quad \alpha \equiv (m\omega/\hbar)^{\frac{1}{2}} \quad . \tag{3}$$

The position and momentum operators can be defined in terms of the annihilation operator a and the creation operator $a^\dagger$ which is its Hermitian adjoint:

$$a = \left(\frac{1}{2m\hbar\omega}\right)^{\frac{1}{2}} \hat{p} - i\left(\frac{m\omega}{2\hbar}\right)^{\frac{1}{2}} \hat{x} \quad . \tag{4}$$

These operators satisfy the usual boson commutation relation,

$$[a,a^\dagger] = I \quad , \tag{5}$$

and have the property of changing the number of quanta by ± 1,

$$\begin{aligned} a|n\rangle &= n^{\frac{1}{2}}|n-1\rangle \quad , \quad a^\dagger|n\rangle = (n+1)^{\frac{1}{2}}|n+1\rangle \quad , \\ a^\dagger a|n\rangle &= n|n\rangle \quad , \quad |n\rangle = (n!)^{-\frac{1}{2}}(a^\dagger)^n\,|0\rangle \quad . \end{aligned} \tag{6}$$

The ordinary coherent states may now be defined in a number of equivalent ways.

2.1. Annihilation Operator Coherent States

The first definition of the ordinary coherent states is that they are eigenstates $|A\rangle$ of the annihilation operator a with complex eigenvalue A,

$$a|A\rangle = A|A\rangle \quad . \tag{7}$$

It is then easy to verify, using Eqs. (6), that these states are the particular combinations of the states $|n\rangle$ of definite number of quanta, given by:

$$|A\rangle = \exp(-\tfrac{1}{2}|A|^2) \sum_{n=0}^{\infty} (n!)^{-\frac{1}{2}}A^n|n\rangle \quad . \tag{8}$$

2.2. Displacement Operator Coherent States

We have already mentioned in Section 1 that the ordinary coherent states are coherent states of the Heisenberg-Weyl group. A unitary realization of this group is obtained by exponentiating the skew-adjoint operators in the algebra. We are thus led to consider the unitary operators,

$$\exp(Aa^\dagger - A^*a + i\phi) \quad ; \quad A \in \mathbb{C} \quad , \quad \phi \in \mathbb{R} \quad .$$

The corresponding unitary representation is then obtained by letting these operators act on a complete set of oscillator states. It is known that this representation is irreducible. By choosing a single representative state from the entire oscillator Hilbert space, and by letting the group elements act on it, we then generate an orbit of the group, whose elements are what a mathematician would call the coherent state of the group. In particular, if we choose the vacuum (ground) state $|0\rangle$ as the representative state, we generate the Glauber or ordinary coherent states. Thus, we define the unitary *displacement operator*,

$$U_1(A) \equiv \exp(Aa^\dagger - A^*a) = \exp(-\tfrac{1}{2}|A|^2)\exp(Aa^\dagger)\exp(-A^*a) \quad , \tag{9}$$

for arbitrary complex A. We note that the displacement operator is so called because of its action on the operators a and $a^\dagger$,

$$U_1(A)aU_1^\dagger(A) = a - A \quad , \quad U_1(A)a^\dagger U_1^\dagger(A) = a^\dagger - A^* \quad . \tag{10}$$

By displacing the vacuum (ground) state $|0\rangle$ we obtain our second definition of the ordinary coherent state $|A\rangle$ as,

$$|A\rangle \equiv U_1(A)|0\rangle = \exp(-\tfrac{1}{2}|A|^2)\exp(Aa^\dagger)|0\rangle \quad . \tag{11}$$

It is trivial to show that the two definitions agree by comparing Eqs. (8) and (11), and making use of Eq. (6).

2.3. Minimum Uncertainty Coherent States

A third definition of the ordinary coherent states arises from the Heisenberg position-momentum uncertainty relation,

$$\Delta x\,\Delta p \geq \tfrac{1}{2}\hbar \quad . \tag{12}$$

The minimum uncertainty coherent states are then defined both to satisfy the uncertainty relation (12) as an equality and to satisfy a certain subsidiary condition which we shall come to presently. It is a simple exercise in quantum mechanics to show that for any three Hermitian operators A, B and C which satisfy the relation,

$$[A,B] = iC \quad , \tag{13}$$

there is an uncertainty relation,

$$\Delta A \Delta B \geq \tfrac{1}{2}|\langle C\rangle| \quad , \tag{14}$$

which is satisfied as an equality by solutions $|\psi\rangle$ to the eigenvalue equation,

$$\Delta B(A-\langle A\rangle)|\psi\rangle = -i\,\Delta A(B-\langle B\rangle)|\psi\rangle \quad . \tag{15}$$

Putting A and B respectively equal to x and p, and using the uncertainty relation (12) as an equality, this equation is easily solved to give the minimum uncertainty coherent states as:

$$\langle x|\psi\rangle = [2\pi(\Delta x)^2]^{-\frac{1}{4}} \exp\left\{-\left(\frac{x-\langle x\rangle}{2\Delta x}\right)^2 + \frac{i}{\hbar}\langle p\rangle x\right\} \quad . \tag{16}$$

We can now see that these particular states are defined in terms of the four parameters $\langle x\rangle$, $\langle x^2\rangle$, $\langle p\rangle$ and $\langle p^2\rangle$, of which three are so far independent, since the uncertainty relation used as an equality imposes one condition between them. However if one now requires that the oscillator ground state $|0\rangle$ is one of these minimum uncertainty coherent syates, then we clearly require by comparison with Eq. (3) that Δx takes the particular value $(2\alpha^2)^{-\frac{1}{2}}$ for this state. Since we have already specified the product $\Delta x\,\Delta p = \frac{1}{2}\hbar$ for these states, the subsidiary condition which enables the oscillator ground state to be one of them can equivalently be expressed by fixing the ratio,

$$\Delta p/\Delta x = m\omega \quad . \tag{17}$$

In this case it is now easy to see that the minimum uncertainty states $|\psi\rangle$ of Eq. (16) subject to the constraint (17) are identical (up to a phase factor) to the annihilation operator uncertainty states $|A\rangle$ of Eq. (8). We do this by writing the defining equation (7) for the states $|A\rangle$ as,

$$\tfrac{1}{2}\left(\frac{\hat{p}}{\Delta p} - i\,\frac{\hat{x}}{\Delta x}\right)|A\rangle \;=\; A|A\rangle \;\equiv\; \tfrac{1}{2}\left(\frac{\langle p\rangle}{\Delta p} - i\,\frac{\langle x\rangle}{\Delta x}\right)|A\rangle \quad , \tag{18}$$

$$\Delta p \;\equiv\; (m\hbar\omega/2)^{\frac{1}{2}} \quad , \quad \Delta x \;\equiv\; (\hbar/2m\omega)^{\frac{1}{2}}$$

where we have used only Eq. (4). We then note firstly that the values Δp and Δx of Eq. (18) are precisely those obtained from the minimum uncertainty condition (12) and the subsidiary condition (17); and secondly that the differential equation (18) is now identical to the equation (15) for $|\psi\rangle$ when $A \to x$, $B \to p$.

We have thus shown that these three definitions of the ordinary coherent states are all equivalent. These states, as is by now well known, have a number of characteristic properties which will be useful later. In the first place it is easy to show that they are not mutually orthogonal,

$$\langle D|A\rangle \;=\; \exp[\,-\tfrac{1}{2}(|A|^2 + |D|^2 - 2D^*A)\,] \;,\; |\langle D|A\rangle|^2 = \exp(-|D-A|^2) \;. \tag{19}$$

Secondly, they form an overcomplete basis, in the sense that they contain more states than necessary for the expansion of an arbitrary state. It is also possible to give a very useful resolution of the identity operator in terms of them. Thus, consider the operator J defined as,

$$J \;\equiv\; \int d^2A\, |A\rangle\langle A| \quad ; \quad d^2A \;\equiv\; d(\mathrm{Re}A)\,d(\mathrm{Im}A) \quad , \tag{20}$$

where the integration extends over the entire complex-A plane. It is not difficult to show that *all* operators $U_1(B)$, defined in Eq. (9), commute with the operator J, by using the definition (11) and the group property

$$U_1(A)U_1(B) \;=\; \exp[\,i\,\mathrm{Im}(AB^*)\,]\; U_1(A+B) \quad , \tag{21}$$

which is readily proven. Since the operators $U_1(A)\exp(i\phi)$ thus form a group (namely the Heisenberg-Weyl group), Schur's lemma immediately gives that J is a constant multiple of the identity operator. By taking the expectation value of J in any state $|B\rangle$, and using the orthogonality relation (19), this constant is readily evaluated, and we find that the identity can be expanded in terms of the ordinary coherent states as,

$$\frac{1}{\pi}\int d^2A\, |A\rangle\langle A| \;=\; I \quad . \tag{22}$$

From the preceding discussion it is clear that the ordinary coherent states should be a useful tool whenever the Heisenberg-Weyl group is a dynamical symmetry of the hamiltonian. In such cases the Heisenberg equations of motion simply become equivalent to the corresponding equations for the classical quantities. The coherent states then remain coherent as time evolves, and the motion in the complex-A plane of the point corresponding to a particular coherent state is just described by its classical path. In such cases one has exactly reduced a quantum problem to a classical problem. An obvious question at this point is whether any of this procedure, and in particular whether any of our definitions of ordinary coherent states, can be generalized to other dynamical symmetry groups. As we asserted already in Section 1, group-theoretical generalizations based both on the annihilation operator definition[5] and on the displacement operator definition,[6] have been given. Our particular motivation to proceed further is now the possibility to understand *pairing correlations* in a similar fashion.

3. PAIRING OPERATORS AND THE SU(1,1) COHERENT STATES

In connection with pairing (of bosons), we naturally consider the operators $a^{\dagger 2}$, a^2, $a^\dagger a$ and $aa^\dagger$. It is trivial to show that these operators are again closed under commutation. More specifically, the three operators

$$K_+ \equiv \tfrac{1}{2}a^{\dagger 2} \quad , \quad K_- \equiv \tfrac{1}{2}a^2 \quad , \quad K_o \equiv \tfrac{1}{2}a^\dagger a + \tfrac{1}{4} \tag{23}$$

are readily seen to satisfy the Lie algebra of SU(1,1), namely

$$[K_o,K_\pm] = \pm K_\pm \quad ; \quad [K_-,K_+] = 2K_o \quad . \tag{24}$$

We note that SU(1,1) is locally isomorphic to the group SO(2,1) of rotations in three-dimensional (2 space + 1 time) Minkowski space. The algebra (24) may be compared with the more familiar Lie algebra of SU(2),

$$[J_o,J_\pm] = \pm J_\pm \quad ; \quad [J_-,J_+] = -2J_o \quad , \tag{25}$$

which is itself locally isomorphic to the group SO(3) of rotations in three-dimensional Euclidean space. Just as in Section 2.2, a unitary realization of the group SU(1,1) may be obtained by exponentiating the skew-adjoint operators in the algebra. We are thus led to introduce the unitary operators,

$$\begin{aligned} U_2(\rho,\theta,\lambda) &\equiv \exp(-\tfrac{1}{4}\rho e^{-i\theta}a^{\dagger 2} + \tfrac{1}{4}\rho e^{i\theta}a^2)\exp(i\lambda a^\dagger a) \; ; \; \rho,\theta,\lambda \in \mathbb{R} \\ U_2^\dagger U_2 &= I \quad . \end{aligned} \tag{26}$$

Using this definition, it is not difficult to show that the mode of action of the operators U_2 on the creation and destruction operators is given by the relations,

$$\begin{aligned} U_2(\rho,\theta,\lambda)aU_2^\dagger(\rho,\theta,\lambda) &= e^{-i\lambda}[\cosh(\tfrac{1}{2}\rho)a + e^{-i\theta}\sinh(\tfrac{1}{2}\rho)a^\dagger] \equiv b \quad , \\ U_2(\rho,\theta,\lambda)a^\dagger U_2^\dagger(\rho,\theta,\lambda) &= e^{i\lambda}[\cosh(\tfrac{1}{2}\rho)a^\dagger + e^{i\theta}\sinh(\tfrac{1}{2}\rho)a] \equiv b^\dagger \quad , \end{aligned} \tag{27}$$

which are the analogues of Eqs. (10) for the (Heisenberg-Weyl) displacement operators. We see that the transformation (27) imposed by the operators U_2 on the boson operators a and $a^\dagger$ is just the usual Bogoliubov transformation to new boson operators b and $b^\dagger$ that still satisfy the boson commutation relations

$$[b,b^\dagger] = I \quad . \tag{28}$$

Furthermore, from Eqs. (27) and the fact that U_2 is unitary, we can trivially prove for any function $f(a,a^\dagger)$ that can be expanded in a power series in its arguments, the relation,

$$U_2 f(a,a^\dagger)U_2^\dagger = f(b,b^\dagger) \Longleftrightarrow U_2 f(a,a^\dagger) = f(b,b^\dagger)U_2 \quad . \tag{29}$$

In terms of the operators U_2, we now introduce the states $|A;\rho\,\theta\,\lambda\rangle$ defined as,

$$|A;\rho\theta\lambda\rangle \equiv U_2(\rho,\theta,\lambda)|A\rangle = U_2(\rho,\theta,\lambda)U_1(A)|0\rangle \quad , \tag{30}$$

which are now our prime objects of study as generalized coherent states. We stress immediately that the representative state that has now been chosen for U_2 to act on, and so to generate an orbit of the group SU(1,1), is just an ordinary coherent state $|A\rangle$. We also note from Eq. (7) that this state $|A\rangle$ is an eigenstate of the operator K_-,

$$K_-|A\rangle = \tfrac{1}{2}A^2|A\rangle \quad , \tag{31}$$

but *not* of the operator K_o. We stress this point only since in the literature, different SU(1,1) coherent states are often defined by letting U_2

(or a similar operator) act on an eigenstate of K_o, and in doing this much of the underlying simplicity is obscured. We also note that Eq. (29) immediately gives that $U_2 a = b U_2$, and hence that the states $|A;\rho\theta\lambda\rangle$ are eigenstates of the destruction operator b,

$$b|A;\rho\theta\lambda\rangle = bU_2(\rho,\theta,\lambda)|A\rangle = U_2(\rho,\theta,\lambda)a|A\rangle = A|A;\rho\theta\lambda\rangle \quad , \tag{32}$$

by making use of Eq. (7).

The special case $A = 0$ is of particular interest since Eq. (32) implies the relation

$$b|0;\rho\theta\rangle = 0 \quad , \tag{33}$$

where we have put $|0;\rho\theta\lambda\rangle \equiv |0;\rho\theta\rangle$ since Eqs. (6) and (26) show that this state is independent of λ. Thus, just as the state $|0\rangle$ was defined by the relation $a|0\rangle = 0$ to be the ground state with respect to the operators a and $a^\dagger$ (i.e., the vacuum for a-type bosons), so the state $|0;\rho\theta\rangle \equiv U_2(\rho,\theta,\lambda)|0\rangle$ obeys Eq. (33), and is thus the vacuum for b-type bosons. Using Eq. (29) again we find,

$$U_2(\rho,\theta,\lambda)\exp(Aa^\dagger - A^*a) = \exp(Ab^\dagger - A^*b)U_2(\rho,\theta,\lambda) \quad , \tag{34}$$

and hence from Eqs. (9), (11) and (30),

$$|A;\rho\theta\lambda\rangle = \exp(Ab^\dagger - A^*b)|0;\rho\theta\rangle \quad . \tag{35}$$

In this way, we see very clearly that the states $|A;\rho\theta\lambda\rangle$ may be viewed as *ordinary* coherent states with respect to the operators b and $b^\dagger$. In other words, they are the Glauber coherent states for the Bogoliubov quasi-particles, or b-type bosons. Before proceeding, we also note at this point a particularly important relation for future use, for the expectation value of the number of b-type bosons in the vacuum $|0\rangle$ for a-type bosons. This quantity is immediately obtained from the definition of Eq. (27) as,

$$\langle 0|b^\dagger b|0\rangle = \sinh^2(\tfrac{1}{2}\rho) \quad . \tag{36}$$

Explicit evaluation of the generalized coherent states of Eq. (30) is not particularly easy with U_2 given by its defining equation (26). This is made much easier by making use of the 'normal-ordering relation',

$$\begin{aligned} &\exp(-\tfrac{1}{2}\rho\, e^{-i\theta}K_+ + \tfrac{1}{2}\rho\, e^{i\theta}K_-) = e^{\sigma K_+}e^{\tau K_o}e^{-\sigma^* K_-} \quad , \\ &\sigma \equiv -e^{-i\theta}\tanh(\tfrac{1}{2}\rho) \quad , \quad \tau \equiv \ln(1-|\sigma|^2) \quad , \end{aligned} \tag{37}$$

which has been given by Perelomov,[6] and which is valid for any operators K_+, K_- and K_o which satisfy the SU(1,1) algebra of Eq. (24). We note parenthetically that it is simplest, and sufficient, to prove Eq. (37) for the most elementary representation $K_o \to \tfrac{1}{2}\sigma_3$, $K_\pm \to \tfrac{1}{2}i(\sigma_1 \pm i\sigma_2)$ of the algebra, in terms of the usual Pauli spin matrices σ_1, σ_2 and σ_3 which obey the SU(2) algebra of Eq. (25). Using the result of Eq. (37), we may write the operator $U_2(\rho,\theta,\lambda)$ of Eq. (26) in the equivalent form:

$$\begin{aligned} U_2(\rho,\theta,\lambda) &\equiv U_2(\sigma,\lambda) \\ &= \exp(\tfrac{1}{2}\sigma a^{\dagger 2})(1-|\sigma|^2)^{\frac{1}{2}a^\dagger a + \frac{1}{4}}\exp(-\tfrac{1}{2}\sigma^* a^2)\exp(i\lambda a^\dagger a) \quad , \end{aligned} \tag{38}$$

where λ is real and σ, given by Eq. (37) is complex and with modulus $|\sigma| < 1$. Similarly, by making trivial use of Eqs. (6) and (8) we can first derive the relation,

$$\exp(\phi a^\dagger a)|A\rangle = \exp[\tfrac{1}{2}|A|^2(e^{\phi+\phi^*} - 1)]\,|Ae^{\phi}\rangle \quad , \tag{39}$$

which is valid for any complex ϕ, and then use this relation to write the generalized coherent states of Eq. (30) in the equivalent form,

$$\begin{aligned} |A;\rho\theta\lambda\rangle \equiv |A;\sigma\lambda\rangle &= U_2(\sigma,\lambda)|A\rangle \\ &= (1 - |\sigma|^2)^{\frac{1}{4}}\exp(-\tfrac{1}{2}\sigma^* A^2 e^{2i\lambda} - \tfrac{1}{2}|\sigma|^2|A|^2)\exp(\tfrac{1}{2}\sigma a^{\dagger 2})|Ae^{i\lambda}(1 - |\sigma|^2)^{\frac{1}{2}}\rangle \;. \end{aligned} \tag{40}$$

In the special case $\lambda = 0 = \sigma$ we simply regain the ordinary coherent states: $|A;00\rangle = |A\rangle$. We also note again that when $A = 0$, the b-type vacuum $|0;\sigma\lambda\rangle \equiv |0;\sigma\rangle$ is independent of λ.

Since, as we have seen, the states $|A;\sigma\lambda\rangle$ may simply be regarded as ordinary coherent states of the operators b and $b^\dagger$, it is clear that for fixed values of σ and λ they therefore obey all of the usual properties of coherent states. For example, the analogues of Eqs. (19) and (22) are respectively,

$$\langle D;\sigma\lambda|A;\sigma\lambda\rangle = \exp[-\tfrac{1}{2}(|A|^2 + |D|^2 - 2D^*A)] \quad ; \tag{41}$$

$$\frac{1}{\pi}\int d^2A\, |A;\sigma\lambda\rangle\langle A;\sigma\lambda| = I \quad . \tag{42}$$

Just as in Eq. (8), the states $|A;\sigma\lambda\rangle$ may also be expressed in terms of the eigenstates $|n;\sigma\lambda\rangle$ of the number operator $b^\dagger b$ for b-type bosons (and compare with Eq. (6)),

$$b^\dagger b|n;\sigma\lambda\rangle = n|n;\sigma\lambda\rangle \quad , \quad |n;\sigma\lambda\rangle = (n!)^{-\frac{1}{2}}(b^\dagger)^n|0;\sigma\rangle \quad , \tag{43}$$

in the form,

$$|A;\sigma\lambda\rangle = \exp(-\tfrac{1}{2}|A|^2)\sum_{n=0}^{\infty}(n!)^{-\frac{1}{2}}A^n|n;\sigma\lambda\rangle \quad . \tag{44}$$

For some purposes it is also useful to compare the ordinary coherent states $|A\rangle$ (with $\sigma = 0 = \lambda$) with our generalized coherent states $|B;\sigma\lambda\rangle$ in the following way. Use of Eqs. (40), (7) and (19) gives the overlap,

$$\begin{aligned} \langle A|B;\sigma\lambda\rangle = (1 - |\sigma|^2)^{\frac{1}{4}}\exp[&-\tfrac{1}{2}|A|^2 - \tfrac{1}{2}|B|^2 + \tfrac{1}{2}\sigma A^{*2} \\ &-\tfrac{1}{2}\sigma^* B^2 e^{2i\lambda} + A^* B\, e^{i\lambda}(1 - |\sigma|^2)^{\frac{1}{2}}] \quad . \end{aligned} \tag{45}$$

We observe that although the states $|A\rangle$ and $|B;\sigma\lambda\rangle$ are not generally orthogonal, they become so in the limit that $|\sigma|$ approaches unity. Finally, use of the completeness relation (22), yields the expansion

$$|B;\sigma\lambda\rangle = \frac{1}{\pi}\int d^2A\, \langle A|B;\sigma\lambda\rangle\, |A\rangle \tag{46}$$

for the generalized coherent states in terms of the Glauber coherent states, and where the overlap integrals $\langle A|B;\sigma\lambda\rangle$ are given explicitly in Eq. (45). A special case of Eqs. (45) and (46) of particular interest is the resolution of the vacuum for b-type bosons,

$$|0;\sigma\rangle = \frac{1}{\pi}(1 - |\sigma|^2)^{\frac{1}{4}}\int d^2A\, \exp(-\tfrac{1}{2}|A|^2 + \tfrac{1}{2}\sigma A^{*2})\,|A\rangle \quad . \tag{47}$$

We now illustrate the use of our generalized SU(1,1) coherent states and their associated Bogoliubov transformation, with an example drawn fron relativistic quantum field theory.

4. A POOR MAN'S BLACK HOLE AND HAWKING RADIATION

In the relativistic case of flat (Minkowski) four-space, it has been known for some time[15-18] that a Bogoliubov transformation analogous to Eq. (27) may be associated with a uniformly accelerated observer moving through the space. Since in many ways this particular example models the behaviour of real black holes, we shall show how our results of Section 3 can now be given a particularly interesting physical interpretation.

In terms of the usual Minkowski time-space coordinates (t,x,y,z), in units with $c = 1$, the inertial (or Minkowski) observer sees the flat space-time line element

$$ds^2 = -dt^2 + dx^2 + dy^2 + dz^2 \quad . \tag{48}$$

One well-known way to introduce a 'horizon' into this space (and hence to model a black hole) is by the transformation to Rindler coordinates (τ,ξ,y,z),

$$\begin{aligned} x &= \xi \cosh \tau \\ t &= \xi \sinh \tau \end{aligned} \quad ; \quad \tau,\xi \in \mathbb{R}(-\infty,\infty) \iff \begin{aligned} \xi &= \pm(x^2 - t^2)^{\frac{1}{2}} \\ \tau &= \tanh^{-1}(t/x) \end{aligned} \quad . \tag{49}$$

We note that this transformation maps only the regions that we now call the right $(x > |t|)$ and left $(x < -|t|)$ 'Rindler wedges' respectively. The forward $(t > |x|)$ and backward $(t < -|x|)$ light cones are unmapped, and the lines $x = \pm t$ (or equivalently $\xi = 0$, $\tau \to \pm\infty$) have become horizons to the Rindler observer. The line element (48) is mapped by the transformation (49) into

$$ds^2 = -\xi^2 d\tau^2 + d\xi^2 + dy^2 + dz^2 \quad . \tag{50}$$

One sees that lines of constant ξ correspond to world-lines of so-called Rindler observers who undergo uniform proper acceleration ξ^{-1} and who see proper time elements $\xi d\tau$.

Our aim is now to do quantum field theory in the Rindler coordinates. For simplicity, we work with a massless scalar (boson) field ϕ, which hence obeys the massless Klein-Gordon (or scalar wave) equation,

$$-\frac{\partial^2\phi}{\partial t^2} + \nabla^2\phi = 0 \iff -\frac{1}{\xi^2}\frac{\partial^2\phi}{\partial\tau^2} + \frac{\partial^2\phi}{\partial\xi^2} + \frac{1}{\xi}\frac{\partial\phi}{\partial\xi} + \frac{\partial^2\phi}{\partial y^2} + \frac{\partial^2\phi}{\partial z^2} = 0 \ . \tag{51}$$

It is easy to show by separation of variables that a complete set of solutions to the classical wave equation (51) in Rindler coordinates is given (for $|x| > |t|$) by,

$$\begin{aligned} u_{\omega\vec{k}}(\tau\xi yz) &= N_{\omega\vec{k}} \exp(-i\omega\tau + ik_2 y + ik_3 z)\, K_{i\omega}(k\xi) \ ; \\ \vec{k} &\equiv (k_2, k_3) \quad ; \quad k = |\vec{k}| \end{aligned} \tag{52}$$

where $N_{\omega\vec{k}}$ is a normalization constant and $K_{i\omega}(\rho)$ is a solution to the modified Bessel equation,

$$\rho^2 \frac{d^2X}{d\rho^2} + \rho\frac{dX}{d\rho} + (\omega^2 - \rho^2)X = 0 \quad . \tag{53}$$

A technical point to note here is that whereas in the right Rindler wedge $(x > |t|)$ the solutions $u_{\omega\vec{k}}$ and $u^*_{\omega\vec{k}}$ represent respectively the positive and negative frequency modes with respect to the operator $i\partial/\partial\tau$, the converse is true in the left Rindler wedge $(x < -|t|)$. The basic reason for this is that in the right Rindler wedge the direction of increasing τ

on lines of constant ξ is in the direction of increasing t, whereas in the left Rindler wedge the direction is opposite. Bearing this in mind, quantization of the massless scalar field in Rindler coordinates is now based on the usual expansion in canonical modes,

$$\phi(\tau\xi yz) = \int_0^\infty d\omega \int \frac{d^2\vec{k}}{(2\pi)^2} \left[u^{(+)}_{\omega\vec{k}}\, b^{(+)}_{\omega\vec{k}} + u^{(-)}_{\omega\vec{k}}\, b^{(-)}_{\omega\vec{k}} + \text{h.c.} \right] , \tag{54}$$

where h.c. indicates the hermitian conjugate; the functions $u^{(+)}_{\omega\vec{k}}$ and $u^{(-)}_{\omega\vec{k}}$ are solutions to the wave equation which vanish respectively in the left and right Rindler wedges but which equal $u_{\omega\vec{k}}$ in the opposite Rindler wedge; and the operators $b^{(\pm)}_{\omega\vec{k}}$ obey the usual canonical commutation relations,

$$\left[b^{(\sigma)}_{\omega\vec{k}} \, , \, b^{(\sigma')\dagger}_{\omega'\vec{k}'} \right] = \delta_{\sigma\sigma'}\, \delta(\omega - \omega')\, \delta(\vec{k}-\vec{k}') \quad ; \quad \sigma = \pm \quad . \tag{55}$$

We now note that the solutions $u^{(\pm)}_{\omega\vec{k}}$ are analytic everywhere except on the horizons $x \pm t = 0$. Indeed it is not difficult to show from Eqs. (52) and (53) that on these horizons, $u_{\omega\vec{k}}$ behaves like a regular function times a singular factor $(x \pm t)^{\pm i\omega}$. However, one can easily verify that the particular linear combinations defined by

$$v^{(\pm)}_{\omega\vec{k}} \equiv \left[1 - \exp(-2\pi\omega) \right]^{-\frac{1}{2}} \left[u^{(\pm)}_{\omega\vec{k}} + \exp(-\pi\omega)\, u^{(\mp)}_{\omega\vec{k}} \right] , \tag{56}$$

are also analytic on the horizons $x = t$ and $x = -t$ respectively, and indeed are analytic respectively in the lower half complex $(t - x)$ plane or the upper half complex $(t + x)$ plane. This fact can then be used to show that the solutions $v^{(\pm)}_{\omega\vec{k}}$ may therefore be used for a Minkowski decomposition of the field ϕ for the inertial observer, and also that in this case the solutions $v^{(+)}_{\omega\vec{k}}$ and $v^{(-)}_{\omega\vec{k}}$ are respectively the positive and negative frequency modes with respect to the operator $i\partial/\partial t$. The Minkowski decomposition analogous to Eq. (54) is therefore

$$\phi(txyz) = \int_0^\infty d\omega \int \frac{d^2\vec{k}}{(2\pi)^2} \left[v^{(+)}_{\omega\vec{k}}\, a^{(+)}_{\omega\vec{k}} + v^{(-)*}_{\omega\vec{k}}\, a^{(-)}_{\omega\vec{k}} + \text{h.c.} \right] , \tag{57}$$

where the operators $a^{(\pm)}_{\omega\vec{k}}$ obey canonical boson commutation relations analogous to Eq. (55).

A comparison of the two decompositions (54) and (57), together with Eq. (56), easily yields the relation,

$$b^{(\pm)}_{\omega\vec{k}} = (1 - e^{-2\pi\omega})^{-\frac{1}{2}} \left[a^{(\pm)}_{\omega\vec{k}} + e^{-\pi\omega}\, a^{(\mp)\dagger}_{\omega\vec{k}} \right] , \tag{58}$$

between the boson operators in the two representations. Defining new operators,

$$a_{\omega\vec{k}} \equiv e^{i\gamma} \left(a^{(+)}_{\omega\vec{k}} + a^{(-)}_{\omega\vec{k}} \right) , \quad b_{\omega\vec{k}} \equiv e^{i\gamma} \left(b^{(+)}_{\omega\vec{k}} + b^{(-)}_{\omega\vec{k}} \right) , \tag{59}$$

where γ is an arbitrary real constant, we find that the relationship between them,

$$b_{\omega\vec{k}} = \left(1 - e^{-2\pi\omega} \right)^{-\frac{1}{2}} \left[a_{\omega\vec{k}} + \exp(-\pi\omega + 2i\gamma)\, a^{\dagger}_{\omega\vec{k}} \right] , \tag{60}$$

is precisely the Bogoliubov transformation (27) with the identification,

$$\tanh(\tfrac{1}{2}\rho) = e^{-\pi\omega} . \tag{61}$$

It is therefore clear that all of our general results may now be applied to this example. In particular, the states $|0\rangle$ and $|0;\sigma\rangle$ now play the role of vacua for our massless scalar meson field as observed respectively by the inertial (Minkowski) and uniformly accelerated (Rindler) observers. In particular, a direct combination of Eqs. (36) and (61) gives the very important result that the inertial vacuum $|0\rangle$ appears to the uniformly accelerated observer, for whom the meson number operator is $b^{\dagger}_{\omega\vec{k}} b_{\omega\vec{k}}$, as a source of mesons with distribution,

$$\langle 0|b^{\dagger}_{\omega\vec{k}}\, b_{\omega\vec{k}}|0\rangle = (e^{2\pi\omega} - 1)^{-1} \quad . \tag{62}$$

We recall that along lines of constant ξ, proper time is marked by the product $\xi\tau$. Hence the Rindler observer interprets a wave varying as $\exp(-i\omega\tau) \equiv \exp[-i(\omega/\xi)\xi\tau]$, as in Eq. (52), as having an angular frequency $w = \omega/\xi$. Comparing Eq. (62) with a black-body radiator, which at temperature T has the usual Planckian distribution $[\exp(\hbar w/k_B T) - 1]^{-1}$ for massless bosons, yields the profound result that the Minkowski vacuum appears to the uniformly accelerated observer as a black-body radiator (at least for massless scalar mesons) of temperature given by,

$$T = \frac{\hbar}{2\pi k_B c\xi} \quad , \tag{63}$$

where we have re-inserted the correct factors of c. Davies[15] seems to have been the first to realize that the Rindler vacuum contains a thermal distribution of quanta relative to the Minkowski vacuum. We note from Eq. (63) that the effect is explicitly both quantum-mechanical and relativistic in origin since T vanishes in both limits $\hbar \to 0$ and $c \to \infty$. We note also that the temperature is directly proportional to the proper acceleration ξ^{-1}.

From our derivation it is clear that the underlying 'cause' of these results is the existence, in the Rindler coordinates, of the horizons at $x = \pm t$. Sciama et al.[18] have stressed how an observer undergoing uniform acceleration may thus profitably be viewed as constituting a "model" or "poor man's" black hole. They have further shown the deep connection between our present result and the thermal properties of real black holes when quantum effects are taken into account. From our own derivation it is no surprise that our result is essentially a poor man's description of Hawking radiation from real black holes, which have been shown[14] to behave as black-body radiators with temperature given as in Eq. (63) but with the proper acceleration ξ^{-1} now replaced by the so-called "surface (acceleration due to) gravity".

5. CONCLUSIONS

We have investigated in this paper the generalized coherent states $|A;\sigma\lambda\rangle = U_2(\sigma,\lambda)|A\rangle$ obtained from the normal or Glauber coherent states $|A\rangle = U_1(A)|0\rangle$. The coherence properties of the Glauber states $|A\rangle$ are already contained in the exponentiation of the one-body operator which defines the displacement operator U_1, where by a one-body operator we mean one linear in the field operators a and $a^{\dagger}$. Similarly the operator U_2 contains exponentials of operators bilinear in a and $a^{\dagger}$, and was thus expected to generate a coherent paired state. We have shown how this generalized coherent state can itself be alternatively viewed as a normal coherent state with respect to new boson (quasiparticle) operators that stand in relation to the original boson (particle) operators by a Bogoliubov transformation. We showed in particular how the Bogoliubov transformation itself arises most naturally in our framework, starting only with the basic concept of building *coherent paired states*. In this respect a number of generalizations immediately suggest themselves.

Firstly, the extension from a single boson (or canonical quantum mode) to a set of n distinct bosons (or modes) is possible. In the case of a single boson treated here, we saw that there is an underlying Lie group, namely SU(1,1) associated with the generalized coherent states. In fact this is not the only appopriate group here, since the three-dimensional Lorentz group SO(2,1) is not only locally isomorphic to SU(1,1) but also to both the group SL(2,R) of real second order matrices with unit determinant and the symplectic group Sp(2,R). Similarly the bilinear products of boson operators $a_i^\dagger a_j^\dagger$, $a_i^\dagger a_j$ and $a_i a_j$ for $i,j = 1, \cdots, n$, also form a basis for a realization of the higher symplectic algebra Sp(2n,R). A unitary realization of this group can similarly be constructed by exponentiating the skew-adjoint operators in this algebra, and it should then be possible to extend in an obvious way the treatment given here to the case of n distinct bosons.

A second generalization concerns the possibility of looking at higher clustering correlations in the same way. Thus, for correlated clusters of m identical bosons, we may envisage taking the various products of order m in the operators a and $a^\dagger$, examining their underlying group structure, and so constructing the appropriate coherent states. In this way one can hope to generalize the Bogoliubov transformation appropriate for (a particular type of) pairing, to transformations appropriate for generating higher clustering and the associated possible new condensed phases. A third possible extension is to repeat the above for clustering correlations in systems of fermions. We are presently interested in each of these extensions.

We have already mentioned in Section 1 various possible physical applications of the pairing coherent states discussed here, and we have illustrated our results in Section 4 in a context perhaps not familiar to condensed matter theorists. We started our discussion in Section 1 with the work of Glauber and others on the ordinary coherent states in the field of quantum optics, and it is perhaps appropriate to return finally to this field. Just as the Glauber states are appropriate for a description of the one-photon coherent states and hence the radiation field from a conventional single-photon laser, so our generalized SU(1,1) paired coherent states have been discussed in quantum optics[10,11] from the viewpoint of two-photon coherent states and the possibility of a two-photon laser. In quantum optics the two-photon coherent states have been called "squeezed states", since although they are evidently minimum uncertainty states in the sense of realizing the uncertainty relation (12) as an equality, it is not difficult to show that the ratio $\Delta p/\Delta x$ for the generalized coherent state is "squeezed" in the sense that it is *not* the value of Eq. (17) for ordinary coherent states, but acquires an extra factor $(1+\sigma)/(1-\sigma)$. There is much present work and excitement in quantum optics on these states. It is our belief that the possible extensions discussed above will have immediate repercussions not only in condensed matter theory but also in quantum optics.

Acknowledgement

We gratefully acknowledge support for this work in the form of a Research Grant from the Science and Engineering Research Council of Great Britain.

REFERENCES

1. E. Schrödinger, Naturwiss. 14:664 (1926).
2. R. J. Glauber, Phys. Rev. Lett. 10:84 (1963); Phys. Rev. 130:2529 (1963) and 131:2766 (1963).
3. E. C. G. Sudarshan, Phys. Rev. Lett. 10:227 (1963);
   R. J. Glauber, Les Houches lectures 1964, in: "Quantum Optics and Electronics," C. DeWitt et al., eds., Gordon and Breach, New York (1965);
   J. R. Klauder and E. C. G. Sudarshan, "Fundamental of Quantum Optics", Benjamin, New York (1968).
4. J. M. Radcliffe, J. Phys. A 4:313 (1971).
5. A. O. Barut and L. Girardello, Commun. Math. Phys. 21:41 (1971).
6. A. M. Perelomov, Commun. Math. Phys. 26:222 (1972); 44:197 (1975); Usp. Fiz. Nauk 123:23 (1977) [translated into English in: Sov. Phys. Usp. 20:703 (1977)].
7. E. Onofri, J. Math. Phys. 16:1087 (1975).
8. M. M. Nieto and L. M. Simmons, Jr., Phys. Rev. Lett. 41:207 (1978); V. P. Gutschick, M. M. Nieto and L. M. Simmons, Jr., Phys. Lett. 76A: 15 (1980); M. M. Nieto and L. M. Simmons, Jr., Phys. Rev. A 19:438 (1979) and Phys. Rev. D 20:1321,1332,1342 (1979); M. M. Nieto, Phys. Rev. D 22:391 (1980); V. P. Gutschick and M. M. Nieto, Phys. Rev. D 22:403 (1980); M. M. Nieto, L. M. Simmons, Jr. and V. P. Gutschick, Phys. Rev. D 23:927 (1981).
9. H. Feshbach and Y. Tikochinsky, Trans. N.Y. Acad. Sci. 38:44 (1977).
10. H. P. Yuen, Phys. Rev. A 13:2226 (1976).
11. D. F. Walls, Nature 306:141 (1983).
12. C. S. Hsue, H. Kümmel and P. Ueberholz, Phys. Rev. D (to be published).
13. A. M. Perelomov and V. S. Popov, Zh. Eksp. Teor. Fiz. 56:1375 (1969) and 57:1684 (1969) [translated into English in: Sov. Phys. JETP 29:738 (1969) and 30:910 (1970)].
14. S. W. Hawking, Commun. Math. Phys. 43:199 (1975).
15. P. C. W. Davies, J. Phys. A 8:609 (1975); Rep. Prog. Phys. 41:1313 (1978).
16. P. Candelas and D. J. Raine, J. Math. Phys. 17:2101 (1976).
17. W. G. Unruh, Phys. Rev. D 14:870 (1976).
18. D. W. Sciama, P. Candelas and D. Deutsch, Adv. Phys. 30:327 (1981).

PREPARING THE GROUND FOR COUPLED CLUSTER CALCULATIONS

H. Kümmel *

University of Manchester
Institute of Science and Technology
Manchester, Great Britain **

I. INTRODUCTION

In this paper it is assumed that we want to calculate the wave function of one or many body systems. If the Hamiltonian H is sufficiently complicated, the standard approach is by using some simple "starting wave functions" ϕ . On top of them one puts those parts which are complicated and typically require sophisticated and/or extended numerical methods. Merely for simplicity of the representation in this paper it is assumed that it suffices to have only one relevant starting wave function, i.e.

$$\Psi = \phi + \text{"corrections"}. \qquad (1.1)$$

Here ϕ may already be a sophisticated wave function, not necessarily eigenfunction to a zero order Hamiltonian H_o. Also, the systems are not necessarily many body systems. Indeed the first example treated in this paper is a one body problem.

The decomposition (1.1) as such is trivial. But it becomes useless if the "corrections" are not small. This invariably implies that there are several or even infinitely many wave functions of about equal importance: Heaving this in mind, it is clear that one is confronted with three general questions:

1. how does one find a good or optimal ϕ ?
2. what is the structure of the "corrections"?
3. which of the terms in the "corrections" are large, which ones are small and therefore hopefully can be neglected?

In this paper possible answers to 1. and 2. will be discussed. There is no general rule for finding optimal ϕ's. Some intuition is still needed. The answer to 2. is fairly unique: it <u>is</u> of the exponential form as used in the coupled cluster method (CCM). In 3. again there is no unique answer. Depending on the problem at hand (Bosons or fermions, hard core force, long range interaction, etc.) very different truncation schemes had to be invented and applied thereafter. Dealing with this aspect would amount to a review article about CCM. Therefore, this point will not be discussed further.

---

* Supported by the Science and Engineering Research Council of Great Britain

** Permanent address: Institut für Theoretische Physik, Ruhr-Universität Bochum, D-463o Bochum 1

II. HOW TO CHOOSE ϕ

Selection of a suitable starting wave function has been a topic in most papers on many body theory. In the present paper only those procedures used in the context of the CCM will be discussed. Moreover, the first model to be considered is the anharmonic oscillator [1,2]. This will be done not only because of it's simplicity, but also because it is a prototype for more general one body, many body or field theoretical problems involving bosons. It is quite surprising that the approach presented here was invented so late.

1. Anharmonic Oscillator

The Hamiltonian is

$$H = \frac{1}{2}p^2 + \frac{1}{2}q^2 + \frac{\lambda}{2} = a^+a + \frac{\lambda}{4}(a+a^+)^4 + \frac{1}{2}\,, \tag{2.1}$$

with

$$a|\phi_a\rangle = 0 \quad \text{and} \quad [a, a^+] = 1. \tag{2.2}$$

Perform now a general (Bogoliubov-)transformation

$$b = Aa + Ba^+ + C\,, \tag{2.3}$$

with

$$b|\phi_b\rangle = 0 \quad \text{and} \quad [b, b^+] = 1. \tag{2.4}$$

It can be shown that for any such transformation and any Hamiltonian

$$|\phi_b\rangle = \exp(\underline{S}_1 + \underline{S}_2)|\phi_a\rangle, \tag{2.5}$$

where

$$\underline{S}_1 = S_1 a^+\,, \quad \underline{S}_2 = S_2\, a^{+2} \tag{2.6}$$

$$S_1 = -A^{-1}C\,, \quad S_2 = -A^{-1}B. \tag{2.7}$$

It is straightforward to replace a by b in H via (2.3). Then it is reasonable to introduce a normal ordering N_b of the operators b. Why this is so will be become clear in a moment. Anyway one naturally arrives at a form

$$H = \omega b^+b + \alpha(b^{+2} + b^2) + \gamma N_b(b+b^+)^4 + E_0. \tag{2.8}$$

Now, due to this normal ordering

$$\langle H\rangle \equiv \frac{\langle\phi_b|H|\phi_b\rangle}{\langle\phi_b|\phi_b\rangle} = E_0. \tag{2.9}$$

One may wish to optimize this by

$$\delta\langle H\rangle = \langle\delta\phi_b|H|\phi_b\rangle = 0. \tag{2.1o}$$

Now,

$$|\delta\phi_b\rangle = b^+|\phi_b\rangle \text{ or } b^{+2}|\phi_b\rangle \tag{2.11}$$

are the smallest components orthogonal to $|\phi_b\rangle$. Therefore necessarily

$$\langle b^+\phi_b|H|\phi_b\rangle = 0 \tag{2.12}$$

and

$$\langle b^{+2}\phi_b|H|\phi_b\rangle = 0. \tag{2.13}$$

This means however that H should not have any terms $\sim$ b or b^+ and b^2 or b^{+2}. From (2.8) it follows

$$\alpha = 0 \quad \text{and} \quad \gamma = 0. \tag{2.14}$$

(Due to the symmetry of the anharmonic oscillator there are no terms $\sim$ b or b^+; so in this special case (2.12) is no condition).

It is clear that these facts are valid not only for the anharmonic oscillator. Therefore the explicit expressions for ω, α and γ in (2.8) are not written down. Furthermore, from (2.9) and (2.5) it follows that (2.12) and (2.13) are equivalent to

$$\frac{\partial E_0}{\partial S_1} = \frac{\partial E_0}{\partial S_2} = 0. \tag{2.15}$$

Returning now to the anharmonic oscillator (where for symmetry reasons $S_1=0$) one finds that

$$\omega - \omega^3 + 6\lambda = 0, \tag{2.16}$$

and

$$E_0 = \frac{1+3\omega^2}{8\omega}. \tag{2.17}$$

One may interpret this result by saying that one has constructed the best basis set with the best ground state $|\phi_b\rangle$. The remarkable feature of this procedure is that for $0 < \lambda < 10^3$ the error of the energy is always smaller than 2%. Also doing a 5x5 diagonalization or low order CCM calculation in the space of the states $(b^+)^n|\phi_b\rangle$ on top of this yields an accuracy as high as all direct diagonalization procedures (in terms of the original harmonic oscillator wave functions) which ever have been done [1]. This method is indeed so efficient that it is practically impossible to compare the computer times. (A programmable pocket computer is all that´s needed). Hsue recently has applied the same ideas to the Lennard-Jones potential [3]. Indeed all single particle potentials with a minimum can be treated in this way.

2. ϕ^4 field theory

The ideas described for the anharmonic oscillator can be generalized quite easily to quantum (boson) field theories. Many types of self-interactions may be considered. Here only the 1+1-dimensional ϕ^4 field theory will be studied as a prototype [4]. This theory has been thoroughly investigated in the past and even some exact results are known. It is particularly worth studying since it breaks a (discrete) symmetry.
There are two forms for the Hamiltonian densities:

$$\mathcal{H}_s = N_m\left\{\frac{1}{2}\dot{\phi}^2 + \frac{1}{2}(\nabla\phi)^2 + \frac{1}{2}m^2\phi^2 + \frac{\lambda}{4}\phi^4\right\}, \tag{2.18}$$

$$\mathcal{H}_{sb} = N_M\left\{\frac{1}{2}\dot{\phi}^2 + \frac{1}{2}(\nabla\phi)^2 - \frac{1}{4}M^2\phi^2 + \frac{\lambda}{4}\phi^4\right\}. \tag{2.19}$$

Classically, the first one has an energy minimum at $\phi = 0$ and oscillator frequencies $\omega_k^2 = k^2 + m^2$. The second is called "symmetry breaking" since it has two classical energy minima at $\phi = \pm M/\sqrt{2\lambda}$ with an excitation spectrum $\omega_k^2 = k^2 + M^2$ in both.

N_m resp. N_M mean normal ordering with respect to the operators a_k and a_k^+ going with the frequencies $\omega_k^2 = k^2 + m^2$, $\omega_k^2 = k^2 + M^2$ respectively;

here

$$\phi(x) = \sum_{\kappa} \frac{1}{\sqrt{2\omega_\kappa}} \left\{ a_\kappa \frac{e^{ikx}}{\sqrt{L}} + a_\kappa^+ \frac{e^{-ikx}}{\sqrt{L}} \right\} \tag{2.2o}$$

(L = normalization volume). This corresponds to the usual conventions. It also guarantee's that no infinite terms occur in 1+1 dimensions. Some exact results will be stated now without proof. They have been known for a long time:

1. The energy is a continuous function of the coupling constant (no first order phase transition, Simon & Griffiths [5]).

2. There is exactly one critical point defined by the vanishing of the energy gap

$$\tilde{m} \underset{\lambda \to \lambda_c}{\to} 0 \tag{2.21}$$

(Lorentz invariance requires that the excitation energy is of the form $\sqrt{K^2 + \tilde{m}^2}$). Since the two point Green's function behaves asymptotically like

$$G(\vec{x}, \vec{x}') \sim e^{-\tilde{m}|\vec{x}-\vec{x}'|} , \tag{2.22}$$

the "correlation length" $\tilde{m}^{-1}$ becomes infinite at $\lambda \to \lambda_c$. This implies that "correlations are strong" [6].

3. $\mathcal{H}_s$ and $\mathcal{H}_{sb}$ are equivalent except for a constant if

$$m^2 + \frac{M^2}{2} - \frac{3\lambda}{4\pi} \ln \frac{M^2}{m^2} = 0. \tag{2.23}$$

One finds that for each $\lambda/m^2 > (\lambda/m^2)_c \approx 9.045$ (for which $\lambda/M^2 = 2\pi/3$) there are <u>two</u> $\mathcal{H}_{sb}$ equivalent to $\mathcal{H}_s$; for $\lambda/m^2 < (\lambda/m^2)_c$ there is none. Since this is the only distinguished point in (2.23) we expect this to be the critical point.

4. The topological charge

$$Q = \sqrt{\frac{\lambda}{2M^2}} \int_0^L \nabla\phi \, dx \tag{2.24}$$

commutes with H and has the eigenvalues

o (" vacuum sector"), ± 1 ("soliton sectors")

(the last ones only if $\lambda/m^2 > 9.045$). These sectors seem to be completely independent (no operators with finite powers of ϕ connect them).

From now on the Bogoliubov transformation followed by a variational principle as described above will be called "Hartree approximation". The generalization to field theories is completely straight forward: A and B in (2.3) become matrices, C is a vector. The "Thouless theorem" (2.5) now involves

$$\underline{S}_1 = \sum_{\kappa} S_{1\kappa} a_\kappa^+ , \quad \underline{S}_2 = \sum_{\kappa\kappa'} \frac{1}{2} S_{2\kappa\kappa'} a_\kappa^+ a_{\kappa'}^+ . \tag{2.25}$$

Inserting b instead of a into H_{sb}, using the expansion

$$\phi(x) = \sum_{\alpha} \frac{1}{\sqrt{2\tau_\alpha}} \left\{ b_\alpha \xi_\alpha(x) + b_\alpha^+ \xi_\alpha^*(x) \right\} , \tag{2.26}$$

one obtains

$$H_{Sb} = \frac{1}{2}\sum_{\alpha\beta}\left\{\tau_\alpha\delta_{\alpha\beta} + \frac{1}{\sqrt{\tau_\alpha\tau_\beta}}\int dx\, \xi^*_\alpha(x)\,\tau_x\,\xi_\beta(x)\right\} b^+_\alpha b_\beta$$
$$-\frac{1}{4}\sum_{\alpha\beta}\left\{\tau_\alpha\delta_{\alpha\beta} - \frac{1}{\sqrt{\tau_\alpha\tau_\beta}}\int dx\, \xi^*_\alpha(x)\,\tau_x\,\xi_\beta(x)\right\} b^+_\alpha b^+_\beta + c.c. \tag{2.27}$$
$$+\int dx\, S_x\, \phi(x) + \lambda N_b \int dx\, C(x)\,\phi^3(x) + \lambda/4\, N_b \int dx\, \phi^4(x) + E_0 .$$

Here

$$\tau_x = -\nabla_x^2 - M^2/2 + 3\lambda C(x) + 3\lambda\beta(x), \tag{2.29}$$

$$S_x = [-\nabla_x^2 - M^2/2 + 3\lambda\beta(x)]\,C(x) + \lambda C^3(x), \tag{2.30}$$

$$E_0 = \int dx\left\{\frac{1}{2}(\nabla C)^2 + \frac{\lambda}{4}C^4 - \frac{3}{4}M^2C^2 + \frac{3}{2}\beta(x)\left(\lambda C^2 - \frac{1}{2}M^2\right) + \frac{3}{2}\lambda\beta^2(x)\right). \tag{2.31}$$

The terms with

$$\beta(x) = \sum_\alpha \frac{1}{2\tau_\alpha}|\xi_\alpha(x)|^2 - \sum_\kappa \frac{1}{2\omega_\kappa}\frac{e^{iKx}}{\sqrt{L}} \tag{2.32}$$

are due to re-normal ordering with respect to b, b^+. They are of purely quantum mechanical origin. As before $\langle\delta\phi_b|H|\phi_b\rangle = 0$ leads to two "Hartree conditions"

$$\langle b^+_\alpha \phi_b|H|\phi_b\rangle = 0 \quad , \tag{2.33}$$

$$\langle b^+_\alpha b^+_\beta\, \phi_b|H|\phi_b\rangle = 0, \tag{2.34}$$

which are generalizations of (2.12) and (2.13). It is evident that now the factors of terms $\sim$ to b_α, b^+_α and $b_\alpha b_\beta$, $b^+_\alpha b^+_\beta$ in (2.27) must vanish. (As before, also $\partial E_0/\partial S_{1\kappa} = \partial E_0/\partial S_{2\kappa\kappa'} = 0$ lead to the same conditions. This also is clear from (2.9) which is still valid).

Eqs (2.33) and (2.34) can be made explicit:

(2.33) becomes $S_x = 0$. (2.35)

(2.34) becomes $\tau_x\,\xi_\alpha(x) = \tau^2_\alpha\,\xi_\alpha(x)$. (2.36)

Before discussing the meaning of these two equations, it can be noticed that (2.36) diagonalizes the oscillator part of the Hamiltonian, such that it now takes the simple form

$$H = \sum_\alpha \tau_\alpha b^+_\alpha b_\beta + \frac{\lambda}{4} N_b \int dx\, \phi^4(x) + \lambda N_b \int dx\, C\,\phi^3(x) + E_0 . \tag{2.37}$$

τ_α are the (low order) excitation energies of the system. Returning to the meaning of (2.35) it is noted that this equation has two types of solutions which are most easily recognized by omitting the quantum terms with β : in this case (2.35) and (2.36) have two sets of solutions, namely

i) $C = \pm M/\sqrt{2\lambda} = \text{const.}$, $\tau^2_\alpha = \tau^2_\kappa = k^2 + \mu^2$, $\xi_\alpha(x) = e^{iKx}/\sqrt{L}$, (2.38)

with the (approximate) physical mass determined by

$$\mu^2 - M^2 - 3\lambda/2\pi \,\ln \frac{M^2}{\mu^2} .$$

This corresponds to two <u>vacuum sectors.</u>

ii) $C(x) = \pm \frac{M}{\sqrt{2\lambda}} \sinh(M(x-x_0))$ ("soliton" or "kink"), (2.39)

τ_α^2, $\xi_\alpha(x)$ = eigenvalues/eigenfunctions of a (well known) soliton stability equation, if one inserts (2.39) into (2.35), (2.36).

This clearly corresponds to the two soliton sectors.

Incorporating the terms with $\beta(x)$ leads to quantum mechanical corrections. In the vacuum sectors this has been done a long time ago by Chang [7]. Only very recently the same could be done in the soliton sectors. This is a typical self consistency problem as known from many body theory. But it is a difficult one since $\beta(x)$ from (2.32) is a finite difference of two infinite terms. For details see Altenbokum´s thesis and a recent paper [8,9].

Some of the results are presented in fig. 1-3. It is no surprise that

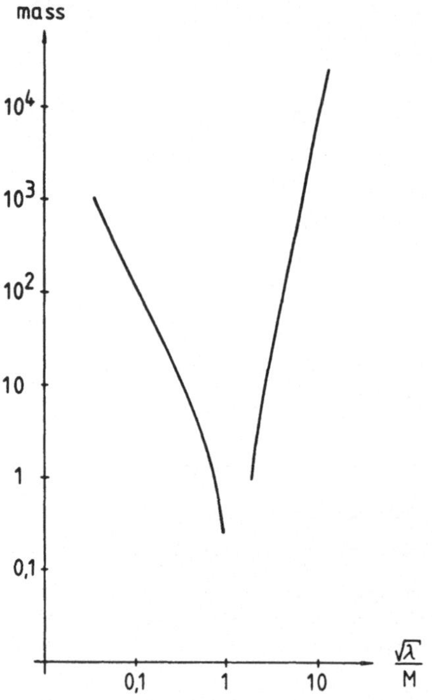

Fig. 1 Soliton mass as a function of coupling constant (for $\mathcal{H}_{sb}$).

near to the critical point as defined above the method breaks down. On the other hand, the soliton mass there goes down to zero, as it must (fig.1). Fig. 2 shows one example for the quantum correction to the soliton wave function, whereas fig. 3 represents the two discrete stability frequencies τ_0 and τ_1 as functions of the coupling constant. It is important to note that $\tau_0 > 0$, except in the limit of $M^2/\lambda \to \infty$ or 0 corresponding to the classical situation. On the other hand, translational invariance requires one $\tau_\alpha = 0$ (and all other $\tau_\alpha^2 > 0$, to have stability). This is closely related to the fact that there is a free parameter x_0 in the soliton wave function C(x) characterizing it´s center: all positions are equivalent and no energy is needed to shift the soliton. But this fact does not mean that a Hartree procedure is impossible: like in the HF approximation for many Fermion systems one may live with this symmetry violation. There are procedures to correct for it, even withim CCM. From (2.26) it is seen that using the Hartree method implies that all τ_α must be $\neq 0$. One also may separate out the center of mass coordinates. The latter method has been explored by Christ and Lee [10,11] who arrive at a form for the Hamiltonian with neatly separated center of mass and relative momenta. The price they

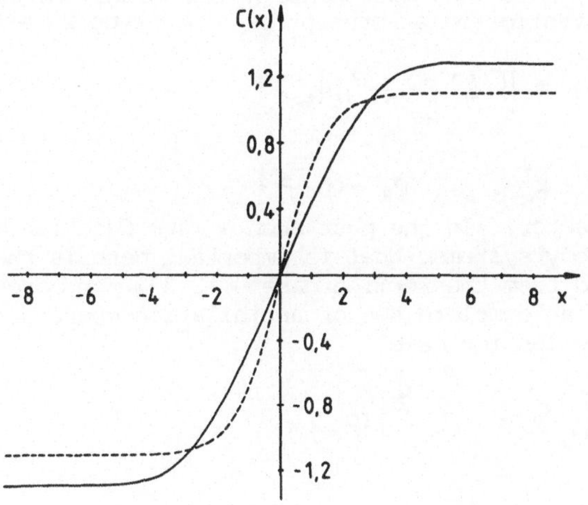

Fig. 2 Soliton solution for $\sqrt{2\lambda}/M = 0.9$.
Solid line: Hartree approximation;
dashed line: classical soliton

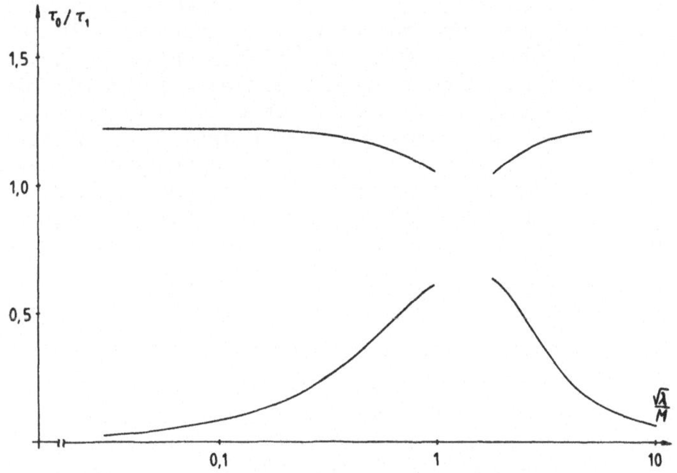

Fig. 3 Discrete stability frequencies as functions of coupling constant

have to pay is the occurrence of a power series in the coupling constant in the Hamiltonian itself. This probably can be handled only by perturbation theory.

Summarizing the Hartree approximation for bosons it is seen that it does not work in the "critical region", i.e. in a region near to what is guessed to be the critical point. Indeed doing a CCM calculation using the Hartree basis [4,12] in both sectors leads to trouble: the techniques working so well outside this region completely fail inside it.

One may still hope to find a way out already using only a starting wave function ϕ, which now must contain high order correlations. One idea which is being investigated at present is a BCS type wave function

$$|\hat{\phi}_b\rangle = \prod_k \left(A_k^+ \frac{v_k}{u_k} + 1\right)|\phi_b\rangle, \tag{2.39}$$

with

$$A_k^+ = b_k^+ b_{-k}^+ , \quad v_k^2 + u_k^2 = 1 \tag{2.40}$$

(for the vacuum sector). In the past such a wave function has often been applied to many body systems. What is important here is the fact that - as distinguished from the fermion case - (2.39) for bosons is no "new vacuum": there is no complete set of annihilation operators. On the other hand - like in the fermion case -

$$|\hat{\phi}_b\rangle = e^{\hat{S}}|\phi_b\rangle, \tag{2.41}$$

with

$$\hat{S} = \sum_{n=0}^{\infty} \hat{S}_{2n} , \quad \hat{S}_{2n} = \frac{1}{n!} \sum_{k_1 \ldots k_n} S^{(n)}_{k_1 \ldots k_n} A^+_{k_1} \ldots A^+_{k_n}, \tag{2.42}$$

and all $S^{(n)}$ of comparable size. Thus $|\hat{\phi}_b\rangle$ is a highly correlated wave function. It is the only one known besides the Hartree wave function leading to an explicit expression for the expectation value of the energy. The procedure proposed here is a Hartree-BCS variational principle, where both the Hartree parameters (here C and μ^2) as well as the BCS-parameters are varied. This has not yet been exploited with the full range of parameters. But, using only C and v_k, it has been found that in some regions of parameter space substantially lower energies are obtained. The minima are still the same as in the Hartree case, however. It is hoped that using all parameters there will be a decrease in energy and that the first order phase transition will vanish. This then would be a better starting wave function near to the critical point. Since $|\hat{\phi}_b\rangle$ is not a "new vacuum" it will be somewhat harder to implement CCM afterwards. But the procedure is well defined even in this case. Future calculations will show whether this is a feasible approach to the critical point problem and whether it can be used in more general and more realistic systems as well.

3. Many Fermion Systems

Many fermion systems - with one exception - in the past have not posed any problems regarding a suitable choice of ϕ.

a) Nuclei: for finite nuclei the oscillator basis works very well. The CCM calculations done with this basis [13] show a very weak dependence on the frequency ω, reflected in the smallness of the S_1. Note that for fermions the Thouless theorem

$$|\phi'\rangle = \exp(\underline{S}_1)|\phi\rangle \tag{2.43}$$

defines the most general determinant not orthogonal to ϕ . ($\underline{S}_1$ creates particle-hole pairs). Small $\underline{S}_1$ therefore implies that ϕ was a good choice. Doing a "maximum overlap" calculation imposing

$$|\langle\psi|\phi\rangle| = \max. \tag{2.44}$$

(leading to $\underline{S}_1$=0) therefore had very little influence in nuclear physics.

b) homogeneous fermi liquids (electron gas [14,15], nucleon matter [16]). Here ϕ necessarily is a Slater determinant of plane waves (momentum conservation). Some of these systems show a phase transition (to crystallization or genuine clustering [17]. Then one has to choose a new ϕ . One possible procedure has been discussed before [18].

c) Atoms and molecules: Here one may safely start from a (typically huge) Hartree-Fock calculation, at least as long as there is no degeneracy problem. CCM calculations on top of this typically have been extremely successful [19].

II. STRUCTURE OF CORRECTIONS

Returning to (1.1) one may ask how the "rest" of the wave functions looks like. A long time ago Hubbard has given a simple answer in terms of perturbation theory for fermions: the exact wave function can be written as

$$|\psi\rangle = \exp(\underline{S})|\phi\rangle, \tag{3.1}$$

with

$$\underline{S} = \sum_{n=0}^{\infty} \underline{S}_n \; , \; \underline{S}_n = \frac{1}{n!}\sum_{\nu_1..\nu_n}\sum_{\rho_1..\rho_n} \langle \rho_1..\rho_n|S_n|\nu_n...\nu_1\rangle a^+_{\rho_1}..a^+_{\rho_n} a_{\nu_n}..a_{\nu_1} \; . \tag{3.2}$$

Here $\langle \rho_1..\rho_n|S_n|\nu_n...\nu_1\rangle = \sum$ all linked (Goldstone)diagrams with

n incoming hole lines (labelled by ν)
n outgoing particle lines (labelled by ρ) (3.4)

One can also well use a more physical approach [19,20]. By definition Ψ contains the Slater determinant ϕ of occupied states. In addition to this there will be contributions where two particles by interacting with each other throw each other out of the Fermi sea. This may be described by $\underline{S}_2|\phi\rangle$, where S_2 creates two "particles" and two "holes". The contribution from two pairs of particles throwing each other out of the Fermi sea independently will be $\frac{1}{2}\underline{S}_2^2|\phi\rangle$. The factor $\frac{1}{2}$ is needed to count each pair only once. Correspondingly the contribution from three pairs doing the same will be $\frac{1}{3!}\underline{S}_2^3|\phi\rangle$. Summing up the contributions from all independent pairs leads to $\sum \frac{1}{n!}\underline{S}_2^n|\phi\rangle = \exp(\underline{S}_2)|\phi\rangle$.

If now triplets of particles are doing the same thing, one has to include all $\underline{S}_3|\phi\rangle$, $\frac{1}{2!}\underline{S}_3^2|\phi\rangle$.. etc., thus $|\psi\rangle = \exp(\underline{S}_2+\underline{S}_3)|\phi\rangle$. Finally, include all quadruplets, etc. and one has $|\psi\rangle = \exp(\underline{S}_2+\underline{S}_3+\underline{S}_4...)|\phi\rangle$.

But it also may well happen that a single particle or several single particles independently are thrown out of the Fermi sea by interaction with the other particles, leading to exp(S_1) in the same way. This is just a physical interpretation of the Thouless theorem (2.43). Summarizing this result it is seen that the exponential form of the wave function is not only exact, but also is "natural". It therefore should be and can be used directly; this is the essence of the CCM. For more details see the reviews [13,20,21,22].

Acknowledgements: The author has greatly benefitted from many discussions with R. Bishop going on now for several years. He wants to thank him and the Mathematics Department of the University of Manchester Institute for Science and Technology, for their generous hospitality and support.

REFERENCES

1. C. S. Hsue, H. Kümmel, Y.Y. Lee and Chyi Lung Lin, unpublished
2. C. S. Hsue and J.L. Chern, Phys.Rev.D29 463(1984)
3. C. S. Hsue, to be published
4. C. S. Hsue, H. Kümmel, P. Ueberholz, Phys.Rev.D., in press
5. B. Simon, R.G. Griffiths, Com.Math.Phys.33 145(1973)
6. J. Glimm, A. Jaffe, Quantum Physics, Springer-Verlag, Berlin-New York 1980
7. J. Chang, Phys. Rev.D12,1071(1973); D13,2778(1976)
8. M. Altenbokum, Thesis 1984
9. M. Altenbokum and H. Kümmel, Phys.Rev.D, in press
10. N. H. Christ and T.D. Lee, Phys.Rev.D12,1606(1975)
11. R. Rajaraman, Solitons and Instantons, North Holland, Amsterdam 1982
12. M. Altenbokum and U. Kaulfuß, Nuov.Cim., in press
13. H. Kümmel, K.H. Lührmann, J.G. Zabolitzky, Phys.Rep.36C 1(1978)
14. R. F. Bishop, K.H. Lührmann, Phys.Rev.B17 3757(1978), B26 5523(1982)
15. K. Emrich, J.G. Zabolitzky, Phys.Rev.B29 2049(1984)
16. B. Day, Phys.Rev.C24 1203(1981)
17. R. Bishop, preprint
18. H. Kümmel, Nucl.Phys.A317 199(1979)
19. K. Szalewicz, J.G. Zabolitzky, B. Jeziorski and H.J. Monkhorst, J.Chem.Phys.81 2723(1984)
20. H. Kümmel, Proc.Int.Summer School of Nucleon-Nucleon Interactions and Nuclear Many Body Problems, Changchun, China 1983, S.S.Wu and T.T.S. Kuo, Editors, World Scientific Publishing Co., Singapore
21. R. Bishop and H. Kümmel, preprint 1985
22. V. Kvasnička, V. Laurinc, S. Biskupič, Phys.Rep.90 160(1982)

THE LMG MODELS AS A MANY BODY PROBE

A.P. Zuker[+], M. Dufour[+] and C. Pomar[++]

[+]Physique Nucléaire Théorique, CRN, 67037 Strasbourg Cedex
France
[++]TANDAR, CNEA Av. del Libertador 8250
1429 Buenos Aires, Argentina

ABSTRACT

A general theory for the ground and low excited states of a many body system is applied to the LMG model. Some valuable indications emerge suggesting how to tackle more complicated cases.

The aim of Linked Cluster Theory[1,2,3] (LCT) is to transform the linear Schödinger problem into a set of coupled non linear equations. In these notes we shall study the solution of the LMG model[4], to illustrate why the truncation of a linear system must involve non linear effects. Emphasis will be put on those features of the calculations whose relevance goes beyond LMG.

Section I reviews very briefly the model and the linear equations are derived. In Section II it is seen how the coupled cluster (expS)[5,6] form of the wavefunction emerges naturally for the ground state (GS), while excited states (ES) demand a degenerate treatment. In Section III we introduce LCT and show how to implement it. In section IV we deal with the critical behaviour of the model. Section V contains some numerical results. Section VI comments on possible extensions. Note : often used notations have double underline when they first appear.

I) THE SU2 AND LMG MODELS

Consider two shells p and h, whose orbits pm and hm have the same quantum numbers, m=1,...,N. The operators

$$S_+ = \sum_m a^+_{pm} a_{hm} \qquad S_- = \sum_m a^+_{hm} a_{pm} \qquad S_o = \sum_m \left(a^+_{pm} a_{pm} - a^+_{hm} a_{hm}\right) \qquad (1)$$

obeys commutation rules

$$[S_+,S_-]= S_o \quad [S_o,S_+]= 2S_+ \quad [S_o,S_-]= \bar{2}S_- \tag{2}$$

The most general, linear plus quadratic, Hermitian Hamiltonian is

$$H= \varepsilon S_o+\eta(S_++S_-)+V_oS_o^2+V_1\left[S_o(S_++S_-)+(S_++S_-)S_o\right]+$$

$$+V_2(S_+^2+S_-^2)+W(\tfrac{1}{2}S_o^2+S_+S_-+S_-S_+) \tag{3}$$

If the system contains N particles, H spans the vector space

$$S_+^k|0>,\ (S_-|0>=0)\ ;\ <0|S_-^kS_+^k|0>=N^{(k)}k!\quad N^{(k)}=N(N-1)\ldots(N-k+1) \tag{4}$$

A general wavefunction can be written as

$$\left[1+A_1S_++A_{-1}S_-+\ldots+A_m\frac{(S_+)^m}{m!}+A_{-m}\frac{(S_-)^m}{m!}+\ldots\right]S_+^k|0>= |\bar{k}> \tag{5}$$

As $|0>$ is the closed h shell, S_+^k creates a k particle-k hole state (kp-kh). The choice of normalization in (5) is intended to privilege this state as a " pivot " from which we shall evolve into the exact " dressed" state $|\bar{k}>$, by means of the " wave operator " $\hat{A}$ (square brackets in eq. (5)). This establishes (forces) a one to one correspondence between the unperturbed an exact wavefunctions. The hope is that few amplitudes A_m with m small will prove sufficient to characterize the state.

In the LMG model only ε and V_2 are different from zero in (3). Only even amplitudes appear in this case and using the value for the norm in eq. (4) it is elementary to obtain the equations,

$$E_kA_\mu=V_2\mu^{(2)}A_{\mu-2}+\left[2(k+\mu)-N\right]\varepsilon A_\mu+V_2\frac{(k+\mu+2)^{(2)}}{(\mu+2)^{(2)}}(\bar{k}-\mu)^{(2)}A_{\mu+2} \tag{6}$$

for $\mu>0$. Here $\bar{k}=N-k$. E_k is the energy.

The equations for $\mu<0$ are obtained by changing signs in the subindeces and interchanging k and $\bar{k}$.

For $\mu=0$ we have

$$E_k= \tfrac{1}{2}V_2\left[k^{(2)}(\bar{k}+2)^{(2)}A_{-2}+\bar{k}^{(2)}(k+2)^{(2)}A_2\right]+(2k-N)\varepsilon=\Delta_k+(2k-N)\varepsilon \tag{7}$$

The amplitudes A_m are a function of k and the notation ${}_kA_m$ will be used if confusion may arise.

II) THE expS BEHAVIOUR FOR GS AND THE TROUBLE WITH ES.

We start by examining eq. (6) for the GS, i.e. k=0. After replacing the value for E in eq. (7) we obtain

$$\left[(N-m)^{(2)}A_{m+2}-N^{(2)}A_2A_m\right]V_2+2m\varepsilon A_m+m^{(2)}V_2A_{m-2}=0 \tag{8}$$

Adopting a perturbative view we neglect at first the terms in brackets. Then

$$A_m=-\frac{m-1}{2}\frac{V_2}{\varepsilon}A_{m-2},\quad A_2=-\frac{V_2}{2\varepsilon}\ \therefore\ A_m=(m-1)!!A_2^{m/2} \tag{9}$$

from which it follows that in this approximation the wave operator is

$$\hat{A} = 1+\sum A_{2n}\frac{(S_+^2)^n}{2n!}=1+\sum A_2^n\frac{(2n-1)!!2^n}{2n!}\left(\frac{S_+^2}{2}\right)^n=1+\sum\frac{A_2^n}{n!}\frac{(S_+^2)^n}{2}=\exp\left(A_2\frac{S_+^2}{2}\right) \tag{10}$$

and a hint of the expS behavior emerges. Analysing the general many body equations one can see why the wave operator for the GS must have this form : to ensure the correct (linear) dependence of E on the number of particles[1]. In LMG, this dependence is trivially obtained since the correlation energy Δ_o (square brackets in eq. (7)) will turn out to be a constant. Therefore the expS form is not necessary for the wave operator to make sense but as we shall see it is quite sufficient. The important point eqs. (9) and (10) bring home is that a naïve truncation never makes sense : the A_{m+2} amplitude may be small enough to be neglected in eq. (8) but eq. (9) says it cannot be null.
Let us now try the general replacement

$$A_4= 3A_2^2+ A_4^L,\qquad A_6= 15A_2^3+ 15A_4^LA_2+ A_6^L,\quad \text{etc.} \tag{11}$$

The superscript L stands for Linked and means irreducibly non factorable. (A_m^L is the same as S_m in expS jargon). It is left as an exercise to show that the coefficients are the correct ones (otherwise see refs.(1) or (3)).

Replacing in eq. (8) for m=2 we obtain immediately the " A_2 " equation

$$(2N^{(2)}-12N+18)A_2^2V_2+2V_2+4\varepsilon A_2+(N-2)^{(2)}V_2A_4^L=0 \tag{12}$$

and after a little more work we obtain the A4 equation

$$\frac{8\varepsilon}{V_2} A_4^L+(8N^{(2)}-96N+264)A_4^L A_2-(48N-192)A_2^3+N-4)^{(2)}A_6^L$$

$$+\ 6A_2\left[(2N^{(2)}-12N+18)A_2^2+2+\frac{4\varepsilon}{V_2}A_2+(N-2)^{(2)}A_4^L\right]=0 \qquad (13)$$

The second line vanishes identically since it is the A_2 equation. To understand what has been achieved we need some dimensional analysis. Now : ε is a constant independent of N and the sensible choice for V_2 is an N^{-1} dependence. Not only because this is what we expect of a two body matrix element in a many particle system but also because it is the only choice that makes eq. (12) dimensionally homogeneous. In other words : if $V_2=0(N^{-1})$, then $A_2=0(N^{-1})$ and the first 3 terms in eq. (12) have the same N-dependence. What about A_4^L ?. Using $A_2=0(N^{-1})$, we see that eq. (13) dictates $A_4^L=0(N^{-3})$. This is an extraordinary result that shows that for large N eq. (12) gives the exact A_2 when A_4^L is neglected. It hinges on the fact that the coefficient of A_2^3 in eq. (13) is linear since the quadratic term cancels. Notice in eq. (13) that the first line contains only terms of $0(N^{-2})$, while the second line contains only terms of $0(N^{-1})$. Fortunately it vanishes. This is the essence of linked cluster ideas : equations must be dimensionally homogeneous.

So far our results coïncide with those of Lührmann[6], and our elementary arguments are meant to be the simplest possible introduction to the coupled cluster method of Coester and Kümmel.

Things get more complicated when we face excited states as can be seen by examining the A_2 and A_{-2} equations for arbitrary k. We have from eqs. (6) and (7)

$$2V_2+4\varepsilon A_2+V_2(\bar{k}-2)^{(2)}\ \frac{(k+4)^{(2)}}{12}\ A_4-\frac{V_2}{2}\left[k^{(2)}(\bar{k}+2)^{(2)}A_{-2}+\bar{k}^{(2)}(k+2)^{(2)}A_2\right]A_2=0$$

$$2V_2-4\varepsilon A_{-2}+V_2(k-2)^{(2)}\ \frac{(\bar{k}+4)^{(2)}}{12}\ A_4-\frac{V_2}{2}\left[k^{(2)}(\bar{k}+2)^{(2)}A_{-2}+\bar{k}^{(2)}(k+2)^{(2)}A_2\right]A_{-2}=0 \qquad (14)$$

The coupling between A_2 and A_{-2} is a slight complication. The true problem comes when we try to make the equations homogeneous.

The correlation energy Δ_k (square brackets in (14)) must be a constant in N. As k grows the equations become increasingly symmetric and by the time $k=N/2$, $E=\Delta=0$ and $A_2=-A_{-2}$. For large k, A_2 and A_{-2} must be $0(N^{-2})$ as can be seen by either requesting a constant Δ or by arguing that if we write the wavefunction in terms of an <u>orthonormal</u> basis the amplitudes must be at most 0(1) (do as exercise). If A_2 or A_{-2} are $0(N^{-2})$ the second term in each of eqs. (14) is negligible and A_4 or A_{-4} must be a <u>negative</u> constant (independent of V and ε) of $0(N^{-4})$ to cancel

$2V_2$. Therefore, in leading order in (large) N we can determine A_4 and A_{-4} quite well, but A_2, A_{-2} and consequently Δ_k are undertermined in the lowest approximation.

For small k the situation is no better except in the very important case of k=1. Then $A_{-2}=0$ and the A_2 equation is (we write ${}_1A_2$ to remind that k=1)

$$(2N^{(2)}-24N+54)\,{}_1A_2^2V_2+2V_2+4\varepsilon\,{}_1A_2+\frac{5}{3}(N-3)^{(2)}\,{}_1A_4^L=0 \tag{15}$$

which becomes identical to (12) for large N ; ${}_1A_2 \simeq A_2$. If A_4^L and the linear and constant coefficients of A_2^2 are neglected the A_2 and ${}_1A_2$ equations are the RPA approximation.

For larger k the factorizations (11) become increasingly meaningless and no simple improvement is available.

The k=1 result suggests that the general many body equations be examined, bearing in mind the Green's function methods in which it is either claimed or proved (the authors are not sure) that knowledge of the GS determines the ES. Equations (14) and (15) say essentially the same thing. (For k=1!).

For future reference we give the solutions for large N

$$NA_2r=-1+(1-r^2)^{1/2} \quad ; \quad r=\frac{NV_2}{\varepsilon}$$

$$E_o=\varepsilon(-N+NrA_2) \tag{16}$$

$$E_1=\varepsilon(-N+2+3NrA_2)=2\varepsilon(1-r^2)^{1/2}+E_o$$

Notice that the correlation energy is 0(1) and hence an unimportant contribution to the total E_o, but a crucial one to the excitation energy. When r=1 the latter goes to zero and for r>1 the equations no longer make sense since A_2 becomes imaginary. More on this in Section IV.

III) IMPLEMENTATION OF LCT

Any formalism for excited states leading to dimensionally homogeneous equations qualifies as LCT. Several variants have been proposed by the Bochum School. The work of Emrich[7] analyses them and contains an important new scheme. Arponen[8] proposes an elegant extension of the exp(S) method to deal with excited states. None of these attempts tackles in a general way the problem of degeneracy which arises when the pivot (i.e. the unperturbed state) is not determinantal state (as in eq. (5) for k≠0). In this case we have a family of determinants, |i> say, which the interaction does not split. To obtain a general theory we start by separating the Hilbert space in two subspaces :

$$|\bar{i}\rangle = |i\rangle + \sum_j A_{ij}|j\rangle \qquad \langle\bar{j}| = \langle j| - \sum_i A_{ij}\langle i|$$
$$\langle i|\bar{i}'\rangle = \delta_{ii'} \qquad \langle\bar{j}|j'\rangle = \delta_{jj'} \qquad \langle\bar{j}|\bar{i}\rangle = 0 \tag{17}$$

We call the i states the model space while the j states are all the others. Now we request that $\langle\bar{j}|H|\bar{i}\rangle=0$ to obtain

$$\varepsilon_{ij}\langle i|\hat{A}|j\rangle=\langle i|V|j\rangle+\langle i|\hat{A}|J\rangle\langle J|V|j\rangle-\langle i|V|I\rangle\langle I|\hat{A}|j\rangle-\langle i|\hat{A}|J\rangle\langle J|V|I\rangle\langle I|\hat{A}|j\rangle$$

$$\text{with} \quad \varepsilon_{ij}=\langle i|H|i\rangle-\langle j|H|j\rangle \quad , \qquad A_{ij}=\langle i|\hat{A}|j\rangle \tag{18}$$

The summation convention is used over repeated capital indeces, V is the non diagonal part of H.

If there is a single i state the last term in (18) reduces to $A_{ij}\Delta$, where Δ is the correlation energy and we have the Schrödinger equation. (e.g. eq. (6)). The dressed $|\bar{i}\rangle$ states are now decoupled from the rest of the space and it only remains to diagonalize in the model space the matrix

$$\langle i|H|\bar{i}\rangle= \langle i|H+H\hat{A}|i\rangle= \langle i|H_{eff}|i\rangle \tag{19}$$

in which the bare Hamiltonial H is replaced by an effective operator containing all the possible contractions of $H\hat{A}$ that stay within the (bare!) model states.

Equation (18) is equation (37) of ref.[1] where more details can be found. To implement it for LMG we start by recognizing that $S_+^k|0\rangle$ is made of $N^{(k)}/k!$ different kp-kh determinants that we choose as model space. The most general wave operator $\hat{A}$ that can take us out of the model space has the form

$$\hat{A}= \sum A_{20}S_r^+S_s^+ + A_{02}S_r^-S_s^- + A_{31}S_r^+S_s^+S_t^+S_u + A_{13}S_r^+S_sS_tS_u \quad \ldots.$$

$$= \sum_{\neq(r,s,t\ldots)} A_{\lambda\mu}\underbrace{S_r^+ \ldots S_s^+}_{\lambda}\underbrace{S_t^- \ldots S_u^-}_{\mu} \qquad \lambda\neq\mu \tag{20}$$

The sum extends over all possible orbits r,s,t,u, which are taken to be different. The operators $S_r^{\pm}$ are the partial jump operators that enter the sums in eq. (1). The coefficients $A_{\lambda\mu}$ are independent of r,s etc. and $\lambda+\mu$=even because of the symmetries of LMG. For the more general H in (3) we must include terms with $\lambda+\mu$=odd.

It only remains to evaluate the matrix elements in eq. (18) and the only difficulty comes from the last term. Let us examine how to write its general contribution. Start from $\langle i|$, which is an arbitrary kp-kh determinant. Then $\langle i|\hat{A}|J\rangle$ can only be of the form A_{pp+2} or A_{p+2p} since otherwise V could not reach back a model state I. These model states are obviously kp-kh determinants that differ in p-i orbits (i=0,1,2) from the $\langle i|$ state. Now the $A_{\lambda\mu}$ matrix element that goes from a given I state to $|j\rangle$ is fixed. The $A_{\ell m}$ equation is the one for which the $|j\rangle$ state is obtained by acting on $\langle i|$ with any of the operators associated with $A_{\ell m}$ in eq. (20). Then, it is not difficult to obtain

$$\langle i|\hat{A}|J\rangle\langle J|V|I\rangle\langle I|\hat{A}|j\rangle= \frac{\ell!m!}{\bar{k}^{(\ell)}k^{(m)}} \sum_{\lambda\mu p} A_{\lambda\mu}(A_{pp-2}C_{\lambda\mu p}+A_{p-2p}D_{\lambda\mu p}) \tag{21}$$

$$C_{\lambda\mu p}= \sum_{\kappa,\rho,i} (1+\delta_{i1}) \frac{\bar{k}^{(p)}(\bar{k}-p+i)^{(\lambda-\kappa)}}{(p-2)!(\lambda-\kappa)!} \frac{(p-i)^{(\kappa)}}{\kappa!} \frac{(p-i)^{(\rho)}}{\rho!} \frac{(k-p+i)^{(\mu-\rho)}}{(\mu-\rho)!} \frac{k^{(p-i)}}{(p-i)!}$$

$$\ell=\lambda+p-i-\kappa-\rho \qquad m=\mu+p-i-\kappa-\rho \qquad i=0,1,2$$

$$\kappa=0\ldots\min(\lambda,p-i) \qquad \rho=0\ldots\min(\mu,p-i)$$

$D_{\lambda\mu p}$= same as $C_{\lambda\mu p}$ but interchange $k \leftrightarrow \bar{k}$ in third and last factor.

The other terms in eq. (18) are totally straightforward to evaluate and we write directly the first few equations.

$$2+\frac{4N}{r}A_{20}+(\bar{k}-2)^{(2)}A_{40}=A_{20}A_{20}\left[\bar{k}^{(2)}+4k(\bar{k}-1)+k^{(2)}\right]$$
$$+6A_{20}A_{02}k^{(2)}+A_{20}A_{31}\,4\left[k(\bar{k}-1)^{(2)}+k^{(2)}(\bar{k}-2)\right]+\ldots \tag{22}$$

$$2-\frac{4N}{r}A_{02}+(k-2)^{(2)}A_{04}=A_{02}A_{20}6\bar{k}^{(2)}+A_{02}A_{02}\left[\bar{k}^{(2)}+4\bar{k}(k-1)+k^{(2)}\right]+A_{02}A_{42}\frac{\bar{k}^{(4)}}{2}$$
$$+\ldots \tag{23}$$

$$2A_{40}(\bar{k}-3)+\frac{4N}{r}A_{31}+A_{51}(\bar{k}-3)^{(2)}=A_{20}A_{20}|6(\bar{k}-1)+6(k-1)|+A_{02}A_{20}12(k-1)$$
$$+\ldots \tag{24}$$

$$2A_{40}+A_{51}4(\bar{k}-4)+\frac{4N}{R}A_{42}+A_{62}(\bar{k}-4)^{(2)}=12A_{20}A_{02}+12A_{20}A_{20}+$$
$$+12\bar{k}^{(2)}A_{20}A_{42}+\bar{k}^{2}A_{02}A_{42}+\ldots \tag{25}$$

For moderate values, k=0(1) and $\bar{k}$=0(N) and we can carry a dimensional analysis. The A_{20} eq. (22) is identical to the A_2 eqs. (12) and (15) to $0(N^{-1})$. Coupling to A_{02} and A_{31} comes to $0(N^{-2})$ and is negligible (the second line is $0(N^{-2})$ while the first is mainly 0(1)). The A_{02} eq. (23) couples strongly with A_{20}. There is also a term in A_{42} in leading order, which can be treated very rigorously through eq. (25) which shows that no new amplitudes appear (in leading order) : the A_{62} amplitude factorizes approximately as $10A_{42}A_{20}+A^{L}_{62}$. The A_{31} es. (24) is trivial to incorporate. For k=0 only A_{20} may appear. For k=1 we add A_{31}. For k=2 we add A_{02} and A_{42}. For k=3 we add A_{13} and A_{53} etc. The rule that emerges is that as we bring in a pair of new amplitudes their calculation in leading order involves only themselves and amplitudes that have appeared before. In spite of the formidale looks of eq. (21) the computations can be practically done by hand.

The effective interaction $H+H\hat{A}$ contains terms in $A_{20}S_-^2S_+^2$, $A_{02}S_+^2S_-^2$, $A_{31}S_-^3S_+^3$ etc. Some care must be taken in the contraction $H\hat{A}$ to respect the condition that all orbits be different in eq. (20). It is left as an exercise to check that using

$$S_-^m|k> = k^{(m)}(\bar{k}+m)^{(m)}|k> \tag{26}$$

$$(N-m)^{(k)} = \sum(-)^{\kappa}\binom{k}{\kappa}(m+\kappa-1)^{(\kappa)}N^{(k-\kappa)} \tag{27}$$

we can derive by induction the correlation energies

$$\Delta_k = \frac{V_2}{2}\bar{k}^{(2)}(k+2)^{(2)}\left[A_{20}+\frac{2}{3!}\,k(\bar{k}-2)A_{31}+\ldots+\frac{2}{(\lambda+2)!\lambda!}k^{(\lambda)}(\bar{k}-2)^{(\lambda)}A_{\lambda+2\lambda}\right]$$
$$+\frac{V_2}{2}k^{(2)}(\bar{k}+2)^{(2)}\left[A_{02}+\ldots+\frac{2}{(\lambda+2)!\lambda!}\bar{k}^{(\lambda)}(k-2)^{(\lambda)}A_{\lambda\lambda+2}\right] \tag{28}$$

This expression should be compared with eq. (7), from which we can identify A_2 and A_{-2} as a function of the new amplitudes. The terms in square brackets are $O(N^{-1})$ except for those in $A_{\lambda+2\lambda}$, λ odd, which are $O(N^{-2})$. It means that as we go up in the spectrum it is increasingly dangerous to neglect amplitudes beyond A_{20} and A_{02}. Fortunately there is an expansion in the coupling constant r (NV/ε) *that can be generalized as an expansion in* A_{20}. Our discussion after eq. (25) shows that $A_{02}=O(A_{20})$, $A_{42}=O(A_{20}^3)$ etc. Since NA_{20} is always smaller than 1 (eq. (16)), it may be expected that for smallish r we can do with few terms in eq. (28). This is an important result that indicates in a very precise sense that LCT (and a fortiori coupled cluster exp(S) generalise Perturbation Theory (PT) while retaining the feature of being (perturbative !) expansions in two small parameters : N^{-1} and NA_2. The crucial point is that in PT one hopes for an expansion in r while now we prove the existence of an expansion in NA_{20}.

The effective interaction obtained in eq. (28) is both non hermitian and many body, i.e., the worst we could expect. In ref.[1] it is extensively discussed why one should not worry about non hermiticity and how to deal with it. H_{eff} will give not only the lowest energy in the kp-kh space (eq. (28)), but all the others. Its many body nature makes it strictly k independent in the model space and the discussion after eq. (25) ensures that to leading order in N^{-1} it is also k-independent. Ref.[2] discusses extensively the way to avoid altogether a full degenerate treatment by dealing with a state dependent H_{eff}. Then one recovers strictu senso exp(S) but the degeneracies show through homogeneous terms in the coupled cluster equations. An interesting prospect worth exploring.

IV) HOW TO DEAL WITH CRITICAL BEHAVIOUR AT $r \geqslant 1$

Eq. (16) shows that for r=1 our equations break down and we are reminded of Thouless theorem stating that imaginary RPA energies signal instability of the Hartree-Fock (HF) solutions. Since Kümmel's contribution in this volume deals with this problem, which is also discussed in ref.[3] we shall be very brief and only indicate how to get excited states and then make some conceptual points.

Consider a canonical transformation

$$a_p^+ \equiv \alpha a_p^+ + \beta e^{i\phi} a_h^+ \qquad a_h^+ \equiv \alpha a_h^+ - \beta e^{-i\phi} a_p^+ \tag{29}$$

$$\alpha^2+\beta^2=1 \qquad \underline{\underline{\alpha^2-\beta^2=a}}\ , \qquad \underline{\underline{2\alpha\beta=b}}\ , \qquad a^2+b^2=1, \qquad \phi=0 \text{ or } \pi/2$$

The second line defines a and b and restricts ϕ to the only interesting choices.

The transformed Hamiltonian is

$$H_{LMG}=V_2\left\{\frac{1}{2}(a^2+1)(S_+^2+S_-^2)-\frac{1}{2}ab\left[S_oS_1+S_1S_o\right]\pm b^2\left(\frac{3}{4}S_o^2-\frac{1}{2}\hat{S}^2\right)+\frac{N}{r}(aS_o\mp bS_1)\right\} \tag{30}$$

upper sign and $S_1=S_++S_-$ if $\phi=0$; lower sign and $S_1=i(S_+-S_-)$ if $\phi=\pi/2$, $\hat{S}=S_+S_-+S_-S_+-\frac{1}{2}\ S_o^2$ is the Casimir of SU2 with eigenvalue N(N+2)/2. This Hamiltonian has the form (3), and equation (6) for k=0 generalizes to

$$A\mu E=\mu^{(2)}A_{\mu-2}+\left[(2\mu-N)\varepsilon-(2\mu-N)^2V_o\right]A_\mu+(N-\mu)^{(2)}V_2A_{\mu+2}+$$
$$\mu A_{\mu-1}\left[\eta+2(2\mu-1-N)V_1\right]+(N-\mu)\left[\eta+2(2\mu+1-N)V_1\right]A_{\mu+1}=0 \tag{31}$$

We can always choose to set $A_1=0$. This is called the maximum overlap condition. It is easily checked that in leading order it is satisfied by

$$\eta-2(N-1)V_1=0=\mp b\ \frac{N}{r}-(N-1)ab \cong bN\left[\frac{1}{|r|}-a\right]=0 \tag{32}$$

where we have used the values in (30) and then resolved the sign ambiguity so that a>0. For $|r|<1$ only b=0 can satisfy eq. (32) i.e. we make no transformation. For $|r|>1$ we need the $a=|r|^{-1}$ solution and a is taken positive to ensure continuity at $|r|=1$. Condition (32) is identical to HF. When enforced, the terms in S_1 in (30) become negligible for the GS calculation and we have a problem of the same form as before except for the appearance of S_o^2 in the diagonal matrix elements (which helps !) and the solution of the A_2 equation is again exact for large N.

For excited states the situation is similar except that to recover the untransformed form we have to cancel A_1 and A_{-1} for the chosen k. To leading order this is achieved by the transformation $a_k = [|r|(N-2k)]^{-1}N$ (left as exercise), which implies that H becomes k dependent but this is no problem.

We end up with a simple recipe to calculate excited states in low order : insert a_k in eq. (30), discard terms in S_1 and proceed as in the previous section.

The conceptual point that deserves attention is that eq. (32) has two solutions since $b = \pm(1-a^2)^{1/2}$. The Hamiltonian is the same to the extent the linear terms in b have been suppressed (an exact result for k=0 and large N). So we conclude that each time we calculate a state, there is another one degenerate with it : e.g. for k=0 we get the two states of eq. (16) at r=1, which will stay degenerate for r>1. For k=1 we get at r=1 the states we had called before k=2 and 3 etc. The numerical examples in next section illustrate the point quite clearly. At some stage the strict degeneracy will be broken : either because $a_k > 1$ at $k = N(1-r^{-1})/2$ and we have to revert to $b_k = 0$, or because the neglected terms in S_1 start playing a role. However, there is an interesting situation where these effects cannot be invoked, in which the degeneracy is still broken : $r=\infty$. Then we have $a=0$, $\alpha=\beta$ and a single transformation (changing the sign in b simply means changing p into h, which does not matter since the orbits are degenerate). Since the diagonal energies go as $(N-2k)^2$, both states k=0 and k=N lie degenerate at the lowest energy. Notice that the choice of phase in eq. (30) ensures that the S_o^2 terms come with a negative definite sign. Otherwise instead of the lowest two states we would obtain the highest two. But precisely : in LMG if E is eigenvalue so is -E (prove as exercise), and if we want to calculate the highest energies we have to choose in eq. (30) the phase ϕ opposite to that of the GS. So the lower end of the spectrum goes up in energy as $-\frac{3}{4}(N-2k)^2 + \frac{1}{4}N(N+2)$ while the upper end goes down as $\frac{3}{4}(N-2k)^2 - \frac{1}{4}N(N+2)$, k=0 corresponds to both the lowest and highest states and each k value represents two states. In the middle of the spectrum we can't have two different H to represent the same levels and we switch back to the a=1, b=0. So we have a region around the centroid where there is no reason whatsoever to have degeneracy.

These qualitative considerations are sufficient to give a broad characterization of the whole spectrum for large N since we have exact or very reliable results at both ends and at the centroid (E=0). For small r the eigenvalues follow a linear pattern that persists up to r=1. Then the linear pattern bends down (up) quadratically for low (high) energy.

V) NUMERICAL RESULTS

The table compares the results of the calculations with exact diagonalizations (Ex). Only absolute energies are given. The notations are A_2=Eqs.(22,23) for $A_{20}+A_{02}$. All other. amplitudes neglected. A_{40} given by eq. (24). After setting $A_{31}=A_{51}=0$.

$A_4 \equiv A_2$ + the $A_{31}+A_{13}+A_{40}$ eqs. included.

$A_6 = A_4$ + the $A_{24}+A_{42}$ eqs. included.

Factorization of higher amplitudes when necessary are derived from their perturbative form (as in eq. (9)).

NT is the H_{LMG} (actually twice the usual definition). T is its transform eq. (30). Only shown when not obvious.

The calculations were done before eq. (21) was derived and before a consistent analysis of the couplings could be done. The discussion after eq. (25) shows (e.g.) that it is dangerous to bring in A_{13} and neglect A_{42}. Therefore the hierarchy chosen is faulty, which explains why for k=3 we get 2 less significant figures than for k=2.

The most interesting aspect of the calculations is not the drive for decimals through higher amplitudes but the remarkable success of A_2 when pushed beyond k=0 in spite of the simplicity of the approximation (it can be solved analytically). It should be noticed that we use the 1/N expansion to neglect higher amplitudes but we do not neglect higher order terms in the coefficients of the leading order equations (e.g. eq. (22) is solved as it stands, we do not set $(N-2)^{(2)}=N^2$etc.). It explains why n=30 behaves much better than N=300 (a number chosen because Lührmann's results indicate that this is where expS does worse). On the basis of 1/N considerations alone the result would be paradoxical.

Emrich's formulation is restricted to k=0 or odd and (consequently !) to r<1. Our lowest approximation gives slightly better results but for k=3 his more consistent treatment is rewarded by an extra decimal.

Arponen and Rankativi do not have these restrictions except for the lack of mechanism to lift degeneracies for k>0. This is a tunneling effect that we should (in principle) be able to solve. Their results are fairly good but open to the objection that their equations are too complicated. For N=30, k=0.8 they have an accuracy of 2.3 % for the excitation energy of k=5 after coupling 8 equations, while our error is 4.5 % in the A_2 approximation.

Table I : Comparison with exact diagonalization (Ex).
See text for notations.

k	r = ∞ A$_2$	r = ∞ Ex	r = 1 NT A$_2$	r = 1 Ex	r = 0.8 A$_2$	r = 0.8 Ex
0	-677.661	-677.666	-30.681	-30.628	-30.366	-30.359
1		"	-29.800	-29.601	-29.006	-28.961
2	-596.036	-596.040	-28.021	-28.173	-27.329	-27.368
3		"	-25.995	-26.537	-25.476	-25.683
4	-520.370	-520.562	-23.851	-24.740	-23.511	-23.799
5		"			-21.473	-21.870
6	-450.760	-451.387			-19.387	-19.866
7		6			-17.270	-17.800
8	-387.341	-388.773			↑	↑
9		53			i.e. Δ=-1.27	vs 1.80
10	-330.315	-333.284				
11		2.992				

N = 30

r = 0.4

k		
0	-30.080	-30.080
1	-28.223	-28.222
2	-26.329	-26.330
3	-24.402	-24.409
	...	
9	-12.389	-12.418

r = .4

A$_2$	Ex	A$_6$
-300.083	-300.083 12	-300.08312
-298.248	-298.247 46	-298.24746
-296.403	-296.408 04	.40806
-294.549	-294.564 90	-294.56382
...		...
-283.256	-283.431	

r = .8

k	A$_2$	Ex	A$_4$	A$_6$
0	-300.396	-300.39485	-300.39484	-300.39484
1	-299.178	-299.16793	-299.169	-299.169
2	-297.755	-297.90896	-297.854	-297.910
3	-296.148	-296.62016	-296.437	-296.482
4	-294.375	-295.30342		
	...			
9	-283.495	-288.35061		

N = 300

k	r = 1 NT A$_2$	r = 1 Ex	T A$_2$	r = 1.3 T A$_2$	r = 1.3 Ex
0	-300.878	-300.813	-300.878	310.51	310.53
1	-300.554	-300.317	-300.878	"	"
2	-298.918	-299.640	Imag.	308.20	308.26
3	-296.496	-298.863		"	"
4				305.86	306.12
5				"	"
6				306.45	304.14
7				"	13
8				300.97	302.37
9					302.33

As in all formulations, if we go to all orders we get all the exact eigenvalues and if we stay in low order we get more than one solution. In our case, since we have tailored the equations for each k we have to discard all solutions except the (unique) one that merges smoothly into the exact perturbative value for small r. In order words, our non-linear equations are designed so that a well defined linearization algorithm leads to the desired solution. *The real non linear problem is HF* (the A_1 equations !). In the LMG model it is very easy to do HF but in general it is not. As of now, we know of no HF calculations that have explored the hidden branch of the non linear couplings that has to be followed systematically when some parameter of the Hamiltonian decides to go critical.

VI) POSSIBLE EXTENSIONS

To comply with editorial space limitations this section is omitted. It will be made available on request. The last paragraph of Section V gives an idea of one of the topics meant to be included.

REFERENCES

1. A. Poves and A.P. Zuker, Physics Reports 71 (1981) 142
2. A.P. Zuker, Springer Lecture Notes 209 (1984) 157
3. A.P. Zuker, Anales de Fisica A81 (1985) 1
4. H.J. Lipkin, M. Meshkov, A.J. Glick, Nucl. Phys. 62 (1965) 188
5. H. Kümmel, K.H. Lührmann and J.G. Zabolitzky, Phys. Rep. 36C (1978) 1
6. K.H. Lührmann, Ann. of Phys. 103 (1977) 253
7. K. Emrich, Nucl. Phys. A351 (1981) 379, 397
8. J. Arponen, Ann. of Phys. 151 (1983) 311
   J. Arponen and J. Rantakivi, Nucl. Phys. A407 (1983) 141

SYSTEMATIC BEHAVIOUR AT LARGE DEGENERACIES IN A SOLUBLE MODEL

M.C. Cambiaggo[+], G.G. Dussel[+], and M. Saraceno

Departamento de Física, Comisión Nacional de Energía Atómica
Av. Libertador 8250
1429 Buenos Aires, Argentina

INTRODUCTION

The use of the Time Dependent Hartree-Fock (TDHF) approximation to describe the time-evolution of one body densities in nuclear systems is by now well understood. It has been tested both in realistic situations involving the collision of heavy nuclear fragments and in model systems where a detailed comparison of exact and TDHF results can be done. It has been stressed[1] that in general TDHF as a zero order approximation lacks an expansion parameter on which to base a systematic analysis of higher order corrections. However in some simple models, namely the Lipkin, Meshkov and Glick[2] (LMG) model and a one dimensional system with attractive delta-interactions such a parameter exists[1]. In this contribution we study the effects that the existence of this large N limit in the LMG model has on the calculation of eigenvalues, its 1/N corrections, and on the evolution of correlations in the exact wave function.

THE MODEL

The LMG model[2] has been extensively used to test various theories in many-body systems and we provide here one more instance of this usage.

The model caricaturizes a closed shell nucleus by distributing N

+ Fellow of the Consejo Nacional de Investigaciones Científicas y Técnicas, Argentina.

fermions in two shells of degeneracy Ω=N with a gap ε between them and interacting with a residual force of monopole type that scatters particle-hole pairs from one shell to the other.

The hamiltonian is

$$H = \varepsilon K_o + \frac{V}{2}(K_+^2 + K_-^2) \tag{1}$$

where K_o, K_+, K_- are the usual quasispin operators

$$K_+ = \sum_{p=1}^{\Omega} a_{p,1}^{+} a_{p,-1} = (K_-)^{+}$$
$$K_o = \frac{1}{2}\sum_{p=1}^{\Omega} (a_{p,1}^{+} a_{p,1} - a_{p,-1}^{+} a_{p,-1}) \tag{2}$$

and $a_{p,\sigma}^{+}$ are the fermion creation operators in level σ (= ± 1) and sublevel p.

The exact eigenstates that couple to the ground state are obtained by diagonalizing the hamiltonian in an SU(2) basis $|K,M\rangle$ where K=N/2 and as usual $-K \leq M \leq K$. The resulting matrices have dimension 2K+1 = N+1. (As states with M of different parity are not connected, they can be further decomposed in two blocks of dimension N/2 and N/2+1). The diagonalization is fast and accurate even for fairly large degeneracies (N$\sim$100).

Once the eigenvalues and eigenvectors are known for a given value of V and N (we take ε as the energy unit) it is easy to compute the exact time evolution of any initial state. For further details and for examples of this evolution we refer to ref.3.

It is clear that this model describes the quantum motion of the quasi spin vector and that the limit N$\to\infty$ corresponds ot the situation where this angular momentum becomes very large and therefore we expect the motion to become classical. This is in fact the case and this classical motion is given by the Time Dependent Hartree-Fock (TDHF) method[1,4].

EIGENGALUES, THE TDHF ACTION AND 1/N CORRECTIONS

We first look at the spectrum in the limit of large N.

In general, we can write the eigenvalues as E_n (N,X) and attempt an expansion in powers of 1/N as follows

$$\frac{E_n(N,X)}{N} = E_n^{(0)}(X) + \frac{1}{N} E_n^{(1)}(X) + \ldots \tag{3}$$

The fact that the energies become proportional to N ensures the existence of a thermodynamic limit where the energy per particle is well defined. The parameter $X = V(N-1)$ is the usual one for this model and is held fixed as $N \to \infty$

To plot the spectrum we define a quantum action by

$$S_n = 2\pi(n - N/2) \tag{4}$$

where n is the integer labeling the eigenvalues for increasing energy. If this model had an $|r\rangle$ representation n would represent the number of nodes of the wave function. Notice that $0 \leq n \leq N$ and therefore $S/\pi N$ is a discrete variable taking values in the range

$$-1 \leq \frac{S}{\pi N} \leq 1 \tag{5}$$

As N becomes very large $S/\pi N$ becomes a continuous variable in its range.

To bring out the existence of the large N limit of the spectrum we plot $S_n/\pi N$ versus E_n/N for increasing values of N. We obtain the result, clearly seen in Fig. 1A that the spectrum crowds along a universal curve -independent of N- which can be interpreted as the action as a function of energy of a classical hamiltonian with one degree of freedom. This classical limit is exactly reproduced by TDHF.

We will not discuss here the TDHF approximation for this model nor its semiclassical requantization. This has been done many times in the literature[1,3-7] and here we will only state the more relevant results.

The TDHF determinant for this model can be parametrized by a complex number Z as

$$Z = \exp(\bar{Z} K_+) |K, -K\rangle \tag{6}$$

where $|K, -K\rangle$ is the unperturbed ground state. This leads to a hamiltonian function

$$\mathcal{H}(p,q) = N\left\{-1/2 + 1/2(p^2+q^2) + X/2(q^2-p^2)[1 - 1/2(p^2+q^2)]\right\} \quad (7)$$

when using canonical variables[8]

$$\omega = Z/\sqrt{1+Z\bar{Z}} = 1/\sqrt{2}(q+ip) \quad (8)$$

The TDHF action is then defined as

$$S = N\,[\oint p\,dq - \pi] \quad (9)$$

where p is obtained as a function of q and E from $\mathcal{H}(p,q) = E$ and the integral is over a periodic solution of the equations of motion. This action is quantized by the usual Bohr-Sommerfeld rule $S=2\pi m$ with $-N/2 \leq m \leq N/2$.

Thus the requantized TDHF levels for a finite but large N are obtained by calculating the classical action (9) as a continuous function of energy in the interval $-1 \leq \frac{S}{\pi N} \leq 1$ and dividing this interval into N equal parts to obtain the discrete levels[8].

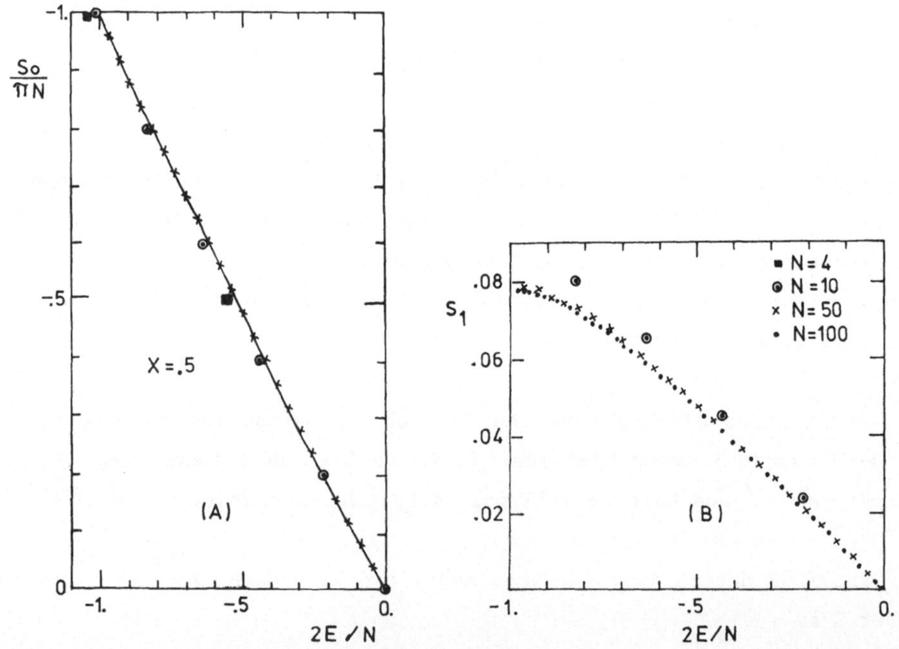

Fig. 1. (A) Action as a function of energy for X=0.5. The full line represents the TDHF result. The exact results correspond to different values of N. (B) 1/N correction to the TDHF action as a function of energy for different values of N.

This procedure could be improved if we could calculate a classical action to the following order in N. To show that this is indeed the case we assume an expansion of the quantum action

$$S = S_0(E,X) + \frac{1}{N} S_1(E,X) + \dots \qquad (10)$$

S is a continuous function of E which provides discrete energy levels via the same procedure as before. $S_0(E,X)$ is the TDHF action given by (9) divided by πN. $S_1(E,X)$ is again a universal curve, and it can be obtained from the exact diagonalization by considering the limit

$$S_1 = N(S-S_0) \quad ; \quad N \to \infty \qquad (11)$$

Fig. 1B shows the curve resulting from this limiting procedure for X=0.5.

Although several methods to extend TDHF have been proposed[9-12] to our knowledge none attempts to identify N as an expansion parameter. Our discussion points to the fact that, for this model, this is indeed a very relevant quantity.

EVOLUTION OF STATISTICAL PROPERTIES

Krieger[3] has investigated the evolution of single particle properties (i.e. the one body density matrix) in the TDHF approximation and compared it with the exact evolution. We show here that again there is much to be learned from looking at this evolution in the large N limit and that a universal (independent of N) behaviour is apparent both in leading order and in the subsequent 1/N terms.

We have chosen as a representative quantity to study the following

$$k = \sqrt{2\,\frac{\mathrm{tr}(\rho^2-\rho)}{\mathrm{tr}\,\rho} + 1} \qquad (12)$$

where $\rho_{\sigma\sigma'}^{(t)} = \langle \Psi(t) | a^+_{\sigma'} a_\sigma | \Psi(t) \rangle$ is the one body density calculated both in the exact or in the TDHF state. For this model it is a 2x2 hermitian matrix with unit trace. It is obvious that for TDHF, $\mathrm{tr}\,\rho^2 = \mathrm{tr}\,\rho$ and therefore k=1 for all time. This can be restated saying that in some single particle basis the occupation of the levels is such that all particles are in one state and the diagonal form of ρ is $\begin{pmatrix} 1 & 0 \\ 0 & 0 \end{pmatrix}$. At the

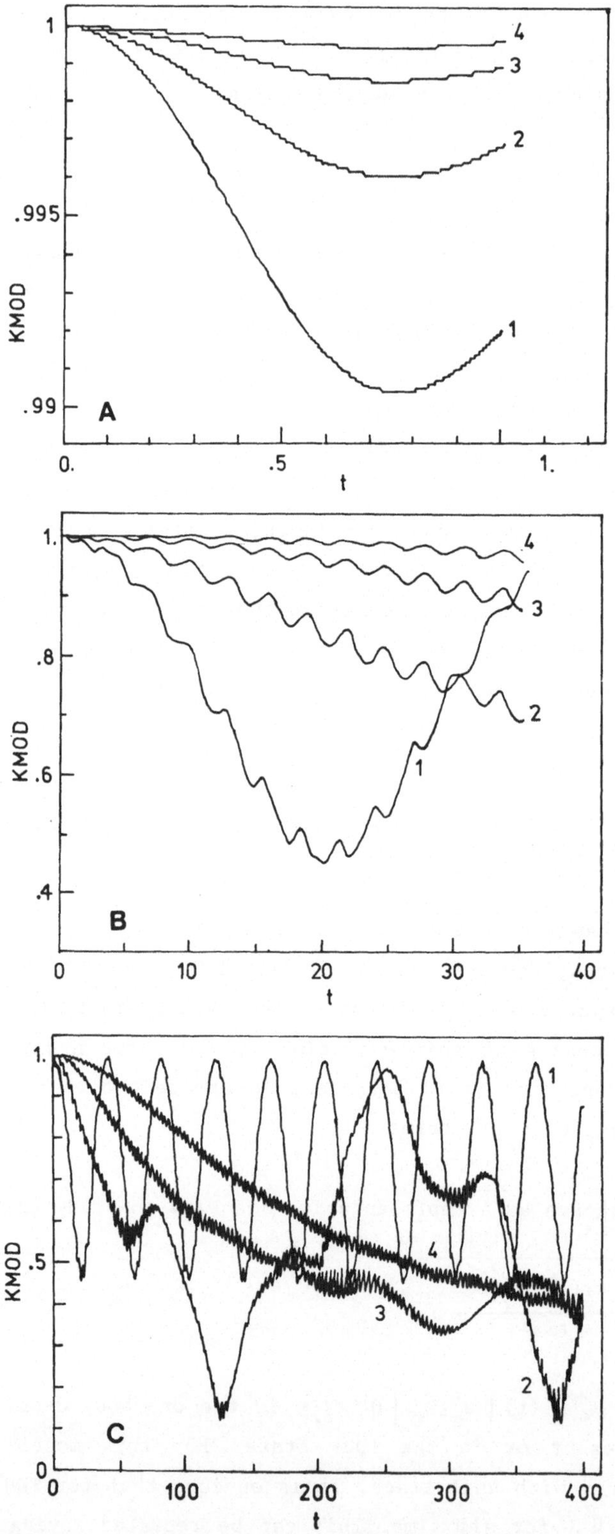

Fig. 2. Quantity k as a function of time for X=.5 and 2E/N=-.15. The numbers 1,2,3,4 correspond to N=4,8,20,50.

other extreme is the totally "thermalized" situation in which the occupation of the levels is equal with a diagonal form $\begin{pmatrix} \frac{1}{2} & 0 \\ 0 & \frac{1}{2} \end{pmatrix}$. For this case $\mathrm{tr}\,\rho^2=\frac{1}{2}$, $\mathrm{tr}\,\rho=1$ and therefore k=0. Thus the way in which k deviates from unity signals the growth of correlations in the evolution and one can check whether and how the evolution of an initial determinant state yields thermalization. k is an entropy-like quantity that measures the disorder of the one particle degrees of freedom[11].

In figure 2 we plot the exact evolution of this quantity for increasing values of N. Fig. 2A shows the behaviour for times short compared to the RPA period. We observe that k contains an oscillation independent of N but that its amplitude decreases with N. This is much more striking in Fig. 2B where one can follow the evolution through many RPA periods. Here, besides the oscillation, a decay is observable with a time which increases with N. From this figure it is quite clear how the evolution of the density comes closer to the TDHF behaviour when N becomes large. In Fig. 2C we look at the behaviour for much longer times. We learn that, besides the atypical behaviour for small N showing oscillations and revivals, the large N behaviour is essentially an oscillation with the RPA frequency independent of N superimposed on a decay which becomes slower as N increases. To investigate the N dependence of this decay we plot the same quantities in Fig. 3 but using $t/\sqrt{N}$ as time coordinate. We then see that the decay becomes independent of N. This is quite apparent in Fig. 3B where the decays corresponding to N=50 and N=100 are almost superimposed. Notice that with the time scaled by $\sqrt{N}$ the RPA oscillations become faster as N increases. In Fig. 3C we show the behaviour for very long times. Again we observe a complicated pattern for low values of N which becomes simpler and universal as N increases. We find empirically that the large N behaviour of k can be well reproduced by

$$\overline{k(t)} = (1-A)\,\exp\left(-\alpha\frac{t^2}{N}\right) + A \qquad (13)$$

when the fast RPA oscillations are averaged. A and α are independent of N (but of course they depend on the energy of the initial state considered).

We conclude that for this quantity the leading order behaviour is k=1 as predicted by TDHF. At order 1/N, instead, a very definite decay is observable with a time of order $\sqrt{N}$. It would be interesting to see if existing extensions of TDHF theory that include collisions can account for this simple behaviour in this model. A detailed comparison with the methods of Ref. 10 and 11 is in progress.

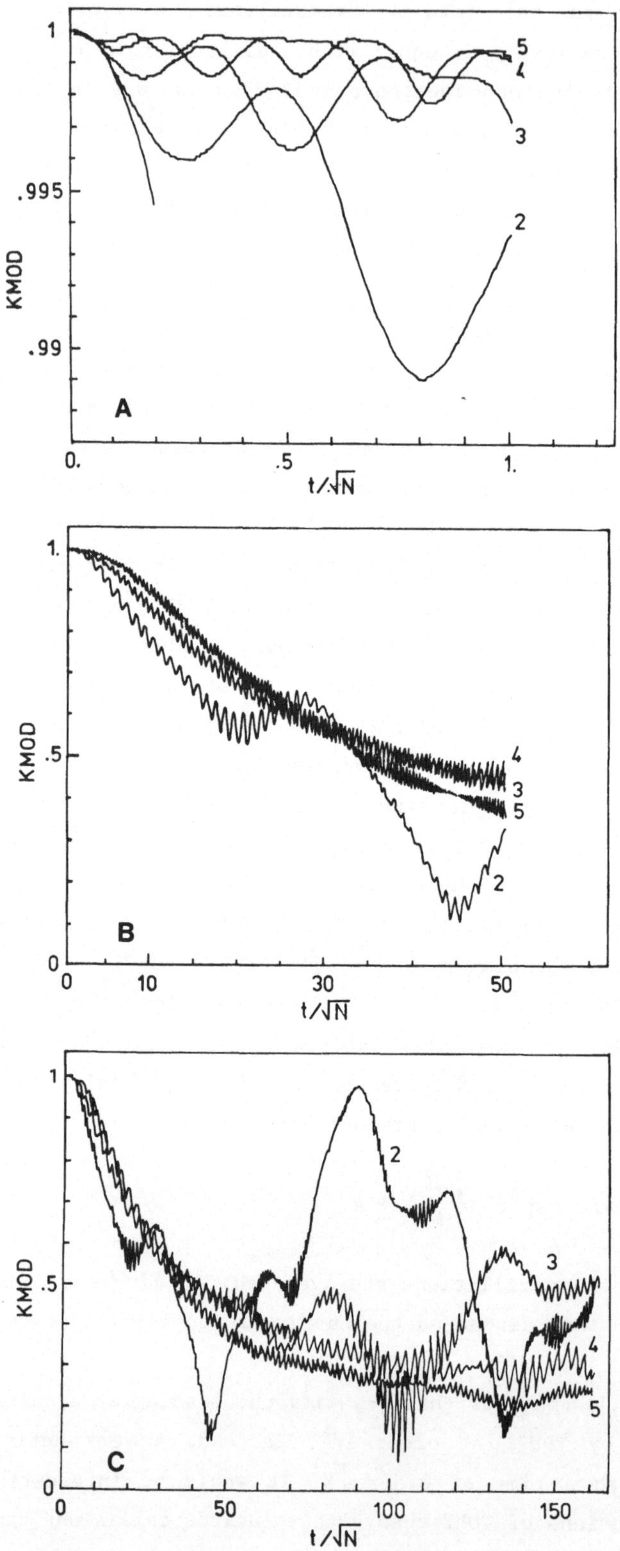

Fig. 3. Quantity k as a function of time/$\sqrt{N}$ for X=.5 and 2E/N=-.15. The numbers 2,3,4,5 correspond to N=8,20,50,100.

References

1) J.W.Negele, Rev.Mod.Phys. 54:913 (1982).

2) H.J.Lipkin, N.Meshkov and A.J.Glick, Nucl.Phys. 62:188 (1965).

3) S.J.Krieger, Nucl.Phys. A276:12 (1977).

4) M.C.Cambiaggio, G.G.Dussel and M.Saraceno, Phys.Lett. B135:10 (1984)

5) K.K.Kan, J.J.Griffin, P.C.Lichtner and M.Dworzecka, Nucl.Phys. A332:109 (1979).

6) S.Levit, J.W.Negele and Z.Paltiel, Phys.Rev. C21:1603 (1980).

7) H.Kuratzuji, Phys.Lett. 108B:367 (1982).

8) M.C.Cambiaggio, G.G.Dussel and M.Saraceno, Nucl.Phys. A415:70 (1984)

9) M.Baranger and I.Zahed, Phys.Rev. C29:1005 (1984).

10) H.S.Köhler, Nucl.Phys. A400:233c (1983).

11) R.Balian and M.Veneroni, Ann.of Phys. 135:270 (1981).

12) M.C.Nemes and A.F.R. de Toledo Piza, Phys.Rev. C27:862 (1983).

IMPROVED METHOD FOR ELIMINATING CENTER-OF-MASS COORDINATES FROM MATRIX ELEMENTS IN OSCILLATOR BASES

Roger H. Richardson
Friends Academy, Locust Valley, Long Island, N.Y. 11560

Joseph Y. Shapiro
Fordham University, Bronx, N.Y. 10458

INTRODUCTION

This work arose out of attempts by the authors and Daniel O. Vona III to develop computer programs that would automatically calculate stationary state properties of nuclei.[1] These use a mixture of configurations constructed from isotropic harmonic oscillator single particle states. As a test of the methods and programs, extensive calculations (up to 350 configurations) were performed on the A = 3 system, primarily using a soft-core Gaussian type potential with tensor and spin-orbit terms due to Eikemeier and Hackenbroich.[2]

These calculations, which have not yet been published, gave two main results. First, convergence of the absolute energies was poor. As configurations were added, the ground state energies kept decreasing, though they were still well above the experimental values. A similar lack of convergence has been found by Majling, et. al.[3] for nuclei with A=2-4, using Gaussian-type central potentials and, for the alpha-particle, by Ceuleneer and Vandepeutte.[4] The latter calculations went up to 2765 configurations (all $10\hbar\omega$ excitations), using the Gogny et.al.[5] (Gaussian-type) and Malfleit-Tjon (central Yukawa-type) potentials.

Second, in spite of this lack of convergence, other calculated properties (^3H - ^3He Coulomb energy difference, nuclear radii, and magnetic moments) were in reasonable agreement with experimental results, so that the method is not without promise.

It became clear that larger model spaces, and hence more efficient computational techniques, were necessary, especially to deal with heavier nuclei. In our programs, almost all of the computer time is spent in a single task, removing the center-of-mass dependence from the matrix elements of the potential energy. The standard method of Brody-Moshinsky brackets[6] is not well suited for large-scale computations, both because it is too slow and because their brackets involve a pair of states coupled to a given total angular momentum L, whereas uncoupled product states enter directly into the calculations. Thus the brackets between uncoupled states, introduced by Talmi, are better. However, in working on methods of evaluating these brackets, we realized that efficiency could be improved by transforming the entire matrix element at once, and directly calculating the coefficients of the Talmi integrals. This work was done in collaboration with F. B. Malik.[8] However, this method, even though it has undergone several modifications which have increased its computational speed by approximately two orders of magnitude, is still too slow. It is also complicated for certain spin-angular operators, e.g., $(\vec{L}\cdot\vec{S})^2$.

In studying the literature, we realized that a much faster method was contained in a brief paper by Chasman and Wahlborn.[9] They proposed calculating the Talmi brackets by first transforming the single particle functions from an isotropic to an axially symmetric oscillator basis. They then transformed to an axially symmetric basis in center-of-mass and relative coordinates. Finally they transform these states back to the isotropic oscillator basis.

The method works because of the simplicity of the transformation brackets to center-of-mass and relative coordinates for the axially symmetric oscillator. These brackets can be written as a product of three factors, which are identical in form. These factors are just the transformation brackets for a one-dimensional oscillator. It follows that the same factorization holds for the transformation brackets for the asymmetric oscillator in Cartesian coordinates. Thus, as a bonus, the method also gives the transformation brackets for the axially symmetric and asymmetric cases.

We here propose an obvious, but nonetheless important, improvement. In the last step of Chasman and Wahlborn's schema, there is no need to transform the center-of-mass states back to isotropic oscillator states, since in all applications they only enter into orthonormality relations.

Thus, instead of the Talmi bracket, we use a modified bracket, in which the center-of-mass coordinates are expressed in an axially symmetric basis, rather than an isotropic oscillator basis.

By storing two small tables of coefficients, we obtain a simple expression for these brackets (Eq.(32)) that involves, besides table lookups, only two summations. One of these tables also gives the brackets for the asymmetric and axially symmetric cases (Eqs.(15) and (21)).

In this paper, we first derive all expressions required to calculate the brackets, using the formulation of second quantization. We then discuss, for the axially symmetric and isotropic oscillator cases, how these brackets can be used to evaluate matrix elements of the potential energy, illustrating with the case of a spin-orbit force.

TRANSFORMATION BRACKETS IN COORDINATE SPACE

In this section we discuss a simple method for deriving the transformation brackets from a product of oscillator states for particles 1 and 2 to a product of states for the center-of-mass and relative motions. For didactic reasons, we first discuss the asymmetric oscillator (Cartesian coordinates), then the axially symmetric oscillator (cylindrical coordinates), and finally the isotropic oscillator (spherical polar coordinates).

The derivations will use the formulation of second quantization. Since this tends to obscure such things as normalization in coordinate space, we begin with a few definitions. For the one-dimensional oscillator, with Hamiltonian

$$H = -\frac{\hbar^2}{2m}\frac{d^2}{dq^2} + \frac{1}{2}m\omega^2 q^2 \tag{1}$$

we introduce a new coordinate x by the scale transformation

$$q = bx, \qquad b = \sqrt{\frac{\hbar}{m\omega}} \tag{2}$$

obtaining

$$H = \frac{\hbar\omega}{2}\left[-\frac{d^2}{dx^2} + x^2\right] \tag{3}$$

We define a and $a^\dagger$, the standard boson annihilation and creation operators, by

$$a = \frac{1}{\sqrt{2}}\left(x + \frac{d}{dx}\right), \qquad a^\dagger = \frac{1}{\sqrt{2}}\left(x - \frac{d}{dx}\right) \tag{4}$$

These satisfy the commutation relations

$$[a,a] = [a^\dagger,a^\dagger] = 0, \qquad [a,a^\dagger] = 1 \tag{5}$$

Acting on a state $|n\rangle$ with n quanta

$$a|n\rangle = \sqrt{n}\,|n-1\rangle, \qquad a^\dagger|n\rangle = \sqrt{n+1}\,|n+1\rangle \tag{6}$$

It follows that

$$|n\rangle = \frac{1}{\sqrt{n!}}\, a^{\dagger n}|0\rangle \tag{7}$$

where $|0\rangle$ is the vacuum state.

For the three-dimensional oscillator we do the transformation of Eq.(2) on all three coordinates, obtaining a Hamiltonian

$$H = \frac{\hbar\omega_x}{2}\left(-\frac{d^2}{dx^2} + x^2\right) + \frac{\hbar\omega_y}{2}\left(-\frac{d^2}{dy^2} + y^2\right) + \frac{\hbar\omega_z}{2}\left(-\frac{d^2}{dz^2} + z^2\right) \tag{8}$$

For the general case of an asymmetric oscillator, with $\omega_x \neq \omega_y \neq \omega_z$, the three scale factors introduced through Eq.(2) are all different, i.e., $b_x \neq b_y \neq b_z$. For the axially symmetric oscillator, we take the z-axis as the axis of symmetry, so that we set $\omega_x = \omega_y = \omega_\perp$, where $\omega_\perp \neq \omega_z$. Consequently $b_x = b_y = b \neq b_z$. For the isotropic oscillator, we set $\omega_x = \omega_y = \omega_z = \omega$ and $b_x = b_y = b_z = b$.

We will normalize all wave functions in the space of the transformed coordinates x, y, and z, not in the original "real" space.

The Asymmetric Oscillator

We first consider oscillators in one dimension. For a two-particle system in which particle 1 has n_1 quanta and particle 2 has n_2 quanta, the product wave function is

$$|n_1,n_2\rangle = \frac{1}{\sqrt{n_1!\,n_2!}}\, a_1^{\dagger n_1}\, a_2^{\dagger n_2}|0\rangle \tag{9}$$

The one-dimensional transformation bracket $\langle n_1,n_2 \mid N,n\rangle$ to a state with N center-of-mass quanta and n relative motion quanta can be obtained by introducing center-of-mass and relative motion annihilation operators A and a, respectively, according to

$$a_1 = \frac{1}{\sqrt{2}}(A + a), \qquad a_2 = \frac{1}{\sqrt{2}}(A - a) \tag{10a}$$

and corresponding creation operators $A^\dagger$ and $a^\dagger$ by

$$a_1^\dagger = \frac{1}{\sqrt{2}}(A^\dagger + a^\dagger), \qquad a_2^\dagger = \frac{1}{\sqrt{2}}(A^\dagger - a^\dagger) \tag{10b}$$

The square roots of two appear because we are using the "symmetric" definition of center-of-mass and relative coordinates

$$\vec{R} = \frac{\vec{r}_1 + \vec{r}_2}{\sqrt{2}} \qquad \vec{r} = \frac{\vec{r}_1 - \vec{r}_2}{\sqrt{2}} \tag{11}$$

Note that we assume that the particles have equal masses. Also that a and $a^\dagger$ (without subscripts), refer to the relative motion, not to a general one-dimensional oscillator as in Eqs.(4)-(7) above.

Substituting Eq.(10b) for $a_1^\dagger$ and $a_2^\dagger$ into Eq.(9), expanding by the binomial theorem, and identifying terms with center-of-mass and relative states using the analogues of Eq.(7) gives

$$\langle n1,n2 \mid N,n\rangle = \delta_{n_1+n_2,N+n}\left[\frac{N!\,n!}{2^{n_1+n_2}\,n_1!n_2!}\right]^{1/2} M_n^{n_1,n_2} \tag{12}$$

where $M_k^{i,j}$ is the coefficient of x^k in the expansion

$$(1+x)^i(1-x)^j = \sum_k M_k^{i,j}\, x^k \tag{13}$$

The Kronecker delta in Eq.(12) just expresses conservation of energy.

The wave function for the three-dimensional oscillator in Cartesian coordinates is just the product of three one-dimensional wave functions

$$|n_x,n_y,n_z\rangle = |n_x\rangle|n_y\rangle|n_z\rangle$$

$$= \frac{1}{\sqrt{n_x!\,n_y!\,n_z!}}\, a_x^{\dagger n_x}\, a_y^{\dagger n_y}\, a_z^{\dagger n_z}\,|0\rangle \tag{14}$$

Introducing two-particle states (as in Eq.(9)) and three-dimensional annihilation and creation operators for the center-of-mass and relative coordinates (as in the Eqs.(10)) and following the same procedure as above, the transformation bracket obviously factors into a product of three one-dimensional brackets,

$$\langle n_{1x},n_{1y},n_{1z};n_{2x},n_{2y},n_{2z} | N_x,N_y,N_z;n_x,n_y,n_z\rangle$$
$$= \langle n_{1x},n_{2x} | N_x,n_x\rangle \langle n_{1y},n_{2y} | N_y,n_y\rangle \langle n_{1z},n_{2z} | N_z,n_z\rangle \tag{15}$$

where the factors on the right are given by Eqs.(12) and (13).

Numerical evaluation of the one-dimensional brackets is straightforward. Expanding the binomials in Eq.(13) gives $M_k^{i,j}$ as a sum. However, it is easier to use recurrence relations. In our computer programs we use

$$\langle n_1,0 | N,n\rangle = \delta_{n_1,N+n} \left[\frac{1}{2^{n_1}}\binom{n_1}{n}\right]^{1/2} \tag{16}$$

which follows from Eqs.(12) and (13), and the recurrence relation on n_2

$$\langle n_1,n_2 | N,n\rangle = \sqrt{\frac{N}{2n_2}}\,\langle n_1,n_2-1 | N-1,n\rangle - \sqrt{\frac{n}{2n_2}}\,\langle n_1,n_2-1 | N,n-1\rangle \tag{17}$$

to prepare a table of one-dimensional brackets. This table, which contains a few hundred entries, is calculated and stored in the computer at the beginning of any run. The above recurrence relation can be obtained by operating to the left and right with a_2 in the matrix element $\langle n_1,n_2-1|a_2|N,n\rangle$, where Eq.(10a) is used to operate to the right.

The Axially Symmetric Oscillator

The transformation brackets for the axially symmetric oscillator can be put into a form identical to Eq.(15) by introducing new annihilation and creation operators by[10]

$$a_+ = \frac{1}{\sqrt{2}}(a_x - ia_y), \qquad a_- = \frac{1}{\sqrt{2}}(a_x + ia_y) \tag{18a}$$

and

$$a_+^\dagger = \frac{1}{\sqrt{2}}(a_x^\dagger + ia_y^\dagger), \qquad a_-^\dagger = \frac{1}{\sqrt{2}}(a_x^\dagger - ia_y^\dagger) \tag{18b}$$

The set of operators $a_+,a_-,a_+^\dagger,a_-^\dagger,a_z,a_z^\dagger$ satisfy the same standard set of commutation relations as the set $a_x,a_y,a_x^\dagger,a_y^\dagger,a_z,a_z^\dagger$, and thus are equally valid. The operators $a_+^\dagger$ and $a_-^\dagger$ create quanta with $L_z=1$ and $L_z=-1$ respectively. A state vector with n_+ quanta, each with $L_z=1$, can be written (in analogy with Eq.(7)) as

$$|n_+\rangle = \frac{1}{\sqrt{n_+!}}\, a_+^{\dagger\, n_+}|0\rangle \tag{19}$$

and a three-dimensional state can be written as

$$|n_+,n_-,n_z\rangle = \frac{1}{\sqrt{n_+!\ n_-!\ n_z!}}\ a_+^{\dagger n_+}\ a_-^{\dagger n_-}\ a_z^{\dagger n_z}|0\rangle \qquad (20)$$

Proceeding precisely as for the case of Cartesian coordinates, one obtains for the transformation brackets to center-of-mass and relative coordinates the factored form

$$\langle n_{1+},n_{1-},n_{1z};n_{2+},n_{2-},n_{2z}\,|\,N_+,N_-,N_z;n_+,n_-,n_z\rangle$$
$$= \langle n_{1+},n_{2+}\,|\,N_+,n_+\rangle\ \langle n_{1-},n_{2-}\,|\,N_+,n_+\rangle\ \langle n_{1z},n_{2z}\,|\,N_z,n_z\rangle \qquad (21)$$

in parallel to Eq.(15). This is Eq.(12) in Chasman and Wahlborn's paper. Note that the above factorization of the brackets does not imply a corresponding factorization of the wave functions.

Most discussions of the axially symmetric oscillator (in either two or three dimensions) use the quantum numbers

$$n_\perp = n_+ + n_-, \qquad \Lambda = n_+ - n_- \qquad (22)$$

rather than n_+ and n_-. Physically, $n_\perp$ is the number of oscillator quanta in the x-y plane, and Λ is the z-component of angular momentum (in units $\hbar$).

In summary, reduction of a product of two axially symmetric oscillator wave functions to center-of-mass and relative coordinates involves nothing new. By introducing the quantum numbers n_+ and n_-, the reduction can be done by Eq.(21), using the same table of one-dimensional transformation brackets as for the asymmetric oscillator (Cartesian coordinates).

The Isotropic Oscillator

To obtain the transformation brackets for the isotropic (spherical) oscillator, we use a modification of the method in Chasman and Wahlborn's paper. As indicated in figure 1, we first transform the wave functions for particles 1 and 2 to cylindrical coordinates. The method for doing this will be discussed below. Next we transform these "axially symmetric" (in quotes because $\omega_\perp = \omega_z$) oscillator wave functions to center-of-mass and relative coordinates, using the method discussed in the above

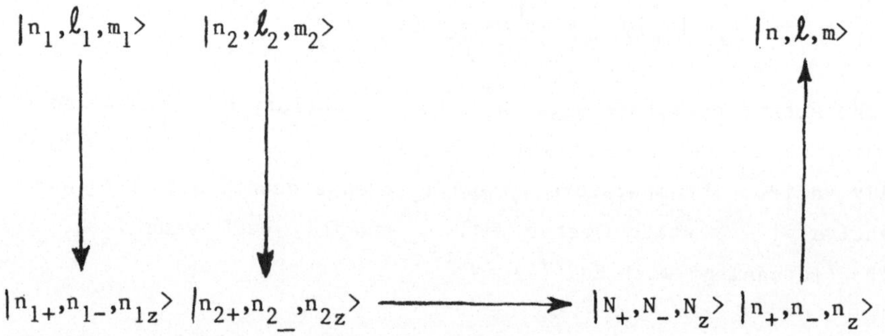

Fig.1 - Scheme for transforming product of two isotropic oscillator wave functions to center-of-mass and relative coordinates

subsection. Finally, we transform the wave function for the relative coordinates back to spherical coordinates. Note that there is no reason to transform the center-of-mass wave function back to spherical coordinates, since in applications it always enters only into orthonormality relations (which are equally valid in cylindrical coordinates) and disappears.

The only new thing involved is the transformation of a single-particle oscillator state from spherical to cylindrical coordinates, i.e., to an axially symmetric oscillator state with all oscillator constants equal. We denote the requisite transformation bracket by $\langle n,\ell,m \mid n_+,n_-,n_z\rangle$. Chasman and Wahlborn give an explicit formula for these brackets as a single sum. We proceed somewhat differently. Consider the special case $m = \ell$. Since both the number of oscillator quanta and the z-component of angular momentum are good quantum numbers, the expansion of the spherical state $|n,\ell,\ell\rangle$ in terms of cylindrical states $|n_+,n_-,n_z\rangle$ must take the form

$$|n,\ell,\ell\rangle = \sum_k c_k |\ell+k,k,2n-2k\rangle \tag{23}$$

since $2n+\ell = n_+ + n_- + n_z$ and $m = n_+ - n_-$.

The coefficients c_k in Eq.(23) are just the desired brackets. They may be evaluated by noting that, for the isotropic oscillator, the angular momentum (in our original "real-space" coordinate system) can be expressed in terms of annihilation and creation operators by

$$L_+ = \sqrt{2}\,(a_- a_z^\dagger - a_+^\dagger a_z) \tag{24a}$$

$$L_- = \sqrt{2}\,(a_-^\dagger a_z - a_+ a_z^\dagger) \tag{24b}$$

$$L_3 = a_+^\dagger a_+ - a_-^\dagger a_- = \mathcal{n}_+ - \mathcal{n}_- \tag{24c}$$

where $\mathcal{n}_+$ and $\mathcal{n}_-$ are the appropriate number operators. The Eqs.(24) follow, after some algebra, from the definition of orbital angular momentum and Eqs.(4) and (18).

Operating with L_+ on Eq.(23) gives zero on the left and two sums on the right (from the right hand side of Eq.(24a)). Identifying coefficients of the same cylindrical state in these two sums gives the recurrence relation

$$C_k \sqrt{k(2n-2k+1)} = C_{k-1} \sqrt{(\ell+k)(2n-2k+2)} \tag{25}$$

which can be iterated to give

$$C_k = \left[\frac{n!\,(\ell+k)!\,\Gamma(n-k+1/2)}{k!\,(n-k)!\,\Gamma(n+1/2)}\right]^{1/2} C_o \tag{26}$$

Substituting Eq.(26) into Eq.(23) gives an expression for the states $|n,\ell,\ell\rangle$ involving a single unknown multiplicative constant C_o. The absolute value of this constant can be obtained by normalizing $|n,\ell,\ell\rangle$, assuming that the states appearing on the right of Eq.(23) are orthonormal. This gives, after performing a sum,

$$|C_o|^2 = \frac{\Gamma(\ell+3/2)\,\Gamma(n+1/2)}{\sqrt{\pi}\,\ell!\,\Gamma(n+\ell+3/2)} \tag{27}$$

We choose the phase of C_o to give the wave function $\langle \vec{r} \mid n,\ell,m\rangle$ its usual definition, i.e., so that

$$\langle \vec{r} \mid n,\ell,m\rangle = N_{n\ell}\, r^{\ell}\, e^{-r^2/2}\, L_n^{\ell+1/2}(r^2)\, Y_{\ell m}(\Omega) \tag{28}$$

where the normalization constant $N_{n\ell}$ is real and positive, the Laguerre polynomials are defined as in Morse and Feshbach[11] and the spherical harmonics as in Condon and Shortley.[12] This gives

$$C_o = (-1)^{n+\ell}\, |C_o| \tag{29}$$

Combining Eqs.(26), (27), and (29) gives the brackets for the case $m = \ell$ as

$$\langle n,\ell,\ell \mid \ell+n_-,n_-,2n-2n_-\rangle$$

$$= (-1)^{n+\ell}\left[\frac{n!\,(\ell+n_-)!\,\Gamma(n-n_-+1/2)\,\Gamma(\ell+3/2)}{\sqrt{\pi}\,\ell!\,n_-!\,(n-n_-)!\,\Gamma(n+\ell+3/2)}\right]^{1/2} \tag{30}$$

The brackets for $m < \ell$ follow from the recurrence relation on m

$$\sqrt{(\ell-m)(\ell+m+1)}\,\langle n,\ell,m \mid n_+,n_-,n_z\rangle$$

$$= \sqrt{2n_-(n_z+1)}\,\langle n,\ell,m+1 \mid n_+,n_--1,n_z+1\rangle$$

$$- \sqrt{2n_z(n_++1)}\,\langle n,\ell,m+1 \mid n_++1,n_-,n_z-1\rangle \tag{31}$$

Eq.(31) may be obtained by substituting Eq.(24b) into the matrix element $\langle n,\ell,m+1 \mid L_+ \mid n_+,n_-,n_z\rangle$ and operating to the left and right.

In our computer programs, Eqs.(30) (actually recurrence relations derived from (30)) and (31) are used to compute and store a table of these brackets (a few thousand entries) at the beginning of each run.

In terms of this bracket, the transformation bracket to center of mass and relative coordinates is given by

$$\langle n_1,\ell_1,m_1;n_2,\ell_2,m_2 \mid N_+,N_-,N_z,n,\ell,m\rangle$$

$$= \sum \langle n_1,\ell_1,m_1 \mid n_{1+},n_{1-},n_{1z}\rangle\,\langle n_2,\ell_2,m_2 \mid n_{2+},n_{2-},n_{2z}\rangle$$

$$\times\ \langle n_{1+},n_{2+} \mid N_+,n_+\rangle\,\langle n_{1-},n_{2-} \mid N_-,n_-\rangle\,\langle n_{1z},n_{2z} \mid N_z,n_z\rangle$$

$$\times\ \langle n,\ell,m \mid n_+,n_-,n_z\rangle \tag{32}$$

The sum is over only two indices, which we take to be n_{1-} and n_{2-}. The other indices are then determined by the laws of conservation of energy and z-component of angular momentum that hold for the individual brackets. Note that we have saved a summation by not transforming the center-of-mass motion back to spherical coordinates. There would be one additional nested summation if we had not stored tables for the single-particle and one-dimensional brackets appearing on the right in Eq.(32). Evaluation of the above transformation brackets is extremely rapid, since all that is involved in Eq.(32) is table lookups and a few multiplications and additions.

REDUCTION OF TWO-BODY MATRIX ELEMENTS

We next discuss the elimination of the center-of-mass coordinates and any spin dependence from the potential energy matrix element. We will skip discussion of the asymmetric oscillator because first, it is seldom necessary for calculations with actual nuclei, and second, the reduction is very similar to that for the axially symmetric case.

We assume that the potential V conserves both parity and total spin S. For the isotropic oscillator basis we also assume rotational invariance, whereas for the axially symmetric oscillator we will assume invariance only for rotations about the axis of symmetry (J_z conservation). This will allow us to eliminate dependence on the azimuthal angle for the axially symmetric oscillator, and on the azimuthal and polar angles for the isotropic oscillator.

The Axially Symmetric Oscillator

We will use a single-particle basis of the form $|n_+,n_-,n_z,m_s\rangle$, where m_s is the z-component of spin (1/2 or -1/2). Then we can expand the two-body matrix element as

$$\langle n'_{1+},n'_{1-},n'_{1z},m'_{1s};n'_{2+},n'_{2-},n'_{2z},m'_{2s} | V | n_{1+},n_{1-},n_{1z},m_{1s};n_{2+},n_{2-},n_{2z},m_{2s}\rangle$$

$$= \sum \langle n_{1+},n_{2+} | N_+,n_+\rangle \langle n_{1-},n_{2-} | N_-,n_-\rangle \langle n_{1z},n_{2z} | N_z,n_z\rangle$$

$$\times \quad \langle n'_{1+},n'_{2+} | N_+,n'_+\rangle \langle n'_{1-},n'_{2-} | N_-,n'_-\rangle \langle n'_{1z},n'_{2z} | N_z,n'_z\rangle$$

$$\times \quad \langle 1/2,m_{1s},1/2,m_{2s} | S,M_s\rangle \langle 1/2,m'_{1s},1/2,m'_{2s} | S,M'_s\rangle$$

$$\times \quad \langle n'_+,n'_-,n'_z,S,M'_s | V | n_+,n_-,n_z,S,M_s\rangle \qquad (33)$$

The sum is over four indices, which we choose as n_+, n_-, n_z, and S. The other indices (n'_+,n'_-,n'_z,N_+,N_-, and N_z) can be expressed in terms of these through conservation laws on the brackets on the right in Eq.(33). The center-of-mass coordinates disappear through orthonormality. The geometrical factors that enter into Eq.(33) are just our tabulated one-dimensional brackets and a few simple Clebsch-Gordan coefficients.

The only complicated step is the evaluation of the matrix element, in relative coordinates, of the potential energy, i.e., of

$$\mathcal{V} = \langle n'_+,n'_-,n'_z,S,M'_s | V | n_+,n_-,n_z,S,M_s\rangle \qquad (34)$$

Here one frequently expresses the states $|n_+,n_-,n_z\rangle$ as a linear combination of states with a definite total angular momentum j. However, because of the two different oscillator constants in the axially symmetric states, this involves an infinite series, which is, in practice, drastically truncated (usually to those states in which the total number of oscillator quanta is unchanged). This makes error estimation very difficult. We would strongly urge that this not be done, and that the potential energy matrix elements be evaluated in the cylindrical basis, i.e., as they stand in Eq.(34). Then the only approximations that need be made are the choice of model, e.g., unified model, and of model space, i.e., what set of axially symmetric oscillator states one is using.

We illustrate the method of evaluation for a spin-orbit force of the form

$$V = V(\rho,z)\,\vec{L}\cdot\vec{S}, \qquad \rho = (x^2+y^2)^{1/2} \qquad (35)$$

Note that this form arises from a potential $V(r)\vec{L}\cdot\vec{S}$ in our original "real" space because of the two different scale factors. We expand

$$\vec{L}\cdot\vec{S} = \frac{1}{2}(L_+S_- + L_-S_+) + L_zS_z \tag{36}$$

noting that though L_z is still given by Eq.(24c), the other components of orbital angular momentum are changed because of the two different scale factors. We now get

$$L_+ = \frac{1+e^2}{\sqrt{2}\,e}(a_-a_z^\dagger - a_+^\dagger a_z) + \frac{1-e^2}{\sqrt{2}\,e}(a_+^\dagger a_z^\dagger - a_-a_z) \tag{37a}$$

$$L_- = \frac{1+e^2}{\sqrt{2}\,e}(a_-^\dagger a_z - a_+a_z^\dagger) - \frac{1-e^2}{\sqrt{2}\,e}(a_-^\dagger a_z^\dagger - a_+a_z) \tag{37b}$$

where

$$e = \sqrt{\frac{\omega_\perp}{\omega_z}} \tag{38}$$

The Eqs.(37) are contained in Bohr and Mottelson's book.[13] For e=1 (the isotropic oscillator), they reduce to the Eqs.(24), as they should. The second terms in the Eqs.(37), which only appear when there is anisotropy, create and annihilate pairs of quanta, so that they couple states where the total number of quanta differs by two.

Substituting the Eqs.(37) into Eq.(36) gives the spin-orbit operator in terms of annihilation and creation operators, which can readily be evaluated when Eq.(35) is substituted into Eq.(34). Since the effect of the spin operators can also be evaluated, one can reduce Eq.(34) to integrals over the coordinates ρ and z. These can always be done numerically. For certain potentials of the Gaussian type, e.g., that of Eikemeier and Hackenbroich, they can be done analytically.

Similar techniques can be used to evaluate terms containing the tensor force. We will not give details here.

The Isotropic Oscillator

We will use j-j coupled single particle states $|n,l,j,m\rangle$. The reduction is standard. We can write

$$\begin{aligned}
&\langle n_1',\ell_1',j_1',m_1';n_2',\ell_2',j_2',m_2' \,|\, V \,|\, n_1,\ell_1,j_1,m_1;n_2,\ell_2,j_2,m_2\rangle \\
&= \sum \langle n_1,\ell_1,j_1,m_1;n_2,\ell_2,j_2,m_2 \,|\, N_+,N_-,N_z;n(\ell,S)j,m\rangle \\
&\times \langle n_1',\ell_1',j_1',m_1';n_2',\ell_2',j_2',m_2' \,|\, N_+,N_-,N_z;n'(\ell',S)j,m\rangle \\
&\times \langle n'(\ell',S)j,m \,|\, V \,|\, n(\ell,S)j,m\rangle
\end{aligned} \tag{39}$$

The sum is over $N_+,N_-,N_z,n,n',\ell,\ell',S,j$, and m, but with conditions which reduce the actual number of indices, namely, conservation of energy and angular momentum in the brackets on the right and properties of particular potentials, e.g., the tensor force vanishes for singlets. We have assumed rotational invariance and that S is conserved. A consequence of the former is that the matrix element of V on the right hand side of Eq.(39) is independent of m.

The geometrical brackets appearing in Eq.(39) can be expanded as

$$
\begin{aligned}
&\langle n_1,\ell_1,j_1,m_1;n_2,\ell_2,j_2,m_2 \mid N_+,N_-,N_z;n(\ell,S)j,m\rangle \\
&= \sum \langle \ell_1,m_1-m_{1s},1/2,m_{1s} \mid j_1,m_1\rangle \langle \ell_2,m_2-m_{2s},1/2,m_{2s} \mid j_2,m_2\rangle \\
&\quad X \quad \langle \ell,m-m_s,S,m_s \mid j,m\rangle \langle 1/2,m_{1s},1/2,m_{2s} \mid S,m_s\rangle \\
&\quad X \quad \langle n_1,\ell_1,m_1-m_{1s};n_2,\ell_2,m_2-m_{2s} \mid N_+,N_-,N_z,n,\ell,m-m_s\rangle
\end{aligned}
\tag{40}
$$

i.e., an expression involving only some simple Clebsch-Gordan coefficients and our modified Talmi bracket of Eq.(32). The summation in Eq.(40) is over the two indices m_{1s} and m_{2s} (note that $m_s = m_{1s} + m_{2s}$) each of which can take at most two values.

One is left with the evaluation of the matrix elements of V in Eq.(39). This is standard. For example, for a spin-orbit force

$$
\begin{aligned}
&\langle n'(\ell',S)j,m \mid V(r)\ \vec{L}\cdot\vec{S} \mid n(\ell,S)j,m\rangle \\
&= \delta_{\ell\ell'}\,\delta_{S1}\,\frac{1}{2}\,[j(j+1) - \ell(\ell+1) - 2]\ \langle n',\ell \mid V(r) \mid n,\ell\rangle
\end{aligned}
\tag{41}
$$

The radial integrals can be evaluated once and for all for a given potential and stored, or they can be reduced to Talmi integrals[7] (which also can be evaluated and stored) by using the series expansion for the radial wave functions.

CONCLUSIONS

We have obtained a concise, efficient method of reducing potential energy matrix elements to relative coordinates, when one is using an oscillator basis. It is especially suited to computer calculations. One nice feature of the method is its modular form, which allows a wide range of calculations. We have written separate Fortran subroutines which calculate and store tables of the one-dimensional brackets of Eq.(12) and the single particle brackets from the isotropic to the axially symmetric oscillator (Eqs.(30) and (31)). These tables are used by other subroutines which calculate our modified Talmi brackets (Eq.(32)) and the brackets with spin of Eq.(40). These subroutines can be combined in various ways for different computations. For example, we have written a program which gives the complete reduction to radial integrals for all terms of the v_{14} potential of Wiringa, et. al.[14]

In our view, the methods developed here are a substantial improvement over what has been done heretofore, and open up many new possibilities for performing nuclear structure calculations.

ACKNOWLEDGMENTS

Part of this work was carried out while one of the authors (J.Y.S.) was visiting the University of Southern Illinois at Carbondale. He would like to thank Dr. F. B. Malik for extending the hospitality of the department. He also benefitted from several stimulating discussions on this work with Dr. Malik and with Dr. Manuel de Llano.

Note added in proof. We are indebted to Dr. de Llano for pointing out that we have ignored some of the early literature. In 1962, Smirnov proposed evaluating the Talmi brackets by transforming to Cartesian coordinates, eliminating the center-of-mass coordinates, and then transforming back to spherical coordinates.[15] Further, he obtained an expression for the brackets of Eq.(12) for the more general case of unequal masses. Pluhar and Tolar have derived Eq.(30) by essentially the method used here.[16] They also have derived expressions for the single particle brackets from spherical to cylindrical and Cartesian coordinates, as have Chacon and de Llano[17] and, for the Cartesian case, Domergue.[18] References 15 and 17 include some limited tables of coefficients. What is new here is 1) pointing out that one need not transform the center-of-mass states back to spherical coordinates, 2) using the cylindrical rather than Cartesian basis (the expressions for the one-dimensional brackets are simpler, involving one summation rather than two[9,17]), and 3) the use of some simple recurrence relations (Eqs. (17) and (31)). The overall effect has been to give an extremely efficient method of eliminating center-of-mass coordinates.

REFERENCES

1. Daniel O'Neal Vona III, Ph.D. Thesis, Fordham University (1979).
2. H. Eikemeier and H. H. Hackenbroich, Nucl. Phys. A169, 407 (1971).
3. L. Majling, J. Rizek, Z. Pluhar, and Yu. F. Smirnov, J. Phys. G: Nucl. Phys. 2, 357 (1976).
4. R. Ceuleneer and P. Vandepeutte, Phys. Rev. C31, 1528 (1985).
5. D. Gogny, P. Pires, and R. de Tourriel, Phys. Lett. 32B, 591 (1970); R. A. Malfleit and J. A. Tjon, Nucl. Phys. A127, 161 (1969).
6. M. Moshinsky, Nucl. Phys. 13, 104 (1959); T. A. Brody, G. Jacob, and M.Moshinsky, Nucl. Phys. 17, 16 (1960).
7. I. Talmi, Helv. Phys. Acta 25, 185 (1952).
8. R. H. Richardson, J. Shapiro, and F. B. Malik, Phys. Rev. C3, 84 (1971).
9. R. R. Chasman and S. Wahlborn, Nucl. Phys. A90, 401 (1967).
10. A. Messiah, "Quantum Mechanics", Wiley, New York (1962) p.454.
11. P. M. Morse and H. Feshbach, "Methods of Mathematical Physics", Wiley, New York (1953) p.784.
12. E. U. Condon and G. H. Shortley, "Theory of Atomic Spectra", Cambridge (1957).
13. A. Bohr and B. Mottelson, "Nuclear Structure, Vol II", Benjamin, Reading, Mass. (1975) p.233.
14. R. B. Wiringa, R. A. Smith, and T. L. Ainsworth, Phys. Rev. C29, 1207 (1984).
15. Yu. F. Smirnov, Nucl. Phys. 39, 346 (1962).
16. Z. Pluhar and J. Tolar, Czech. Journ. Phys. B14, 287 (1964).
17. E. Chacon and M. de Llano, Revista Mexicana de Fisica 12, No.2, 57 (1963).
18. M. L. Domergue, Comptes Rendues B264, 403 (1967).

MICROSCOPIC CALCULATIONS OF ALPHA-NEUTRON SCATTERING

J. Carlson, K. E. Schmidt, and M. H. Kalos
Courant Institute of Mathematical Sciences
251 Mercer St, New York, NY 10012

INTRODUCTION

In recent years, many calculations[1,2,3] have been reported on three and four body nuclear systems with realistic interactions. The most recent calculations indicate that a microscopic model with realistic two and three nucleon interactions can accurately predict many properties of both ^{3}He and ^{4}He. Calculations of p-shell nuclei will be very valuable as a further test of these interactions. As a first step toward this goal, we have calculated the low energy alpha nucleon phase shifts for the J=1/2 and J=3/2 states. These initial results are for the Reid V8[4] interaction plus the Urbana Model V[1] three nucleon interaction (TNI).

An accurate treatment of the ^{5}He system is required in order to successfully perform variational calculations of the ^{6}He and ^{6}Li systems. These six body nuclei are very loosely bound, and previous calculations[5] of these nuclei as three body (alpha plus two nucleon) systems interacting with phenomenological alpha nucleon interactions have been rather successful. Therefore, one would expect that a successful treatment of ^{5}He would allow one to extend the calculations to heavier systems.

The ^{5}He system has been studied previously by several authors.[6,7] These authors employ relativistic optical models in order to explain the magnitude of the splitting between the J=1/2 and J=3/2 states of alpha nucleon scattering. An important goal of this calculation is to determine to what extent the splitting can be explained by realistic two and three nucleon interactions alone.

The phase shifts are calculated using the variational Monte Carlo method described in reference 8. Several enhancements to the method are presented. These improvements substantially reduce the statistical error in determining the phase shifts. They also allow us to accurately determine other quantities, including the contribution of various terms in the interaction to the splitting.

INTERACTION AND WAVE FUNCTION

The hamiltonian for a nuclear system may be taken to be

$$\sum_i \frac{-\hbar^2}{2m}\nabla_i^2 + \sum_{i<j} V_{ij} + \sum_{i<j<k} V_{ijk}. \tag{1.1}$$

The two nucleon interaction V_{ij} is given by

$$V_{ij} = \sum_k v^k(r_{ij})\, O^k_{ij}. \tag{1.2}$$

where the sum over the operators k extends to 8 for the Reid V8 interaction or 14 for the Urbana,[9] Argonne,[10] or Paris[11] interactions. The three nucleon interaction of reference 1 may be written as a sum of a two pion exchange term ($V^{2\pi}$) which is attractive in light nuclei, and an intermediate range repulsive term (V^R).

The variational wave function used for the ^{4}He and ^{5}He systems is given by the expression

$$\Psi = S\,[\prod_{i<j} F_{ij}]\,\Phi. \tag{1.3}$$

S indicates a symmetrization of the pair correlations, and Φ is a slater determinant of one body states coupled to the correct total angular momentum

$$\Phi = A \prod_i \,[\,\chi_\sigma(i)\ \chi_\tau(i)\ \phi(r_{i\alpha})]. \tag{1.4}$$

A is an anti-symmetrization operator and $r_{i\alpha}$ is the distance from particle i to the system's center of mass in ^{4}He, and the distance to the center of mass of the four s-state particles in ^{5}He. The center of mass of the alpha particle in ^{5}He is different for different terms in the anti-symmetrized Φ.

The single particle states $\phi(r)$ are taken to be solutions of the Schroedinger equation in a Woods-Saxon well with the desired boundary condition. The s-wave solutions are bound states and the p-wave solutions are scattering states. The depth, radius and skin thickness of the well are variational parameters, and may be different for the two states. This parametrization is different from earlier s-shell calculations.[1,3,12] In previous calculations, Φ had no spatial dependence and the asymptotic form of the wave function as one particle separates from the rest was incorporated into the pair correlation operators.

The pair correlation operators are obtained by solving the differential equations:

$$-\frac{\hbar^2}{m}\left[\tilde{\phi}_{S,T}\nabla^2 f_{S,T} + 2\nabla\tilde{\phi}_{S,T}\cdot\nabla f_{S,T}\right] \tag{1.5}$$

$$+\left[\lambda_{S,T} + v_{S,T}\right] f_{S,T}\,\tilde{\phi}_{S,T} = 0$$

in the singlet channels and the coupled equations :

$$-\frac{\hbar^2}{m}\left[\tilde{\phi}_{S,T}\nabla^2 f_{S,T} + 2\nabla\tilde{\phi}_{S,T}\cdot\nabla f_{S,T}\right] \tag{1.6}$$

$$+\left[\lambda_{S,T} + v_{S,T}\right] f_{S,T}\tilde{\phi}_{S,T} + 8\ v_{t,T}\ f_{t,T}\tilde{\phi}_{S,T} = 0$$

$$-\frac{\hbar^2}{m}\left[\tilde{\phi}_{S,T}\nabla^2 f_{t,T} + 2\nabla\tilde{\phi}_{S,T}\nabla f_{t,T} - \frac{6}{r^2} f_{t,T}\tilde{\phi}_{S,T}\right] \tag{1.7}$$

$$+\left[\lambda_{S,T} + v_{S,T} - 2v_{t,T} - 3v_{b,T}\right] f_{S,T}\tilde{\phi}_{S,T} + \left[v_{t,T}\right] f_{S,T}\tilde{\phi}_{S,T} = 0$$

The functions $f_{S,T}$ and $f_{t,T}$ are then recast into operator form so that

$$F(r_{ij}) = f^c(r_{ij})\left[1 + u_3 \sum_k u^k_{ij}\ O^k_{ij}\right], \tag{1.8}$$

where f^c and u^k are obtained from 1.5-1.7, and u_3 is the three body correlation given in reference 12.

The boundary conditions require that f^c and u^σ go to zero at a distance d, while the tensor correlation has a long range. The functions $\tilde{\phi}_{S,T}$ in the two body equations are s and p wave radial functions in a harmonic oscillator potential. The strength of the oscillator and the distance d are variational parameters. These correlations do not include $L\cdot S$ terms, and these terms may be important for ^5He. We plan to introduce them in future calculations.

METHOD

The method used to calculate the phase shifts is presented in detail in reference 8. We give a brief review of the original method and discuss the enhancements introduced in these calculations.

The low energy phase shifts in a one channel scattering problem may be determined by converting it into an equivalent bound state problem. The energy of this bound state may then be calculated with standard variational techniques. If the two systems are confined such that the distance between them is less than a distance R_a, and the Schroedinger equation is solved with the boundary condition that the wave function be zero when the systems are separated by R_a, the phase shift δ_l is given by:

$$\tan(\delta_l) = \frac{j_l\,(kR_a)}{n_l\,(kR_a)}, \tag{2.1}$$

where

$$k = [2\,\mu\,E_{sep}/\hbar^2]^{1/2}. \tag{2.2}$$

In these equations, μ is the reduced mass and j_l and n_l are spherical bessel functions. E_{sep} is the separation energy, defined as the difference between the total energy of the complete system and the sum of the energies of the separated systems. This procedure requires R_a to be large enough so that the potential acting between the systems is zero when they are separated by R_a. Boundary conditions other than $\Psi(R_a)=0$ may be employed, and equation 2.1 altered appropriately.

We use the Metropolis Monte Carlo Method to calculate the energies of light nuclei.[12] This method involves sampling a set of configurations in coordinate space with a probability density proportional to the square of the wavefunction. For light nuclei, all spin-isospin states of the wavefunction are calculated at every step of the random walk. The expectation value of the hamiltonian is determined by summing over all spin-isospin states at each point and then taking the average over all configurations.

The quantity of interest in these calculations is the separation energy E_{sep}, which is the difference between the ^5He and alpha particle energies. The energies of the two systems are determined by the variational Monte Carlo approach which has an inherent statistical error. The statistical error in E_{sep} would be dominated by the error in the energy of the alpha particle if two separate Monte Carlo calculations were undertaken and the energies subtracted. In order to avoid this, we write

$$E(^4He) = \frac{\int dr_1 \cdots dr_5 \ \Psi_4^\dagger(1,2,3,4) \ H_4 \ \Psi_4(1,2,3,4) G(\vec{r}_{5,\alpha})}{\int dr_1 \cdots dr_5 \ \Psi_4^\dagger(1,2,3,4) \Psi_4(1,2,3,4) G(\vec{r}_{5,\alpha})}, \tag{2.3}$$

where Ψ_4 is the alpha particle wave function and H_4 is the Hamiltonian acting on nucleons 1 through 4. As long as G is a function of the vector from r_5 to the center of mass of the alpha particle, the integrals over r_5 cancel and we are left with the usual expression for the energy of the alpha particle.

Rewriting the energy in this way is useful because it allows us to use the same set of configurations obtained in calculating the energy of ^5He to calculate the energy of the alpha particle. If we have a set of points R_i distributed with probability density $W(r_1..r_5) = \Psi_5^\dagger(r1..r5)\Psi_5(r1..r5)$, both the numerator and denominator of equation 2.3 may be multiplied by W, and the energy of the alpha particle is given by

$$E(^4He) = \frac{\sum_i \frac{\Psi_4^\dagger H_4 \Psi_4}{\Psi_4^\dagger \Psi_4} \frac{\Psi_4^\dagger \Psi_4 G(\vec{r}_{5,\alpha})}{W(R_i)}}{\sum_i \frac{\Psi_4^\dagger \Psi_4 G(\vec{r}_{5,\alpha})}{W(R_i)}} \tag{2.4}$$

In order to minimize the variance, we choose

$$G(\vec{r}_{5,\alpha}) = \phi(r_{5,\alpha})^2 \,\Theta(r_{5,\alpha} - R_0) \tag{2.5}$$

where ϕ is the single particle wave function in the p-wave state. This choice leads to a low variance because each term in the denominator of equation 2.4 is near 1. For the extreme case of two non-interacting particles, this method will give the correct answer with zero variance. The Θ function is included in $G(\vec{r}_{5,\alpha})$ so that those configurations in which all the particles are close together do not contribute to the expectation value in ^4He. This is necessary to prevent large fluctuations when particle 5 is near one of the other particles.

The energy of ^5He may be determined from the same set of configurations in the usual manner:

$$E(^5He) = \frac{\sum_i \frac{\Psi_5^\dagger H \Psi_5}{W(R_i)}}{\sum_i \frac{\Psi_5^\dagger \Psi_5}{W(R_i)}} \tag{2.6}$$

The statistical error in the difference $E(^5He) - E(^4He)$ is significantly lower when calculated in this manner than when the two energies are determined independently. A reduction in computer time by a factor between 5 and 25 is obtained, depending upon the state being calculated and the value of R_a. It is also very useful to minimize the variational energy of ^5He by reweighting a set of configurations rather than performing independent calculations. Energy differences may easily be determined within 0.1 to 0.2 MeV in this manner.

RESULTS

The results of our calculations are summarized in table 1 and figures 1 and 2. The first step in determining the phase shifts for ^5He is to calculate the binding energy of ^4He. The results of this calculation are presented in the first column of table 1. The alpha particle is overbound by about 1 MeV with this interaction, and the three nucleon interaction contributes approximately 8 MeV. Without the TNI, we obtain a minimum energy of -23.6 ± 0.3 MeV, which is statistically indistinguishable from the energy published previously[12] ($-22.9 \pm .5$ MeV) using the different parametrization. Therefore it seems plausible that our wave function provides an accurate description of the alpha particle.

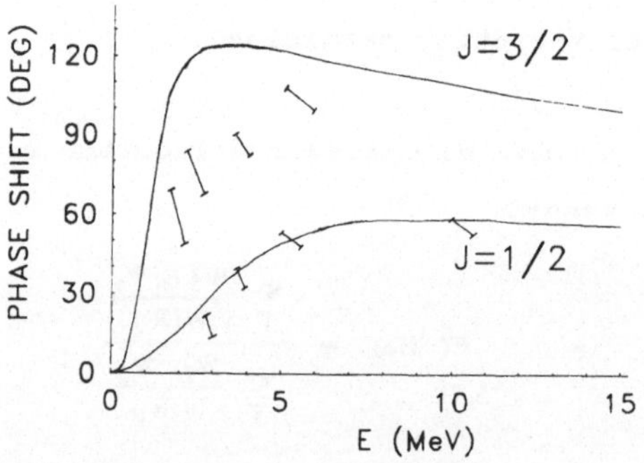

Figure 1) Phase shifts (calculated and experimental) versus energy for J=1/2 and J=3/2 alpha nucleon scattering.

TABLE 1. Potential Energy Contributions (R_n=5.5 fm)

term	^{4}He	^{5}He (J=1/2)	^{5}He (J=3/2)
V6	−151.0 ± 2.0	−17. ± 2.0	−23. ± 3.0
V8-V6	3.7 ± 0.2	1.6 ± 0.2	−1.2 ± 0.3
$V^{2\pi}$	−8.3 ± 0.2	0.05 ± 0.08	−1.3 ± 0.2
V^R	1.4 ± 0.04	0.24 ± 0.06	0.5 ± 0.1
energy	−29.9 ± 0.2	10.3 ± 0.3	5.5 ± 0.4

Figure 1 presents the calculated and experimental[13] phase shifts for both the J=1/2 and 3/2 states, and the equivalent information is presented in figure 2 as a plot of the separation energy E_{sep} versus R_n. In this figure, the dashed line represents the energy for a free nucleon confined within R_n, the solid lines gives the experimental results, and the points with error bars are the results of our calculations.

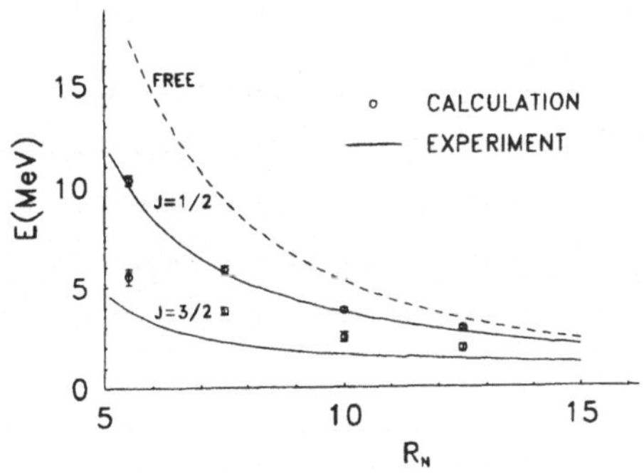

Figure 2) Energy versus radius (R_n) for ^{5}He. The dashed line indicates the energy of a free particle confined within R_n.

As can be seen in figure 2, the J=1/2 energies are consistent with experiment, while the J=3/2 energies are too high by roughly 1.5 MeV. Thus, we obtain a splitting between the two states which is somewhat smaller than the experiment would indicate. Nevertheless, we obtain 85 per cent of the difference between the free particle solution and the experimental J=3/2 result at R_n=5.5 fm. This represents approximately three fourths of the experimental splitting between the J=1/2 and J=3/2 states.

The total energy and the contributions of various potential terms are presented in Table 1. The ^{5}He results are given for R_n=5.5 fm. Column 1 gives the expecation values for the alpha particle, and columns two and three give the additional contribution in the J=1/2 and 3/2 ^{5}He states, respectively. These differences are obtained with the same subtraction technique used for the total energy. It is apparent from the table that significant contributions to the energy splitting arise from both the L·S and the three nucleon interaction. These terms contribute approximately 3 and 1 MeV, respectively.

CONCLUSIONS

These calculations suggest that a substantial fraction of the splitting between the J=1/2 and J=3/2 states in alpha neutron scattering may be explained in a microscopic non-relativistic treatment with realistic interactions. In particular, we obtain accurate results for the J=1/2 state using the Reid V8 plus TNI model. The J=3/2 state is somewhat too high (≈1.5 MeV) in energy, however. This discrepancy will make it difficult to perform variational calculations on ^{6}He and ^{6}Li, since they are very loosely bound.

These calculations will be repeated with a more realistic two nucleon interaction in order to determine how strongly our results depend upon the interaction. We also plan to enhance the variational wave function by introducing L·S pair correlations. Results with this more general wave function will help us to determine the accuracy of our variational calculations.

In summary, this microscopic model provides a good qualitative picture of low energy alpha neutron scattering. However, discrepancies do remain, and the source of these discrepancies should be more fully investigated. In addition, we believe that the subtraction techniques employed in this work will prove very useful in many low energy scattering problems.

This work was supported by the Division of Nuclear Physics of the Office of High Energy and Nuclear Physics, U. S. Department of Energy under contract DE-AC02-79ER10353.

REFERENCES

1. J. Carlson, V. R. Pandharipande and R. B. Wiringa, Nucl. Phys. A401 (1983) 59.
2. C. R. Chen, G. L. Payne, J. L. Friar, and B. F. Gibson, Los Alamos preprint LA-UR-85-1472 (1985).
3. R. Schiavilla, V. R. Pandharipande and R. B. Wiringa, University of Illinois preprint ILL-(NU)-85-#40 (1985).
4. R. V. Reid, Ann. Phys. (NY) 50 (1968) 411.
5. D. R. Lehman and W. C. Parke, Phys. Rev. Lett. 50 (1983) 98.
6. L. G. Arnold, B. C. Clark, R. L. Mercer, D. G. Ravenhall, and A. M. Saperstein, Phys. Rev. C14 (1976) 1878.
7. L. G. Arnold, B. C. Clark and R. L. Mercer, Phys. Rev. C19 (1979) 917.
8. J. Carlson, V. R. Pandharipande and R. B. Wiringa, Nucl. Phys. A424 (1984) 427.
9. I. E. Lagaris and V. R. Pandharipande, Nucl. Phys. A355 (1981) 331.
10. R. B. Wiringa, R. A. Smith and T. L. Ainsworth, Phys. Rev. C29 (1984) 1207.
11. M. Lacombe, B. Loiseau, J. M. Richard, R. Vinh Mau, J. Côté, P. Pirès, R. de Tourreil, Phys. Rev C21 (1980) 861.
12. J. Lomnitz Adler, V. R. Pandharipande, and R. A. Smith, Nucl. Phys. A361 (1981) 339.
13. R. A. Arndt and L. D. Roper, Nucl. Phys. A209 (1973) 447.

GREEN'S FUNCTION MONTE CARLO CALCULATIONS OF EFFECTIVE PI-ELECTRON HAMILTONIANS

Michael A. Lee, S. Klemm, and S. Risser

Department of Physics and Liquid Crystal Institute
Kent State University
Kent, Ohio 44242

INTRODUCTION

Effective electronic Hamiltonians have long been popular in both physical and chemical studies of systems with delocalized outer valence electrons. We will report here on a Green's Function Monte Carlo (GFMC) method which we have employed to calculate the ground state energy and polarizability of a class of organic molecules with extended pi-electron systems. We have used the most popular and accurate effective pi-electron Hamiltonian for these conjugated molecules, the Pariser-Parr-Pople[1] Hamiltonian. This Hamiltonian includes long range coulomb interactions between electrons in addition to the usual Hückel hopping term.

THE EFFECTIVE HAMILTONIAN

We will consider here the simplest class of molecules which can be adequately described by effective pi-electron Hamiltonians, linear hydrocarbon chains. In these chains, which are shown in Figure 1, each carbon atom has a p-orbital, and contributes one loosely bound electron to the electronic system. Many of the physical and chemical properties of these molecules, such as their polarizability and reactivity are attributed to the delocalized pi-electron system. This suggests describing the loosely bound pi-electrons in terms of a Hamiltonian which includes "effective" coulomb interactions.

A particularly simple effective Hamiltonian is the tight-binding, or Hückel[2] Hamiltonian. It neglects the two-body interactions and describes the system in terms of a "hopping" Hamiltonian

$$H = - \sum_{i,\sigma} t_i (a^\dagger_{i\sigma} a_{i+1,\sigma} + a^\dagger_{i+1,\sigma} a_{i\sigma})$$

The transfer integral t_i characterizes the probability of an electron to hop from site i to i+1 and vice versa. $a^\dagger_{i+\sigma}$ ($a_{i,\sigma}$) creates (destroys) an electron of spin σ in a p-orbital at atomic site i. The value of the transfer integral depends on the atomic species (carbon here) and the separation between atoms i and i+1. Exact analytic and numerical solutions for the Hückel Hamiltonian can be found in elementary texts on quantum chemistry.[2] In the worst case it requires the diagonalization of an NxN matrix, where N is the number of carbon atoms.

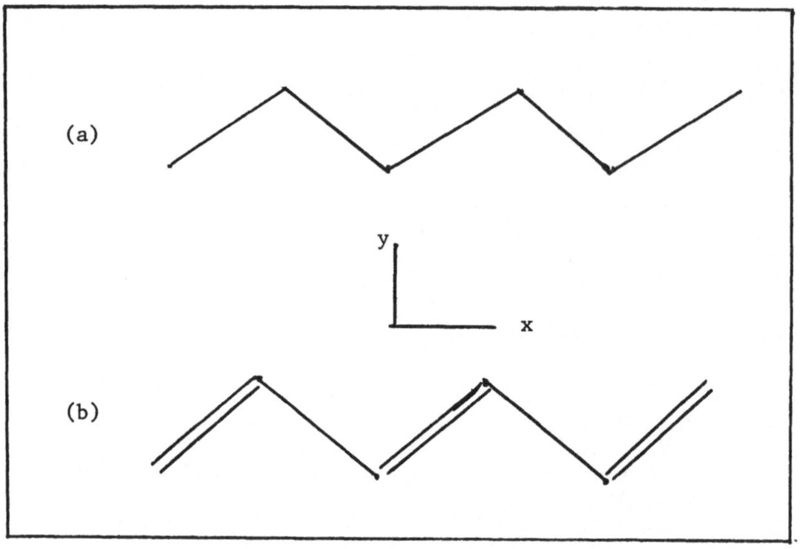

Fig. 1. Nonalternating (a) and alternating (b) bond length models of a six carbon chain.

Hubbard[3] proposed a model Hamiltonian, originally for d-band transition metals, which extended the Hückel-type Hamiltonian to include a repulsive interaction when two electrons occupied the same atomic site, i.e.

$$\frac{1}{2}\sum_i U n_i(n_i-1).$$

The number operator

$$n_i = \sum_\sigma a^\dagger_{i\sigma} a_{i\sigma}$$

takes on the value 0, 1 or 2. This model Hamiltonian is particularly attractive for one-dimensional Hamiltonians which attempt to describe organic conductors and charge-transfer salts.[4] In an infinitely long one-dimensional system, the Hamiltonian can be solved analytically[5] when $t_p = t$ = constant and $U = 4t$ and can be solved numerically[6] when U/t is arbitrary. Unfortunately, neither the Hückel or Hubbard forms of the Hamiltonian are accurate enough to characterize quantitatively the chemical and physical properties of conjugated hydrocarbon molecules.

The Pariser-Parr-Pople Hamiltonian includes long range coulomb electron-electron, electron-nuclear, and nuclear-nuclear interactions, by adding to the Hubbard Hamiltonian the term

$$\sum_{i<j} V_{ij}\,(n_i-z_i)(n_j-z_j)$$

where z is the screened nuclear charge and equals the number of electrons contributed by the ith atom.

Much effort has gone into obtaining the best values for the effective interaction constants of the PPP Hamiltonian. Ohno[7] has arrived at a form for V which is coulomb-like at large distances, $V_{ij} \sim e^2/r$, and equals U when two electrons occupy the same atomic site.

$$V_{ij} = \frac{U}{(1+(r_{ij}/a)^2)^{1/2}}$$

The constant a is determined by the large r behavior once the onsite interaction is chosen.

The parameters which remain to completely specify the PPP Hamiltonian are t_i at each site, the onsite interaction U, and the positions of the atoms. We have chosen these constants to be consistent with literature values and in particular to be the same as those chosen by Soos and Ramasesha.[8] They have reported calculations of the ground state energy and other electronic properties of carbon rings and chains of up to 14 atoms. Their solution method involves a choice of a basis set that is different from ours but one which allows the reduction of the matrices to a more manageable size. A direct diagonalization approach to the PPP Hamiltonian for N electrons on N atomic sites would require a matrix of rank $(N!/(N/2)!^2)^2$. This is 400 for N=6 and 853,776 for N=12. This is a clear motivation to find alternative methods.

The full PPP Hamiltonian which we have used is

$$H = - \sum_{i,\sigma} t_i(a^\dagger_{i\sigma}a_{i+1,\sigma} + a^\dagger_{i+1,\sigma}a_{i\sigma}) + \frac{U}{2}\sum_i n_i(n_i-1) + \sum_{i<j} V_{ij}(n_i-z_i)(n_j-z_j) \tag{1}$$

with $z_i = 1$, $U = 11.26$ ev and $t_i = 2.40$ ev $(1-\delta(-1)^i)$. The bond alteration parameter δ takes on the value $\delta=0$ when all carbon atoms have a bond length of 1.40 Å (i.e. nonalternating carbon chains) and $\delta=.07$ when the chain exhibits nominally alternating single and double bonds of length 1.45 Å and 1.35 Å respectively. We will furthermore assume ideal 120° angles between the bonds in all cases.

THE GFMC METHOD

The GFMC method was originally developed by Kalos and coworkers[9] and used primarily to study quantum fluids. Its presentation in a form suitable for treating Hubbard Hamiltonians has been given elsewhere.[10] We will briefly review the method here with particular emphasis on the complications arising from the more complex geometry of finite molecules.

From the theory of hermitian matrices, we know that if one multiplies the Hamiltonian matrix with an arbitrary vector repeatedly, then the resulting vector will eventually converge to the eigenvector of the system that has the largest (in absolute magnitude) eigenvalue. Similarly, repeated multiplication of an arbitrary vector by the inverse of the Hamiltonian yields the eigenvector with the smallest (most near zero) eigenvalue. In the latter case, the eigenvector will be the ground state when the eigenvalue spectrum is positive definite. This can always be achieved by simply adding a constant $W > -E$ to the Hamiltonian and defining the Green's function as the inverse of H+W, i.e.

$$(H+W)G = I \tag{2}$$

The above procedure is actually more generally applicable to excited states when implemented in matrix operations, but the GFMC method is particularly suited to ground state properties and hence we will always pick $W > -E$. This also has the benefit that the Green's function itself is positive definite in the basis set that we have chosen.

The GFMC method employs a random walk process which moves particles on lattice sites in real space. We will thus define a basis set for $N = N\uparrow + N\downarrow$ spins in terms of an N-tuple,

$$J = (j_1, j_2, \ldots, j_N)$$

where j_1 to $j_{N\uparrow}$ specifies the sites occupied by the up spins while the remaining indices identify the locations of the down spins. Because of the antisymmetry of the fermion wavefunction, not all such basis vectors are linearly independent, unless we add the restriction that $j_\ell < j_{\ell+1}$ when particles ℓ and $\ell+1$ have the same spin.

If we denote by ψ_J the amplitude of the ground state wavefunction on basis state J, and the Green's function matrix by $G_{IJ} = \langle I|G|J\rangle$, then the equation we will use to obtain ψ is

$$\psi_I = (E+W) \sum_J G_{IJ}\psi_J \tag{3}$$

The iterative solution of the above relation can be described in terms of a random walk process. Consider a population of configurations where the probability of finding any one configuration on the basis state J is proportional to the value of the ground state wavefunction ψ_J (not $\psi^2{}_J$). The quantity $(E+W)G_{IJ}$ describes the likelihood of the movement of a configuration from state J to state I during a random walk. Precisely, it is the expected density of endpoints or terminations of a random walk beginning from state J. The actual details of each step of the random walk are contained in the definition of G in Eq. (2) when expressed in terms of matrix elements of G. Substituting the PPP Hamiltonian into Eq. (2) yields

$$-\sum_{nn} t_{Jnn}\, G_{I,J+nn} + (W+V_J)G_{IJ} = \delta_{IJ} \tag{4}$$

The notation $V_J = \langle J|V|J\rangle$ refers to the value of the potential or non-hopping part of the Hamiltonian. J+nn is a configuration that differs from J by moving a single particle one site left or right. t_{Jnn} is the value of t_i for that pair of sites involved in the move. The sum is over all such nearest neighbor states of J excluding those which would result in two particles of the same spin occupying the same site. To obtain this form of the equation we have used the symmetry $G_{IJ} = G_{JI}$. Rearranging Eq. (4) to obtain $(E+W)G_{IJ}$ yields

$$(E+W)G_{IJ} = \delta_{IJ}\frac{(E+W)}{(W+V_J)} + \frac{S_J}{(W+V_J)}\sum_{nn}\frac{t_{Jnn}}{S_J}(E+W)G_{I,J+nn}.$$

where $S_J \equiv \sum_{nn} t_{Jnn}$.

Since $(E+W)G_{IJ}$ is the density of terminations on basis element I, the first term tells us that the density of terminations in exactly one step on state J=I is $(E+W)/(W+V_J)$. The second term says that the density of moves to an adjacent configuration is $S_J/(W+V_J)$, and with probability t_{Jnn}/S_J the move is to the particular neighboring state specified by nn. t_{Jnn}/S_J is a normalized probability. Once zero, one, or more configurations are moved, they are propagated from their new locations J+nn by repeating the above procedure, this time for $(E+W)G_{I,J+nn}$. The locations of all terminations (there may be none from any particular configuration in the population) are the starting points for a new set of random walks in the next iteration of Eq. (3). Since the true ground state energy is not known, a trial value E_T must be used in the random walks. The number (density) of configurations will on the average increase or decrease depending on whether E_T is higher or lower than the true ground state energy. By adjusting E_T until a stable population size is achieved, a value of E_T arbitrarily close to the correct ground state energy may be obtained.

The above prescription with a population of a few hundred configurations and iterating the process a few hundred times will give a ground state energy to less than a percent uncertainty in about 15 minutes on a Ridge 32C computer. In practice, an order of magnitude increase in efficiency can be achieved if one has a good analytic trial wavefunction, $\psi_T(J)$, to use with importance sampling in the random walk. This is a technical point, and it is implemented in the same fashion for a lattice model as it is in the continuous space GFMC developed by Kalos.[9] We will, therefore, not describe it here. Suffice it to say that the importance sampling scheme allows us to evaluate the energy as an average of local energies

$$E \approx \left\langle \frac{H\psi_T(J)}{\psi_T(J)} \right\rangle$$

taken over fewer random walks.

RESULTS

We would first like to quote the results for the ground state energy of alternating and nonalternating carbon chains with $N \leq 30$. These are shown in Table I. The statistical uncertainty of our results is better than one part in a thousand, and agrees with the results Soos and Ramasesha[8] for the $N \leq 12$ chains. An indication of the accuracy of the GFMC results can be obtained by closely comparing one chain, the N=6 nonalternating chain with two choices of parameters. The results quoted in Ref. 8 are for a bond length of $r_o = 1.397$ Å. The bond length used in our work is 1.40 Å. Soos[11] calculates the energy of these two systems to be -1.9519 ev and -1.9504 respectively. In the latter case, we obtain -1.9505 ± .00015. These numbers serve to demonstrate not only that our results agree with numerically exact matrix diagonalization methods, but that these quantum simulation methods can provide answers that are accurate enough to distinguish between proposed model Hamiltonians which may not be soluable by traditional theoretical methods.

The ground state energy of an infinite PPP chain can be estimated by plotting energy per particle E/N vs. 1/N. The primary differences between the energy per particle of a finite chain and the infinite system will be end effects and should scale like N^{-1} for large N. Our estimate for the $N=\infty$ limit is E/N = -2.135 ± .001 for the nonalternating chains and -2.202 ± .002 for the alternating chains. These estimates are in agreement with the predictions of Soos and Ramasesha[8] which were -2.14 ± .02 and -2.20 ± .02 ev respectively. We are currently in the process of extending these calculations to systems with periodic boundary conditions. Such rings have no chemical reality, but should show a change in the energy per particle which scales as N^{-2}, thus allowing a much more accurate $N=\infty$ extrapolation.

One of the qualitative features attributed to conjugated molecules is the high polarizability of the pi-electron cloud. While it is possible

Table I. Ground State Energies[a] for Alternating (δ=.07) and Nonalternating (δ= 0) Carbon Chains in units of t = 2.40 ev .

N	δ = 0	δ = .07
4	-.7782 ± .0001	-.8358 ± .0001
6	-.8128 ± .0001	-.8621 ± .0002
8	-.8309 ± .0001	-.8754 ± .0002
10	-.8421 ± .0001	-.8837 ± .0002
20	-.8652 ± .0004	-.9005 ± .0002
30	-.8740 ± .0003	-.9061 ± .0003

[a]Energy per particle.

to do perturbation theory using GFMC techniques,[12] it is also possible to apply an external field $\vec{F}$ to the Hamiltonian

$$- \sum_i e\vec{F}\cdot\vec{r}_i$$

and directly calculate the energy as a function of field strength. Figure 2 shows a plot of the change in total energy of a nonalternating chain as a function of a field applied along the long axis (x-axis) of the molecule. Also plotted in this graph are the energy changes of the same system as calculated by directly diagonalizing a 400 by 400 Hamiltonian matrix including the field. For comparison, we have also calculated the energy change for the Hückel Hamiltonian.

The first and most obvious observation is that the Hückel Hamiltonian predicts a larger polarizability. This is clearly due to the fact that

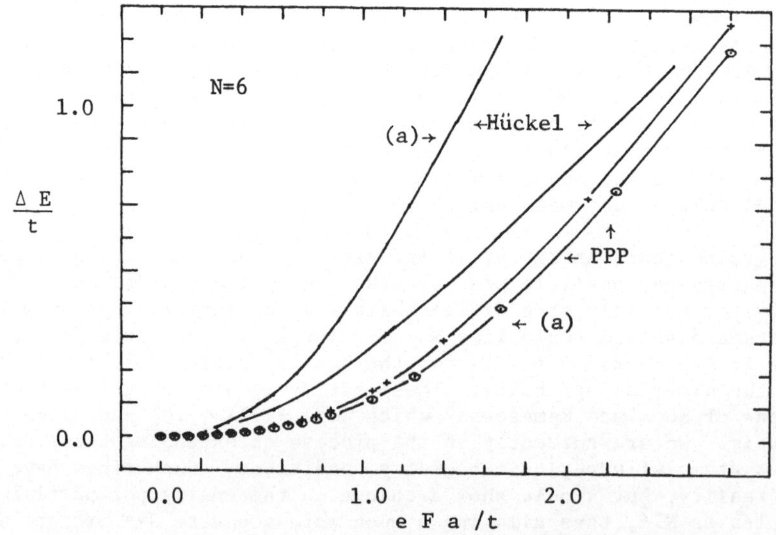

Fig. 2. Change in energy per particle vs field strength F in dimensionless units. t=2.40 ev, a=1.40A. Field is in the x direction. Solid lines are diagonalization results. (a) means alternating.

the absence of coulomb repulsion in the Hamiltonian allows the electrons to be pushed more easily to one end of the molecule. Secondly, at high fields the energy versus field becomes linear. This merely reflects the fact that one element of the basis set, that with all the electrons on one-half of the molecule, dominates the ground state in an extreme field. The slope of this line is simply the dipole moment of that basis state.

The primary reason for plotting the energy change versus field is to extract the molecular polarizability. This was done by fitting the small field region of the plot to a sixth order polynomial. The range of field strengths which were fit was decreased until the second and fourth order coefficients were independent of that range. In Table II we have listed the molecular polarizability of each molecule for a field along the long molecular axis. The clear trend is towards increasing molecular polarizability with increasing chain length. Elementary arguments would suggest that the polarizability should increase as N^2. Although this behavior is qualitatively correct, these elementary arguments are unable to predict the actual magnitude of the polarizability. Another feature evident from the table is that the alternating chains are less polarizable than the nonalternating chains. Having observed this behavior, one might be willing to say that this reflects the more localized nature of the electrons in the double bonds of the alternating chain. This observation does not relate particularly to experimental properties of nonalternating even carbon chains because they do not exist. Rather, it says that models which do not include the alteration are not quantitatively accurate.

CONCLUSIONS

Even carbon chains are the simplest systems which can be treated by effective pi-electron Hamiltonians. The accuracy of the PPP Hamiltonian and the capabilities of quantum simulation methods encourage the investigation of the properties of the many other aromatic molecules, including those with hetero atom substituents and more complicated geometries. The properties of carbon rings with 4N+2 atoms is a straightforward extension of this method. The full molecular polarizability tensor is calculable in this model and is currently under investigation. With the addition of an accurate core-core interaction between the shielded nuclei the ground state configuration of the molecule can be calculated using GFMC methods. A currently popular subject, the structure and energy of a soliton in a polyacetylene chain is presumably also within the scope of this method.

We cannot end this discussion without making two points clear. Firstly, the accuracy of the lattice GFMC method, and its ability to predict

Table II. Calculated Dimensionless Molecular Polarizabilities[+]

N	Nonalternating	Alternating
4	.294 ± .002	.295 ± .004
6	.609 ± .004	.530 ± .014
8	1.024 ± .024	.841 ± .02
10	1.485 ± .04	1.321 ± .06

[+]The dimensionless polarizability α is defined as

$$\Delta E/t = -\alpha\,(eFa/t)^2$$

physical and chemical properties, depends on the accuracy of the Hamiltonian. Secondly, the method as presented here relies on the positive definite nature of the Green's function on the ordered basis which we chose. This requirement is not met by ring structures like naphthalene because it is possible for two particles to interchange positions during the random walk without occupying the same site in the process. The chemical importance of multiply-connected ring molecules encourages attempts to resolve this limitation.

Given the above caveat, the GFMC method has demonstrated its ability to accurately characterize certain molecular properties that are inaccessible to traditional matrix methods.

ACKNOWLEDGEMENTS

The authors would like to acknowledge the benefit of discussions with Z. Soos. The preliminary calculations and program development were done on a Ridge 32C, made available to us through the special acquisitions program of Ridge Corporation. The final number crunching was possible because of a generous allotment of computer time by the Kent State University Department of Computer Services. The material presented here is based in part upon work supported under a National Science Foundation Graduate Fellowship held by one of us (S.K.).

REFERENCES

1. R. Pariser and R.G. Parr, J. Chem. Phys. 21:446 (1953), and 21:767 (1953); J.A. Pople, Trans. Faraday Soc. 42:1375 (1953).

2. L. Salem, "The Molecular Orbital Theory of Conjugated Systems," W.A. Benjamin, Inc. (1966).

3. J. Hubbard, Phys. Rev. B 17:494 (1978).

4. "Highly Conducting One-Dimensional Solids," J.T. Devrees, R.P. Evrard, and V.E. Van Doren, ed., Plenum Press, New York (1979).

5. E.H. Lieb and F.Y. Wu, Phys. Rev. Lett. 20:1445 (1968).

6. H. Shiba, Phys. Rev. B 6:930 (1972).

7. K. Ohno, Theor. Chim. Acta 2:219 (1964).

8. Z.G. Soos and R. Ramasesha, Phys. Rev. B 29:5410 (1984), and Phys. Rev. Lett. 51 (1983).

9. M.H. Kalos, D. Levesque, and L. Verlet, Phys. Rev. A 9:2178 (1974); M.H. Kalos, M.A. Lee, P.A. Whitlock, and G.V. Chester, Phys. Rev. B 24:115 (1981).

10. M.A. Lee, K.A. Motakabbir, and K.E. Schmidt, Phys. Rev. Lett. 53:1191 (1984).

11. Z.G. Soos, personal communication.

12. K.E. Schmidt, personal communication.

LONG-RANGE AND ELEMENTARY CONTRIBUTIONS FOR

QUANTUM FLUIDS AT ZERO TEMPERATURE

S. Rosati, M Viviani, and E. Buendìa*
Dipartimento di Fisica, Università di Pisa, Pisa, Italy
Istituto Nazionale di Fisica Nucleare, Sezione di Pisa
Pisa, Italy

A.Fabrocini+
Istituto di Fisica Generale, Facoltà di Ingegneria
Università di Pisa, Pisa, Italy
Istituto Nazionale di Fisica Nucleare, Sezione di Pisa
Pisa, Italy

ABSTRACT

The effect of the optimal correlation function on the energy and momentum distribution is analized in ^{4}He and normal ^{3}He liquids. The interpolating integral equation is used for approximating the elementary contribution. The energies obtained in both cases are in good agreement with the results furnished by the scaling approximation. The value estimated for the ^{4}He condensate fraction is close to that one calculated by PUOSKARI and KALLIO by using a mixture formalism, but differs to some extent from the prediction of the scaling technique. The inclusion of a long-range tail in the distribution function results to have a small effect on the normal ^{3}He momentum distribution.

1. INTRODUCTION

Many variational approaches to the study of quantum fluids at zero temperature are based on trial correlated wave function of the form

$$\Psi_v(1,\dots,A) = F(1,\dots,A)\,\Phi(1,\dots,A) = \left[\prod_{i<j=1}^{A} f(i,j)\right]\left[\prod_{i<j<k=1}^{A} f(i,j,k)\right]\dots\,\Phi(1,\dots,A). \tag{1}$$

A is the number of the particles enclosed in a volume Ω with a constant ϱ density in the thermodynamical limit $(A,\Omega\rightarrow\infty)$. The model function Φ takes into account the statistics (and the excitation properties) of the system. For the ground state of a Bose system it is $\Phi=1$, whereas for a Fermi system the function Φ is usually chosen as the Slater determinant of the lowest single-particle wave functions of the Fermi sea. In eq.(1), the correlation operator F(1,...,A) is written as a product of terms involving two-particle, three-particle,..., correlations. The two-body correlation factor f(i,j) between the i and j particles is frequently chosen to depend only on the interparticle distance r_{ij}. The inclusion of correlations involving three or more particles requires a relevant numerical effort, and only for very accurate calculation the three-body correlations have been also included.

We first consider the problem of calculating the radial distribution function (r.d.f.) in the case of a Bose system described by a trial wave function containing two-body correlations only. An important question is how the best variational correlation function f(r) or, equivalently, the corresponding r.d.f. could be determined. In this case g(r) is the solution of a Euler integro-differential equation. In solving such an equation, hovewer, some approximations must necessarily be introduced in order to estimate the terms of "elementary" type which contribute to the coefficients of the Euler equation. The procedure adopted to approximate such elementary contributions is discused in the first part of sect.2. The remaining part of the section is devoted to the calculation of the momentum distribution n(k) and to the discussion of some numerical results obtained for the energy per particle and for n(k).

For Fermi systems the determination of the best variational correlation function f(r) is quite a difficult numerical problem. Due to this, an alternative approximate approach based on the underlying fictitious Bose system is presented at the end of sect.3. The advantages of the method, as well as possible improvements, are discussed in section 4.

2. BOSE SYSTEMS AT T=0

We limit our discussion to the case where the trial wave function is a product of two-body correlation factors depending only on the interparticle distances, i.e.

$$\Psi_v = \prod_{i<j=1}^{A} f(r_{ij}). \tag{2}$$

The r.d.f. g(r) can be written [1,2] as

$$g(r) = 1 + N(r) + X(r) \tag{3}$$

where the nodal function N(r) satisfies the Orstein-Zernike equation

$$N(r_{12}) = \varrho \int d\vec{r}_3 \, X(r_{13}) \left[X(r_{23}) + N(r_{23})\right] \tag{4}$$

and the function X(r) is given by

$$X(r) = f^2(r) \exp[N(r) + E(r)] - 1 - N(r). \tag{5}$$

The function E(r) is the sum of all the so-called elementary contributions and an (approximate) evaluation of the E(r) is necessary for solving eqs.(4) and (5).

A). The method of interpolating integral equation

Two approximations often used to replace the exact expression (5) of X(r), are referred to as the HNC/0 approach and the PY (Percus-Yevick) [3] equation. In the HNC/0 approach the elementary function E(r) appearing in eq.(5) is assumed to be zero. The PY approximation is obtained by linearizing eq.(5) to get

$$X_{PY}(r) = f^2(r)\,[1 + N(r)] - 1 - N(r). \tag{6}$$

The PY equation would be exact if the elementary function E(r) had the form

$$E_{PY}(r) = - N_{PY}(r) + \ln[1 + N_{PY}(r)]. \tag{7}$$

Since in most of cases the PY equation results to be fairly accurate, eq.(7) can be utilized to get an approximate estimate of the elementary function. In a previous paper [4] two of us have derived an integral equation which interpolates between HNC/0 and PY approximations, by adopting for X(r) the expression

$$X_\alpha(r) = f^2(r)\left[\left(\exp\left(\alpha(r)N_\alpha(r)\right)-1\right)/\alpha(r)+1\right] - 1 - N_\alpha(r) \tag{8}$$

where the function $N_\alpha(r)$ satisfies eq.(4). The interpolating function $\alpha(r)$ is expressed in terms of the solution of the PY equation, namely

$$\alpha(r) = 1 + \alpha_o\, E_{PY}(r)/N_{PY}(r) \tag{9}$$

where α_o is a constant parameter. As it is discussed in refs.[4] and [5], the parameter α_o can be determined as it follows. Eqs.(3)-(5) apply exactly to a classical real gas at temperature T, interacting through a two-particle potential

$$v(r) = - KT \ln f^2. \tag{10}$$

The isothermal compressibility of a classical gas is given [6] by

$$\chi = (\partial P/\partial \varrho)_T/KT = 1 - \varrho/3 \int d\vec{r}\, \left(g(r) + \varrho/2(\partial g/\partial \varrho)_T\right) r v'(r)/KT \tag{11}$$

where P is the thermodynamical pressure, or, alternatively, it can be obtained from the relation

$$1/\chi = 1 + \varrho \int d\vec{r}\, \left(g(r) - 1\right). \tag{12}$$

When g(r) is calculated in an approximate way, the last two equations give in general different estimates for χ. The parameter α_o can then be fixed so as to make eqs.(11) and (12) compatible. Explicitely, such "compressibility consistent condition" (CCC) is written in the following form:

$$\left(1+\varrho\int d\vec{r}\,(g_{\alpha}(r)-1)\right)^{-1} = 1-2\varrho/3\int d\vec{r}\; r\left(g_{\alpha}(r)+\varrho/2(\partial g_{\alpha}/\partial\varrho)_{T}\right)f'(r)/f(r). \quad (13)$$

This methods of interpolation allows for very accurate results [4] in the case of correlation functions (or r.d.f.) with short-range tails. Hovewer, if the function g(r) has a tail given as r^{-2}, eq.(13) is no longer useful [7] since both its members are infinite and a different criterion to fix α_o must be devised.

B). The Euler equation for g(r)

The best variational choice of g(r) is fixed by the condition

$$\frac{\delta\langle\Psi_v\,|\,H\,|\,\Psi_v\rangle/A}{\delta g} = 0 \quad (14)$$

where H is the Hamiltonian of the system. As a consequence, the following Schröedinger-type equation can be derived [8,9]

$$-\hbar^2/m\,\nabla^2 g^{1/2}(r) + g^{1/2}(r)\left[v(r) + w_o(r) + w_E(r)\right] = 0. \quad (15)$$

In the eq.(15), v(r) is the two-particle potential, $w_o(r)$ and $w_E(r)$ can be interpreted as potentials induced by the medium. They are given by

$$\hat{w}_o(k) = -\hbar^2/(4m)\left(k^2(S(k)-1)^2\right)\left(2S(k)-1\right)/S^2(k) \quad (16)$$

and

$$w_E(r) = \hbar^2/(4m)\left(\nabla^2 E(r) + E^*(r)\right) \quad (17)$$

$$E^*(r) = \int d\vec{s}\;\nabla^2 g(s)\,\frac{\delta E(s)}{\delta g(r)} \quad (18)$$

where

$$S(k) = \varrho\int d\vec{r}\,(g(r)-1)\exp(i\vec{k}\cdot\vec{r}) \quad (19)$$

is the static structure function. The notation

$$\hat{x}(k) = \varrho\int d\vec{r}\,x(r)\exp(i\vec{k}\cdot\vec{r}) \quad (20)$$

has been adopted. To solve the Euler equation (15), first of all it is necessary to calculate the functions E(r) and E*(r) which appear in eq.(17). The procedure adopted for obtaining E(r) is the following. As a preliminary step, a variational calculation is performed by adopting a short-ranged correlation factor f(r). The corresponding interpolating function

$$\alpha^{(sr)}(r) = 1 + \alpha_o E_{PY}^{(sr)}(r)/N_{PY}^{(sr)}(r) \tag{21}$$

is derived and the constant α_o is fixed via the CCC. As a result, one can obtain an accurate estimate $E^{(sr)}(r)$ of the elementary function. In a further step, the functions

$$\alpha(r) = 1 + \alpha_o^* E_{PY}^{(sr)}(r)/N_{PY}^{(sr)}(r) \tag{22}$$

$$E(r) = \ln\left[\left(\exp\left(\alpha(r)N(r)\right) - 1\right)/\alpha(r) + 1\right] - N(r), \tag{23}$$

are adopted an the Euler equation is solved. The constant α_o^* is fixed in such a way as the relation

$$E(0) = E^{(sr)}(0) \tag{24}$$

holds. This condition can be justified on the ground that the inclusion in g(r) of a long-ranged tail should very scarcely modify the behaviour of the elementary function at small values of r. Moreover, such a feeling has been confirmed by numerical checks on the elementary contributions involving only four and five particles. In order to calculate the function E*(r) from eq.(18), we notice that E*(r) corresponds to the sum of all the elementary diagrams containing once the $\nabla^2 g(r)$ function and it can be estimated by means of an interpolating procedure. The details of the calculation are given elsewhere [11]. Here we simply point out that, as a rule, the contribution from E*(r) to the energy per particle is small (as an example, $\simeq$0.02 K for the ground state energy per particle of liquid ^{4}He at the equilibrium density).

C). Momentum distribution

The one-body density matrix

$$\varrho(\vec{r}_{11'}) = \varrho\int d\vec{r}_2\ldots d\vec{r}_A \Psi(1,\ldots,A)\,\Psi^*(1',\ldots,A)/\int d\vec{r}_1\ldots d\vec{r}_A|\Psi(1,\ldots,A)|^2 \tag{25}$$

is related to the momentum distribution by the relation

$$n(k) = \int d\vec{r}\,\varrho(r)\,e^{i\vec{k}\cdot\vec{r}} = An_o\,\delta(k) + \int d\vec{r}\left(\varrho(r) - \varrho(\infty)\right)\,e^{i\vec{k}\cdot\vec{r}} \tag{26}$$

where n_o is the condensate fraction. The calculation of $\varrho(r)$ can be performed adopting the procedure developed by FANTONI in ref.[10]. As a result, $\varrho(r)$ is expresed [10] in terms of the following functions

$$g(r) \qquad N(r) \qquad E(r) \tag{27}$$

$$g_{wd}(r) \qquad N_{wd}(r) \qquad E_{wd}(r) \tag{28}$$

$$g_{ww}(r) \qquad N_{ww}(r) \qquad E_{ww}(r) \tag{29}$$

through the relations

$$\varrho(r) = \varrho\, n_o \exp\left(N_{ww}(r) + E_{ww}(r)\right) \tag{30}$$

Table I. Energy per particle in K of liquid ^{4}He at T=0 calculated with the L-J potential. The estimates $E_{sr/\alpha}$ and $E_{EUL/\alpha}$ correspond to the short-ranged correlation factor of eq.(36) and to the solution of the Euler equation (15), respectively. Numbers in parentheses are from ref. [14]. σ= 2.556 Å.

$\varrho\sigma^3$	b	$E_{sr/\alpha}$	$E_{EUL/\alpha}$
0.330	1.15	-5.92	-6.01 (-5.93)
0.365	1.15	-5.74	-5.84 (-5.85)
0.401	1.15	-5.28	-5.42 (-5.55)

$$n_o = \exp(2R_w - R_d) \tag{31}$$

where

$$\begin{aligned} R_w &= \varrho \int d\vec{r} \left(g_{wd}(r) - 1 - N_{wd}(r) - E_{wd}(r) \right) \\ &\quad - 1/\left(2\varrho(2\pi)^3\right) \int d\vec{k} \left(\hat{g}_{wd}(k) - 1\right)\left(\hat{N}_{wd}(k) + 2\hat{E}_{wd}(k)\right) + E_w . \end{aligned} \tag{32}$$

R_d is obtained by replacing in the last equation the subscripts w by d. The quantities E_w and E_d represent (small) contributions from certain elementary diagrams [10]. The functions in eq.(28) and those in eq.(29) are related by equations similar to those valid for g(r), N(r) and E(r). As it is discussed with more details in ref. [11], in order to solve these equations we can utilize two new interpolating functions of the form

$$\beta(r) = 1 + \beta_o E_{PY}^{(wd)}(r)/N_{PY}^{(sr)}(r) \tag{33}$$

$$\gamma(r) = 1 + \gamma_o E_{PY}^{(ww)}(r)/N_{PY}^{(ww)}(r). \tag{34}$$

The parameters β_o and γ_o are fixed so as to satisfy the conditions

$$\varrho(0)/\varrho = 1, \qquad T_{MD} = \hbar^2/2m \int d\vec{k}\, k^2 n(k)/(2\pi)^3\varrho = T_{JF} \tag{35}$$

Table II. The same as the Table I, but with the HFDHE2 potential.

$\varrho\sigma^3$	b	$E_{sr/\alpha}$	$E_{EUL/\alpha}$
0.328	1.191	-6.02	-6.14 (-6.12)
0.365	1.196	-5.71	-5.90 (-5.94)
0.401	1.199	-4.96	-5.20 (-5.46)

Table III. Condensate fraction n_o for ^4He at T=0. The values in colunms 2 and 4 have been obtained by the interpolating procedure, Those in columns 3 and 5 are from refs. |15| and |16|, respectively. The upperscript * means that the value has been obtained by linear interpolation of the avaible estimates.

$\varrho\sigma^3$	n_o			
	LJ potential		HFDHE2 potential	
0.365	0.134	0.139	0.114	0.098
0.380	0.120	0.124	0.102	0.087*
0.401	0.105	0.095*	0.086	0.071

where T_{JF} is the Jackson-Feenberg expression of the mean value of the kinetic energy.

D). Numerical results for liquid ^4He at T=0

Many calculations on the properties of liquid ^4He at T=0 have been performed by using the Lennard-Jones (LJ) 6-12 potential, or the HFDHE2 potential proposed by AHLRICHS et al. |12|, with values of parameters as determined by AZIZ et al. |13|. The energy per particle calculated for these potentials is given in tables I and II, for a few density values, in correspondence to a short-ranged function of the form

$$f(r) = \exp\left(-(b\sigma/r)^5/2\right) \tag{36}$$

and to the solution of the Euler equation (15). The numbers reported in parentheses were obtained in ref. |14| where (i) a scaling approximation for the elementary contributions has been adopted and (ii) the behaviour of the correlation factor f(r) has been fixed to the form $(1+\alpha/r^4)$ for sufficiently large values of r. The agreement between the interpolating and scaling procedures is remarkable at the equilibrium density, small differences appairing at larger density values.

The values obtained for the condensate fraction n_o of liquid ^4He at T=0 by using the interpolating procedure are given in columns two and four of Table III in correspondence to three density values. For the L-J potential the present estimates are in close agreement with those of PUOSKARI and KALLIO |15| reported in column three of the table. The results in column five are from ref. |16| and they differ to some extent from the corresponding ones given by the interpolating procedure.

3. FERMI SYSTEMS AT T=0.

Here we will discuss the case of a Fermi system described by the trial wave function

Table IV. Energy per particle in K of normal liquid ^{3}He at T=0. The meaning of various quantities is specified in the text.

$\varrho\sigma^3$	$E_{sr/\alpha}$	$E_{EUL/\alpha}$
0.237	-1.41	-1.52
	(-1.37)	(-1.46)
0.277	-1.12	-1.25
	(-1.08)	(-1.28)
0.301	-0.79	-0.93
	(-0.76)	(-1.06)
0.330	-0.12	-0.29
	(-0.16)	(-0.60)

$$\Psi_v = \prod_{i<j=1}^{A} f(r_{ij})\, \Phi(1,\ldots,A) \tag{37}$$

which is a particular case of the general form given in eq.(1). It is possible to derive [17] an Euler equation to obtain the best variational choice of f(r) but the numerical solution of the equation is a difficult problem. As an alternative [4,5], we can first study the underlying fictitious system of Bose particles -with the same interaction and mass as for the Fermi particles- described by the function

$$\Psi_v^{(BOSE)} = \prod_{i<j=1}^{A} f(r_{ij}). \tag{38}$$

For such Bose system we can solve the corresponding Euler equation as it

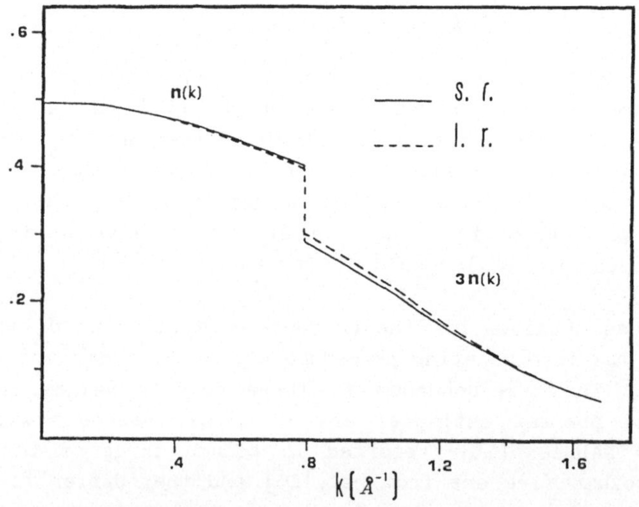

Figure 1. Momentum distribution n(k) for liquid ^{3}He at T=0 in correspondence to the HFDHE2 potential and a density value $\varrho\sigma^3$ = 0.277. The full line corresponds to a correlation factor as in eq.(36) with b=1.15. The dashed line corresponds to a correlation factor including the correct long-ranged tail.

is discussed in sect.2, so as to get the interpolating function $\alpha(r)$ (see eq.(21)) and the r.d.f.. The functions $\alpha(r)$ and f(r) obtained in such a way are used to interpolate [4,5] between the FHNC/0 and FPY equations. The resulting FHNC/α equations can be found elsewhere [5,11] and will not be explicitely written here.

The results obtained for normal liquid ^{3}He by adopting the HFDHE2 potential are presented in Table IV. $E_{sr/\alpha}$ corresponds to the short-ranged correlation factor f(r) specified by eq.(36) with the value b=1.15 for all the density values considered. The estimate $E_{EUL/\alpha}$ has been obtained by utilizing the solution of the FHNC/α equations with a correlation factor extracted from the solution of the Euler equation correspondent to the underlying Bose system. The results reported in parentheses in table IV are those given in ref. [18], where the scaling approximation has been adopted. The agreement between these estimates and the ones obtained by using the interpolation technique is quite satisfactory for density values close to the experimental equilibrium one.

The calculation of the one-body density matrix and momentum distribution can be performed as in ref. [11] by utilizing the interpolating functions $\alpha(r)$, $\beta(r)$ and $\gamma(r)$ determined for the underlying Bose system. The momentum distribution n(k) calculated for $\rho\sigma^3 = 0.277$ is plotted in fig. 1. It is apparent that the inclusion of a long-ranged tail in the r.d.f. has a negligible effect on the momentum distribution.

4. DISCUSSION AND CONCLUSIONS

In the present paper we have discussed two important problems which appear in the study of strongly interacting quantum fluids. The first problem concerns the accurate evaluation of the elementary contributions which rapidly increase with the density of the system. To this end, two tecnhiques widely adopted in the last years consist in an explicit numerical calculation of terms involving only few particles, or in adopting the so-called scaling approximation [14]. The second problem is related to determining the best variational two-body correlation factor f(r) by solving the appropiate Euler equation.

In this paper the first topic is dealt with by means of a procedure based on integral equations which interpolate the ones avaible from the HNC/0 and PY approaches. This procedure is utilized in solving the Euler equation, so as to take properly into account both elementary contributions and long-range effects.

The energy per particle and the momentum distribution obtained in this way for liquid ^{4}He and ^{3}He at T=0 compare favourably with other available estimates.

The trial wave functions considered contain only two-body correlations. However, for dense quantum systems effective correlations interesting more than two particles are important. The extension of the method of interpolating integral equations to include such more involved correlations will be discussed in a subsequent paper [11].

ACKNOWLEDGMENTS

The authors wish to thank S.Fantoni for many fruitfull discussions. One of us (E.B.) acknowledges the Istituto Nazionale di Fisica Nucleare and the Spanish Comisiòn Asesora Cientìfica y Técnica for financial support.

* On leaves from Departamento de Fìsica Nuclear, Facultad de Ciencias de la Universidad de Granada, Granada, Spain.

+ Present address: Departament of Physics and Material Research Laboratory, University of Illinois of Urbana-Champaign, Urbana, Illinois 61801 USA.

REFERENCES

(1) J.M.J. VAN LEEWEN, J.GROENEVELD and J.DE BOER: Physica, 25, 792(1959).
(2) T.MORITA: Progr.Theor.Phys., 20, 920(1958).
(3) J.K.PERCUS and G.J.YEVICK: Phys.Rev., 110, 1(1958).
(4) A.FABROCINI and S.ROSATI: Nuovo Cimento, D1, 615(1982).
(5) A.FABROCINI and S.ROSATI: Nuovo Cimento, D1, 567(1982).
(6) See, for instance, A.MÜNSTER: Statistical Thermodynamics (Springer, New York, 1974).
(7) M.VIVIANI and S.ROSATI: Anales de Fìsica, 81, 121(1984).
(8) L.LANNTO, A.D.JACKSON and P.J.SIEMENS: Phys.Lett.,68B, 311(1977).
(9) R.A.SMITH, A.KALLIO, M.PUOSKARI and P.TOROPAINEN: Nucl.Phys., A328, 186(1979).
(10) S.FANTONI: Nuovo Cimento, 44A, 191(1978).
(11) M.VIVIANI, E.BUENDIA and S.ROSATI: Preprint 1985.
(12) R.AHLRICHS, P.PENCO and G.SCOLES: Chem. Phys., 19, 119(1976).
(13) R.A.AZIZ,V.P.S.NAIN,J.S.CARLEY,W.L.TAYLOR and G.T.McCONVILLE: J.Chem. Phys., 70, 4330(1979).
(14) Q.N.USMANI, B.FRIEDMAN and V.R.PANDHARIPANDE: Phys.Rev., B25, 4502(1982).
(15) M.PUOSKARI and A.KALLIO: Phys.Rev., B30, 152(1984).
(16) E.MANOUSAKIS, V.R.PANDHARIPANDE and Q.N.USMANI: Phys.Rev., B31, 7022(1985).
(17) J.C.OWEN: Phys.Lett., 89B, 303(1980).
L.J.LANTTO and P.J.SIEMENS: Nucl.Phys., A317, 55(1979).
E.KROTSCHECK: Phys.Rev., A15, 397(1977).
(18) E.MANOUSAKIS, S.FANTONI, V.R.PANDHARIPANDE and Q.N.USMANI: Phys.Rev., B28, 3770(1983).

VARIATIONAL THEORY OF IMPURITIES IN LIQUID ^{4}He

K. E. Kürten and M. L. Ristig

Institut für Theoretische Physik, Universität zu Köln
D-5000 Köln, West Germany

J. W. Clark

McDonnell Center for the Space Sciences
and Department of Physics
Washington University, St. Louis, Missouri 63130, U.S.A.

INTRODUCTION

Impurities immersed in bulk liquid ^{4}He yield important information on the response and the induced intrinsic spatial structure of the perturbed medium.[1 2] The dynamical properties of the impurity-helium system are attracting experimental attention, notably through studies of the fate of H,D and T atoms during and after the recombination process at the helium surface.[3] In particular, one is interested in knowing if atoms of hydrogen or one of its isotopes is clustered together with itself, and if hydrogen is bound to the helium surface or dissolves into bulk liquid helium.

In this contribution we shall be concerned with the static, ground-state properties of a system consisting of atomic and molecular hydrogen impurities, or heavy atomic impurities, embedded in liquid He. We adopt a Jastrow variational description, the optimization being performed within a paired-phonon analysis (PPA) in conjunction with the hypernetted chain (HNC) approximation for a binary boson mixture.[4,5] The HNC results are supplemented by results from a variational Monte Carlo procedure,[6] tailored to the special circumstances produced by the foreign particle.[7] After introducing the explicit expressions for the physical quantities of immediate relevance, we examine the behavior of hydrogen-isotopic atoms and molecules in liquid ^{4}He. Results are also presented for a single Xe or Cs impurity in liquid ^{4}He.

FORMALISM

We consider a homogeneous system of N-1 identical background bosons of mass m_1 and one foreign particle of mass m_2. The behavior of the foreign particle can be described by a Hamiltonian which takes the form

$$H = T + V \quad , \tag{1}$$

where

$$T = -\frac{\hbar^2}{2m_1}\sum_{i=1}^{N-1} \Delta_i - \frac{\hbar^2}{2m_2}\Delta_N \quad .$$

and

$$V = \sum_{i<j}^{N-1} v_{11}(r_{ij}) + \sum_{i=1}^{N-1} v_{12}(r_{iN}) \quad .$$

The quantities $v_{11}(r)$ and $v_{12}(r)$ represent realistic interaction potentials between two background particles and between a background particle and the foreign particle respectively. As a spatially correlated trial function of Jastrow type we choose the symmetric form

$$\Psi = \exp\left\{\frac{1}{2} \sum_{i<j}^{N-1} u_{11}(r_{ij}) + \sum_{k=1}^{N-1} u_{12}(r_{kN})\right\} \quad . \tag{2}$$

Since a single representative of a given particle species experiences no exchange correlation, such a wave function is suitable whether the impurity particle is a boson or a fermion. The pseudopotentials $u_{11}(r)$ and $u_{12}(r)$ may be obtained by functional minimization of the energy expectation value

$$\langle H \rangle = \frac{\langle \Psi | H | \Psi \rangle}{\langle \Psi | \Psi \rangle} \quad . \tag{3}$$

This minimization is well accomplished through a paired-phonon analysis.[8] This procedure rests on the variational principle for the ground state expectation value and leads in the thermodynamic limit to a set of two Euler-Lagrange equations for the radial distribution functions $g_{11}(r)$ and $g_{12}(r)$ of the background particles and of a background and the foreign particles, respectively. Adopting the hypernetted-chain approximation (HNC/0), which relates the pseudopotentials $u_{11}(r)$ and $u_{12}(r)$ to the radial distribution functions $g_{11}(r)$ and $g_{12}(r)$, the Euler-Lagrange equations can be cast into a simple form and solved by standard methods involving the paired-phonon analysis.[9] Greater accuracy in the actual evaluation of (3) can be achieved by implementing Monte Carlo integration[6] in place of the HNC/0 approximation.

The behavior of the foreign particle in the interior of the ^4He liquid is determined by the chemical potential μ, defined by the zero concentration limit of the partial derivative of the energy expectation value (3) per particle with respect to the concentration x of the foreign particle,

$$\mu = \left.\frac{\partial(\langle H \rangle/N)}{\partial x}\right|_{x=0} \quad . \tag{4}$$

Equation (4) may be cast into the following form

$$\mu = 2\left(E_{12} - E_{11}\right) + \partial E_{11}/\partial x\Big|_{x=0} \quad , \tag{5}$$

the quantities E_{11} and E_{12} being written out explicitly in Ref. 9. For a finite number of particles one has to deal instead with the expression

$$\mu = N\left(E_i - E_b\right) \quad . \tag{6}$$

Here, the quantity E_b is the energy expectation value of N background particles, while E_i is the energy expectation value of (N-1) background particles and one foreign particle. Thus, μ represents the change in energy upon replacing one background particle of the liquid by an impurity. Another salient physically measurable quantity is the volume coefficient α, defined by

$$\rho(p,x) = \rho(p,0)/(1 + \alpha x) \tag{7}$$

in terms of the density of a dilute mixture relative to the density of the pure background medium at fixed pressure p. This quantity may be directly extracted from the partial optimal structure function $S_{12}(k)$ via

$$-\alpha = 1 + S_{12}(k=0) \quad . \tag{8}$$

The exact chemical potential μ and volume coefficient α associated with the true ground-state wave function are related by[10]

$$\frac{\partial\mu}{\partial\rho} = \frac{\alpha}{\rho}\frac{\hbar^2 k^2}{4mS_{11}(k)} \quad . \tag{9}$$

where $S_{11}(k)$ is the static structure function for the ^4He background atoms.

Using an approximate ground-state wave function such as (2), the expressions on the left- and right-hand sides of equation (9) differ from each other. The magnitude of this deviation provides some indication of the influence of multiparticle correlation factors on the density dependence of the chemical potential difference μ.

For the interaction between background He atoms we adopt the HEDHE2 potential of Aziz *et al.*, and for the interaction between background atoms and the hydrogenic impurities we use a Lennard-Jones 6-12 form suggested in Refs. 5,11. Actual parameters are given in Table 1.

Table 1. Parameters of Lennard-Jones Models of He-He and He-Impurity Potentials.

	E (K)	α (Å)
^3He, ^4He	10.22	2.556
H,D,T	6.60	3.200
H_2,HD,T_2	15.56	3.010
Xe	25.18	3.697
Cs	1.34	6.896

NUMERICAL RESULTS FOR HYDROGEN ATOMS

In earlier theoretical discussions of binary mixtures of atomic hydrogen, deuterium, and tritium with helium, it has been found that the boson systems H-^4He and T-^4He completely phase separate at zero temperature, and that H, D, and T atoms do not penetrate the surface of liquid ^4He since their chemical potentials are positive.[12 13] Adopting an improved theoretical approach, we are particularly interested in examining the density dependence of the chemical potential difference within the PPA/HNC/0 approximation.

Figure 1 shows our results for the chemical potential corresponding to the replacement of one ^{4}He atom in the bulk by one H, D, or T atom, as a function of density. In all three cases, this chemical potential is positive and increases rapidly with increasing density. Thus, it is energetically more favorable for the hydrogen atoms to reside in regions of low density. More pointedly, isotopic hydrogen atoms cannot penetrate the helium surface.

Results on the volume coefficients α for both atomic and molecular hydrogen isotopic impurities are depicted in Fig. 2. For H and T atoms α is positive at all densities considered; it increases as the density is lowered and becomes singular at about ρ = 0.016 Å^{-3}. It is well known that at this density the optimal ground state of the ^{4}He liquid is unstable against density fluctuations. The quantity $\partial\mu/\partial\rho$, as calculated for H, D, or T atoms via Eq. (9), turns out to be positive, and increases with increasing density. This behavior is consistent with the results for the slope of the chemical potential $\mu(\rho)$ which may be derived from Fig. 1.

NUMERICAL RESULTS FOR HYDROGEN MOLECULES

Recently, Silvera presented results from an experiment which furnishes direct evidence that, after recombination of individual atoms, hydrogen penetrates the surface of liquid ^{4}He either in the form of single molecules or molecular clusters.[3]

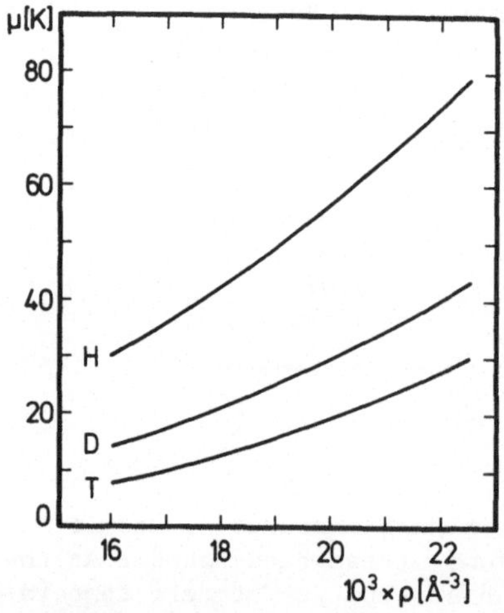

Figure 1. Chemical potential μ for replacing one He atom in the liquid at density ρ by one H, D, or T atom (HNC/0 results).

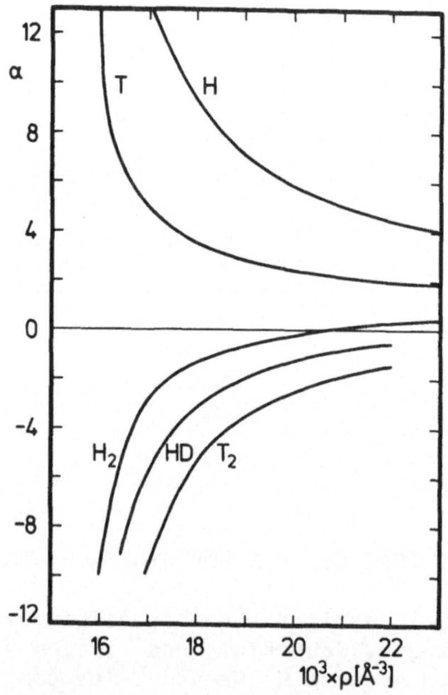

Figure 2. Theoretical volume coefficients α of H and T atoms and H_2, HD, H_2 molecules in liquid helium, as functions of density ρ, based on the HNC/0 approximation.

Figure 3 displays the various hydrogenic chemical potentials μ as functions of the mass ratio m_2/m_1. Except at hypothetically small impurity masses, the chemical potentials are negative at experimental saturation density and decrease monotonically with increasing molecular mass. Thus, in contrast to the situation for hydrogen atoms, hydrogen molecules penetrate the surface of liquid ^{4}He and tend to dissolve into the interior. Monte Carlo evaluation yields only small negative corrections to the hypernetted-chain results, supporting this prediction.

Figure 4 allows us to study the penetration of the liquid by H_2 molecules in more detail. The chemical potentials μ, in hypernetted-chain approximation, are negative at all densities considered, increasing slightly at higher densities. On the other hand, the numerical results on the volume coefficient of the H_2 molecules are negative at densities $\rho < 0.0205$ Å^{-3}, whereas they are positive for higher densities (Fig. 2). This behavior signals the existence of a relative minimum of the chemical potential as a function of density. Comparison of the HNC/0 and Monte Carlo results in Fig. 4 shows that the magnitude of the elementary diagrammatic contribution is small everywhere and decreases with decreasing density. As expected, the Monte Carlo results display a shallow, minimum (at a density $\rho = 0.0185$ Å^{-3}). Provided the chosen trial function (2) is sufficiently accurate, these findings suggest that H molecules penetrate the surface of the liquid but reside in regions with density below the saturation value.

However, we should be aware that inclusion of multiparticle factors in the ground-state ansatz will correct the detailed density dependence of the approximate chemical potentials. Even so, we do not believe that these backflow corrections will lead to significant changes in the essential features of the variational results based on our trial wave function.

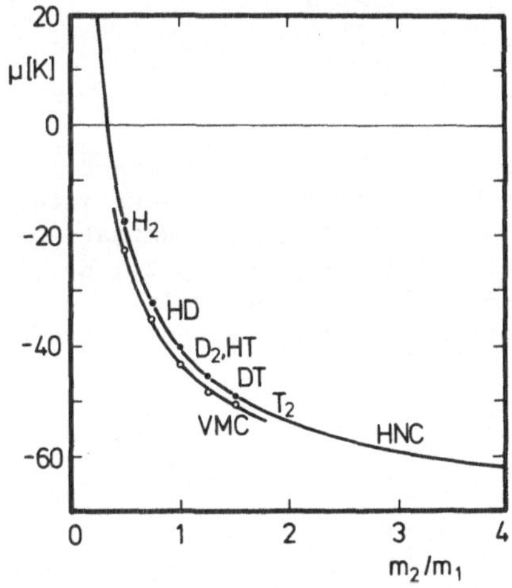

Figure 3. Chemical potential μ for replacing one ^{4}He atoms of the liquid at density $\rho = 0.02185$ Å^{-3} by one diatomic hydrogen molecule, as a function of the impurity/^{4}He atom mass ratio, calculated by variational Monte Carlo (VMC) and HNC/0 approximation.

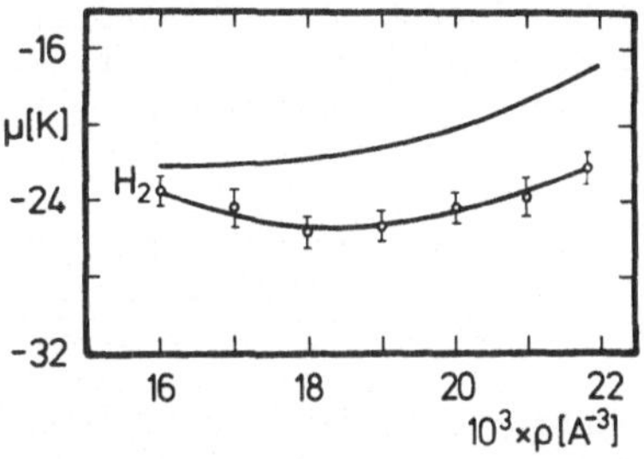

Figure 4. Chemical potential μ for replacing one ^{4}He atom in the liquid by one H_2 molecule at density ρ. The lower curve with dots and error bars represents the Monte Carlo results; the upper curve is from the HNC/0 calculation.

NUMERICAL RESULTS FOR Xe AND Cs ATOMS

These two examples, Xe and Cs, represent opposite extremes, as may already be discerned from inspection of the potential parameters[14] in Table 1. The strong attraction of Xe acts to concentrate the density of He atoms in the vicinity of the Xe atom, whereas the repulsive component of the Cs-He interaction is so overwhelming that a net enhancement of host atoms in the vicinity of the impurity is strongly discouraged.

The numerical results for the chemical-potential difference μ of an Xe or Cs impurity are listed in Table 2. The Xe atom is strongly bound in the ^{4}He liquid with a large binding energy of about -290 K. Figure 5 shows a comparison of variational Monte Carlo and HNC/0 results for the radial distribution function $g_{12}(r)$ and the associated structure function $S_{12}(k)$. The strong oscillatory behavior of the radial distribution function $g_{12}(r)$ produces large-scale cancellations within and among the elementary diagrams neglected in the HNC/0 scheme. Thus the HNC/0 approximations suffers remarkably small errors, relative to the Monte Carlo evaluation. Relative to the Monte Carlo result the HNC/0 approximation for μ is in error by only about 1%. The results for the mixed structure function $S_{12}(k)$ also agree quite well, except at smaller wave numbers where the disagreement may be ascribed to the finite box size of 15 Å for the 64 particles of the Monte Carlo treatment. As in the Monte Carlo treatment of the pure ^{4}He system, the size dependence of the Monte Carlo estimate of μ is rather weak: no substantial differences were observed between systems with $N = 32$ and $N = 108$ particles.[15]

Table 2. Results for Chemical-Potential Differences Associated with Heavy Atomic Impurities

Impurity	μ (Monte Carlo)	μ (HNC/0)
Xe	-290.0 ± 7.0	-287.0
Cs	250.6 ± 9.6	277.0

As anticipated, the numerical results for the Cs impurity stand in marked contrast to those for Xe. Due to the fat core of the Cs-He interaction, the size dependence of the corresponding Monte Carlo estimate of μ is considerable. Satisfactory convergence is not achieved until N reaches about 150. We observe that the Cs-^{4}He interaction leads to a net repulsion: in reality the Cs atom would be propelled to the surface of a finite drop.

Figure 6 presents some Monte Carlo results for the partial radial distribution function of the Cs impurity problem. The converged $g_{12}(r)$ shows milder oscillations than in the Xe impurity problem, as expected. Consequently the aforementioned large cancellation within and among the elementary diagrams, attributed to strong oscillations of the distribution functions, especially $g_{12}(r)$, do not occur here. Thus, the HNC/0 result for μ misses the Monte Carlo result by about 10%. Moreover, we note the important property (due to the large core) that the peak of $g(r_{12})$ lies much farther out in r than does the peak of the Xe-He distribution function.

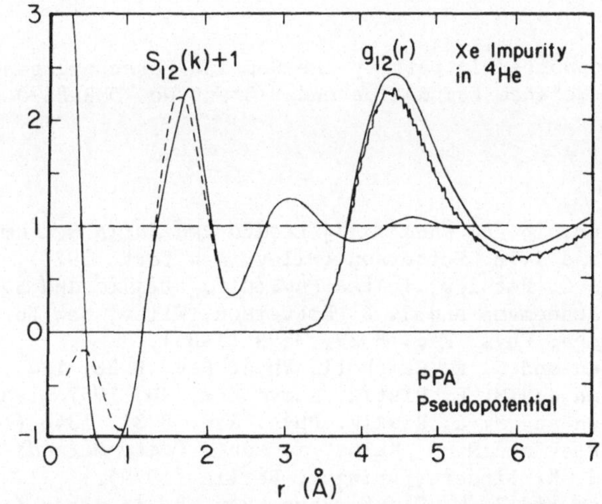

Figure 5. Comparison of Monte Carlo (dashed or jagged curves) and HNC/0 (solid curves) results for a single Xe atom embedded in liquid ^{4}He. The partial distribution function $g_{12}(r)$ and partial structure function $S_{12}(k)$ are plotted against radial distance r and wave number k, respectively. Pseudopotentials $u_{11}(r)$ and $u_{12}(r)$ determined optimally within the PPA-HNC/0 scheme are assumed. The Monte Carlo calculation refers to N = 64 particles.

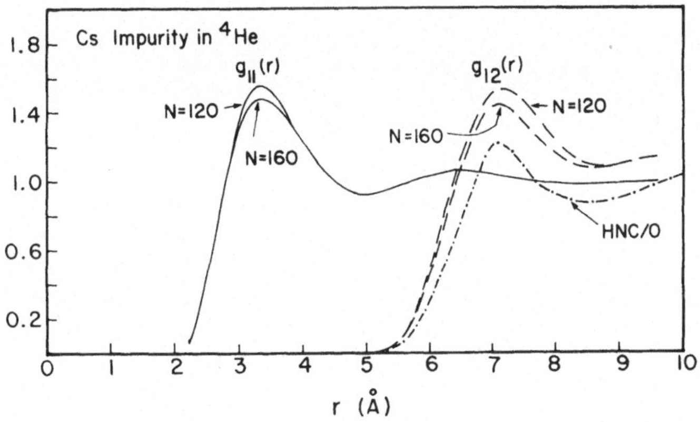

Figure 6. Monte Carlo result for partial radial distribution functions $g_{11}(r)$ and $g_{12}(r)$ in the case of a Cs impurity in liquid ^{4}He, assuming PPA-HNC/0 optimal pseudopotentials. The substantial size dependence in this problem is illustrated by the difference between results at N = 120 and N = 160, convergence having been reached by the latter value. The HNC/0 version of $g_{12}(r)$ (dot-dash curve) is included for comparison.

ACKNOWLEDGMENTS

Research supported in part by the Deutsch Forschungsgemeinschaft and by the National Science Foundation under Grant No. DMR-83-04213.

REFERENCES

1. A. L. Fetter, in *The Physics of Liquid and Solid Helium*, ed. K. H. Bennemann and J. B. Ketterson (Wiley, New York, 1976).
2. G. Baym and C. Pethick, in *The Physics of Liquid and Solid Helium*, ed. K. H. Bennemann and J. B. Ketterson (Wiley, New York, 1976).
3. I. F. Silvera, Phys. Rev. **B 29**, 3899 (1984).
4. K. E. Kürten and C. E. Campbell, Phys. Rev. **B 26**, 124 (1982).
5. K. E. Kürten and M. L. Ristig, Nuovo Cim. **2D**, 1057 (1983); K. E. Kürten and M. L. Ristig, Phys. Rev. **B 31**, 1346 (1985).
6. D. M. Ceperley and M. H. Kalos, in *Monte Carlo Methods in Statistical Physics*, ed. K. Binder (Springer, Berlin, 1979).
7. K. E. Kürten and J. W. Clark, Phys. Rev. **B**, in press (August 1, 1985); K. E. Kürten, in *Recent Progress in Many-Body Theories*, ed. H. Kümmel and M. L. Ristig (Springer, Berlin, 1983).
8. C. E. Campbell and E. Feenberg, Phys. Rev. **188**, 396 (1969).
9. K. E. Kürten and M. L. Ristig, Phys. Rev. **B 27**, 5479 (1983).
10. G. B. Baym, Phys. Rev. Lett. **17**, 952 (1966).
11. R. Guyer and M. D. Miller, Phys. Rev. Lett. **42**, 1754 (1979).
12. J. B. Mantz and D. O. Edwards, Phys. Rev. **B 20**, 4518 (1979).
13. M. D. Miller, Ann. Phys. (N.Y.) **127**, 367 (1980).
14. W. Buck, Adv. Chem. Phys. **30**, 368 (1975); J. Gspann (private communication).
15. M. H. Kalos, M. A. Lee, P. A. Whitlock, and G. V. Chester, Phys. Rev. **B 24**, 115 (1981).

SPIN-POLARIZED DEUTERIUM

H. R. Glyde

Department of Physics
University of Delaware
Newark, DE 19716

and

S. I. Hernadi

Department of Physics
University of Ottawa
Ottawa, Canada K1N 6N5

ABSTRACT

Several ground state properties of (electron) spin-polarized deuterium ($D\downarrow$) such as the energy, single quasiparticle energies and lifetimes, Landau parameters and sound velocities are evaluated. The calculations begin with the Kolos-Wolneiwicz potential and use the Galitskii-Feynman-Hartree-Fock (GFHF) approximation. The deuteron nucleas has spin I = 1, and spin states $I_z = 1,0,-1$. We explore $D1\downarrow$, $D2\downarrow$ and $D3\downarrow$ in which, respectively, one spin state only is populated, two states are equally populated, and three states are equally populated. We find the GFHF describes $D1\downarrow$ well, but $D2\downarrow$ and $D3\downarrow$ less well. The Landau parameters, F_L, are small compared to liquid ^3He and very small for doubly polarized $D1\downarrow$ (i.e. the F_L decrease with nuclear polarization).

I. INTRODUCTION

We explore the properties of a fluid of deuterium atoms. The deuterium atom consists of a single electron and a deuteron nucleas having nuclear spin I = 1. We assume that the spin of each electron is aligned, downward antiparallel to an applied field B ($D\downarrow$). In practice[1] this requires special production of atoms plus large B and low temperature to maintain alignment. The electron spin alignment prevents D_2 molecule formation. To date $D\downarrow$ gas at modest density, $n \sim 10^{14}$ atoms/cm^3, has been produced.[2]

With their electron spins aligned, two $D\downarrow$ atoms interact weakly via the $b^3\Sigma_u^+$ potential.[3] This has a well depth $\varepsilon \sim 6.4$ K at a separation ~ 4.2 Å. This $D\downarrow - D\downarrow$ ($H\downarrow - H\downarrow$) potential, which is weaker than the He-He potential, is shown in Fig. 1. The deuterium atom is a composite Fermion.[4] It consists of an odd number of Fermions and interchange of

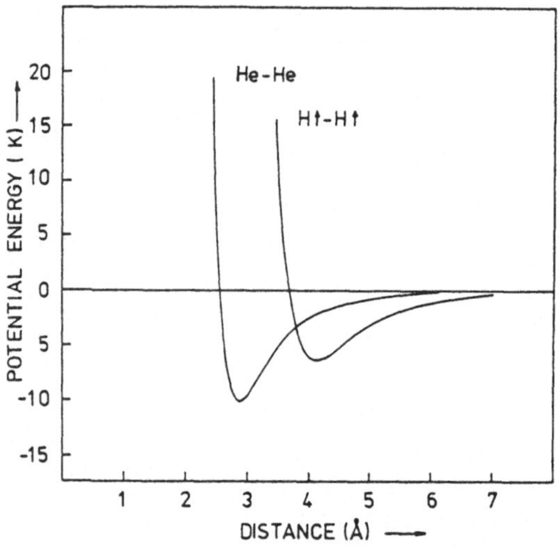

Fig. 1. The $H^{\downarrow}$ - $H^{\downarrow}$ (= $D^{\downarrow}$ - $D^{\downarrow}$) $b^3\Sigma_u^+$ potential compared to the He-He potential (from Ref. 10).

each identical Fermion in a pair of D atoms shows that the total wave function of two atoms must be antisymmetric. With the electron spins frozen downward only the nuclear spin I = 1 is variable. Thus $D^{\downarrow}$ is spin 1 Fermion.

A fluid of $D^{\downarrow}$ atoms has great intrinsic interest as an example of a weakly interacting Fermi Fluid of spin 1. With weak interaction, microscopic calculations may be more successful than, say, in liquid ^{3}He or nuclear matter. A central purpose of the present paper is to test microscopic Green function methods in $D^{\downarrow}$. Also interesting generalizations to spin 1 are required. In atomic hydrogen, the hyperfine coupling between the nuclear and electron spins can assist flipping of electron spins and H_2 atom formation. With this mechanism the $H^{\downarrow}$ atoms having their proton spins ($\Uparrow$) aligned opposite to the electron combine with other $D^{\downarrow}$ atoms to form H_2 molecules. This leaves a gas of doubly ($|\downarrow\Downarrow>$) polarized $H^{\downarrow}$ atoms. A similar mechanism operating in $D^{\downarrow}$ (and $T^{\downarrow}$) could provide a method of producing nuclear spin aligned D and T. Nuclear spin aligned D and T is of great practical interest in laser fusion since DT fusion reactions[5] are enhanced if the spins are aligned.

Here we simply assume electron spin alignment but take B = 0. We also ignore the hyperfine interaction between electron and nuclear spins. The Hamiltonian of the fluid is then

$$H = \sum_i \frac{P_i^2}{2m} + \sum_{i<j} v(r_{ij})$$

with v(r) depicted in Fig. 1. With I = 1, three nuclear spin states I_z = 1,0,-1 are possible in each atom. As a model system, we study $D_1^{\downarrow}$ in which only one nuclear spin state (say I_z = -1) is allowed. This is doubly polarized D . We also, as in previous work, consider $D_2^{\downarrow}$ and $D_3^{\downarrow}$ in which two and three spin states, respectively, are assumed to be equally occupied. The $D_1^{\downarrow}$ is directly analogous to fully nuclear spin polarized ^{3}He ($^3He^{\uparrow}$). $D_2^{\downarrow}$ is analogous to normal ^{3}He while $D_3^{\downarrow}$ is a new three spin state Fermi fluid. We consider high densities in the range $n \sim 1\text{-}5 \times 10^{21}$ atoms/cm^3. In practice, $D^{\downarrow}$ will appear in a finite field B having partial nuclear polarizations. $D_1^{\downarrow}$ best represents realizable $D^{\downarrow}$ since it is in a pure state $|\downarrow\Uparrow>$. It will be the state remaining after $D^{\downarrow}$ atoms in other states have participated in D_2 molecule formation.

In their pioneering studies, Etters and co-workers[6] explored the ground state properties of H, D and T using Monte Carlo methods. For $D_3^{\downarrow}$ they evaluated the ground state energy, E, the pair correlation function, g(r), the compressibility, κ and the pressure p. Stwalley and Nosanow,[7] drawing upon the Quantum Theory of Corresponding States developed by Nosanow et al[8] and Miller et al,[9] established the highly quantum nature of $H^{\downarrow}$, $D^{\downarrow}$ and $T^{\downarrow}$. $D^{\downarrow}$ is a highly quantum fluid because the mass is light and the interaction is weak. This may be expressed in a single quantum parameter $\eta = \hbar/m\varepsilon\sigma^2$ where ε is the well depth of v(r) and σ defined by $v(\sigma) = 0$ ($\sigma \sim 3.7$ Å). Miller and Nosanow[11] and Nosanow[12] developed the quantum properties, such as E, of fluid $D^{\downarrow}$ as a function of η. They showed that $H^{\downarrow}$ ($\eta = 0.547$) is "too quantum" to form a self bound fluid ($E > 0$). However, it was not clear whether $D^{\downarrow}$ ($\eta = 0.274$) would have $E < 0$ or not. Also, the Fermi wave vector, k_F, of a Fermi liquid ($k_F^3 = (6\pi n)/n_s$) depends upon the number of spin states occupied, $n_s = 1,2,3$. Thus, with $\varepsilon_F^o = \hbar^2 k_F^2/2m$ and a zero order kinetic energy

$$<KE> = \frac{3}{5}\varepsilon_F^{\,o} \tag{1}$$

we expect the KE of $D_1^{\downarrow}$, $D_2^{\downarrow}$ and $D_3^{\downarrow}$ to be approximately in the ratio $1/2^{-2/3}/3^{-2/3}$ and to find $E_1^{\downarrow} > E_2^{\downarrow} > E_3^{\downarrow}$.

Clark et al[13] and Krotscheck et al[13] evaluated E for $D_1^{\downarrow}$, $D_2^{\downarrow}$ and $D_3^{\downarrow}$ using the method of correlated Basis Functions (CBF). They find $D_3^{\downarrow}$ has the lowest energy and generally $E_1^{\downarrow} > E_2^{\downarrow}$ as expected. None of the upper bounds were negative. Panoff et al[14] using variational Monte Carlo methods, have established that $D_3^{\downarrow}$ will form a self bound liquid ($E_3^{\downarrow} < 0$). This calculation provides a bench mark for other calculations. Lim,[15] employing lower bound methods, predicts $D^{\downarrow}$ will have $E < 0$.

Leggett[16] and Modawi[17] have investigated superfluidity in $D^{\downarrow}$. A transition temperature $\lesssim 10^{-6}$ K at densities $n \sim 10^{21} - 10^{22}$ is predicted. Modawi[17] has developed a microscopic theory of $D^{\downarrow}$ including the hyperfine interaction. Particularly, he suggests a superfluid state may be possible at much higher temperatures if it can be assisted by a magnetic field dependent resonance scattering in which $D^{\downarrow} - D^{\downarrow}$ couples to the attractive D_2 molecular state through the hyperfine interaction. In this case a $T_c \sim 1$ mK at $n \sim 10^{18}$ atoms/cm^3 is possible.

Bedell and Quader[18] have determined the scattering amplitude and Landau parameters of $D_1^{\downarrow}$ using their semi-microscopic Fermi liquid model. Buckle[19] has extended Fermi liquid theory to spin 1 and set out the regions of temperature and density in which Fermi liquid effects will be important in $D^{\downarrow}$.

In section II we survey the general interaction between quasiparticles in a spin 1 Fermi fluid making contact with Buckle's[19] formulation. The elements of the Galitskii-Feynman-Hartree-Fock (GFHF) theory are also sketched. A purpose here is to explore how well the GFHF theory can describe $D^{\downarrow}$. Values of the ground state energy are presented and compared with previous values in section III. Single particle energies, Landau parameters effective masses and sound velocities are presented in sections IV and V.

II. SPIN 1 FERMI LIQUID THEORY

A. The Interaction

We seek the form[20] of the interaction, $f_{1234}(p_1p_2;p_3p_4)$, between two spin 1 quasiparticles (qp), a and b, in a fluid. Here p1(1) is the incoming momentum (spin projection, ↑, ⇸, ↓) of qp a before interaction and (p_3) (3) is its momentum (spin) after interaction. The p_2(2) and p_4(4) are the corresponding labels for qp b. We may think of the interaction as a scattering event (see Fig. 2). The nuclear spin operator $\hat{I}(a)$ of atom a satisfies

$$\hbar^2\hat{I}^2(a) = I_a(I_a + 1)\hbar^2 \tag{2}$$

$$\hbar\hat{I}_z(a) = I_z(a)\hbar \qquad (I_z = 1,0,-1)$$

The total spin of the pair is $\vec{I} = (\vec{I}_a + \vec{I}_b)$ and the pair state is denoted $|I,I_z\rangle$.

The interaction between a and b depends entirely on the electron wave functions. It is intrinsically independent of the nuclear spin state. However, the symmetry of the electron wave functions depend upon the symmetry of the nuclear spin state $|I,I_z\rangle$ of the interacting pair. There are two possiblilites: the pair nuclear spin state $|I,I_z\rangle$ is symmetric (space state antisymmetric) or $|I,I_z\rangle$ is antisymmetric (space state symmetric). Thus there are two independent interactions,

$$f_{\text{spin symm.}} = 2a_o \tag{3}$$

$$f_{\text{spin anti.}} = 2a_e \tag{4}$$

where a_o means that we retain only odd (antisymmetric) space angular momentum components in the interaction. The nuclear spin affects the interaction only by dictating which angular momentum components of the interaction must be included. All forms of f must be expressible in terms of (3) and (4).

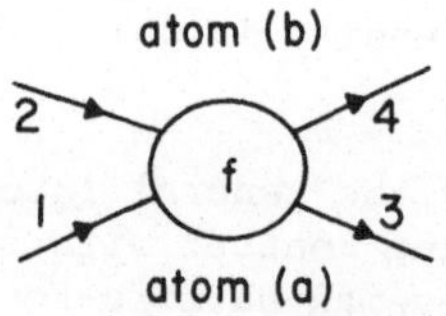

Fig. 2. The interaction of two quasiparticles having initial momentum and spin 1 and 2.

By relating the pair states $|II_z\rangle$ to single particle states, $|\Uparrow\Downarrow\rangle$, we find that $I = 0,2$ states are spin symmetric while $I = 1$ is spin antisymmetric.

We assume zero magnetic field. In this case there is no preferred direction and f can depend upon $\hat{I}_a$ and $\hat{I}_b$ only in the scaler form $(\hat{I}_a \cdot \hat{I}_b)$. For a 3×3 matrix $\hat{I}_a$, the most general form for f_{1234} is,[21]

$$f_{1234} = f_1 \delta_{13} \delta_{24} + f_2 \langle 12|(\hat{I}_a \cdot \hat{I}_b)|34\rangle + f_3 \langle 12|(\hat{I}_a \cdot \hat{I}_b)^2|34\rangle. \qquad (5)$$

All higher powers $(\hat{I} \cdot \hat{I})^n$ can be related[21] to $n = 0,1,2$. The f_1, f_2 and f_3 must now be determined.

From $\vec{I}^2 = (\vec{I}_a + \vec{I}_b)^2 = I(I+1)$ we have $(\hat{I}_a \cdot \hat{I}_b) = \frac{1}{2} I(I+1) - 2$. Thus f in (5) depends only upon I as expected (independent of I_z) and

$$\begin{aligned} \langle 00|f|00\rangle &= f_1 - 2f_2 + 4f_3 = 2a_o \\ \langle 1I_z|f|1I_z\rangle &= f_1 - f_2 + f_3 = 2a_e \\ \langle 2I_z|f|2I_z\rangle &= f_1 + f_2 + f_3 = 2a_o. \end{aligned} \qquad (6)$$

This shows that $f_3 = f_2$ and that

$$\begin{aligned} f_1 &= 2a_o \\ f_2 &= (a_o - a_e). \end{aligned} \qquad (7)$$

If we have a microscopic model for the interaction, f_1 and f_2 are now determined in terms of the odd (a_o) and even (a_e) angular momentum components of this microscopic interaction.

As in Landau theory for spin ½ Fermions it is convenient to define a spin-symmetric (f^s) spin-antisymmetric (f^a) interaction. The appropriate definition for spin 1 is[19]

$$\begin{aligned} f^s &= \frac{1}{3} (f_{\Uparrow\Uparrow} + f_{\Uparrow\nrightarrow} + f_{\Uparrow\Downarrow}) \\ f^a &= \frac{1}{3} (f_{\Uparrow\Uparrow} - f_{\Uparrow\Downarrow}). \end{aligned} \qquad (8)$$

Here $f_{\Uparrow\Uparrow} \equiv f_{\Uparrow\Uparrow\Uparrow\Uparrow} = \langle\Uparrow\Uparrow|f|\Uparrow\Uparrow\rangle$ and $f_{\Uparrow\Downarrow} \equiv f_{\Uparrow\Downarrow\Uparrow\Downarrow} = \langle\Uparrow\Downarrow|f|\Uparrow\Downarrow\rangle$. As in the spin ½ case, the Landau f^s, f^a are defined in terms of the no spin flip scatterings. The f_{1234} may be readily evaluated by expressing the $|\Uparrow\Downarrow\rangle$ states in terms of $|II_z\rangle$ states and (6). There are 3 independent interactions

$$\begin{aligned} f_{\Uparrow\Uparrow\Uparrow\Uparrow} &= f_{\nrightarrow\nrightarrow\nrightarrow\nrightarrow} = f_1 + 2f_2 = 2a_o \\ f_{\Uparrow\Downarrow\Uparrow\Downarrow} &= f_{\Uparrow\nrightarrow\Uparrow\nrightarrow} = f_1 + f_2 \\ f_{\Uparrow\Downarrow\Downarrow\Uparrow} &= f_{\nrightarrow\Uparrow\Uparrow\nrightarrow} = f_2. \end{aligned} \qquad (9)$$

The last interaction corresponds to a spin flip scattering. All other interactions can be related to one of (9) by rotation or vanish because spin is not conserved. Thus

$$f^s = (f_1 + (\tfrac{4}{3})f_2) = \frac{1}{3}(4a_o + 2a_e)$$

$$f^a = \frac{1}{3} f_2 = \frac{1}{3} (a_o - a_e). \tag{10}$$

Eq. (10) relates $f^{s,a}$ to the angular momentum components of a microscopic or model interaction (e.g. see (17)). In terms of $f^{s,a}$,

$$f_{1234} = (f^s + \frac{4}{3} f^a)\delta_{13}\delta_{24} + 3f^a[(I \cdot I) + (I \cdot I)^2]. \tag{11}$$

This specifies the pair interaction quite generally for a spin 1 Fermi fluid in zero magnetic field.

B. Fermi Liquid Theory

A straightforward generalization[19] to three spin states gives the compressibility, κ, effective mass m^* and magnetic susceptibility as

$$\begin{aligned} (n\kappa)^{-1} &= n\left(\frac{dn}{d\varepsilon}\right)^{-1} [1 + F_o{}^s] \\ \chi &= \frac{2}{3}\left(\frac{dn}{d\varepsilon}\right)\mu_B{}^2/[1 + F_o{}^a] \\ m^*/m &= [1 + F_1{}^s/3]. \end{aligned} \tag{12}$$

Here the dimensionless Landau parameters are, as usual,

$$F = \left(\frac{dn}{d\varepsilon}\right) f. \tag{13}$$

The density of states per unit volume at ε_F for n_s = 1,2 or 3 spin states is

$$\left(\frac{dn}{d\varepsilon}\right) = n\left(\frac{3}{2\varepsilon_F}\right) = \frac{n_s}{2} \frac{m^* k_F}{\pi^2\hbar^2} \tag{14}$$

where $k_F{}^3 = (6\pi n/n_s)$ or $n = n_s(k_F{}^3/6\pi^2)$. A factor of 2/3 appears in χ because only two of the three spin states $|\Uparrow\rangle$ and $|\Downarrow\rangle$ enter the magnetization, M. The state $|\Leftrightarrow\rangle$ does not contribute to M. Otherwise (12) are the usual expressions as for spin ½.

C. Galitskii-Feynman-Hartree-Fock

The GFHF theory has been discussed in detail in previous applications to normal[22] and spin-polarized[23] ^{3}He. The only new feature here is the separate forms of the spin-symmetric and spin-antisymmetric interactions for $D_1{}^\downarrow$, $D_2{}^\downarrow$, $D_3{}^\downarrow$ given in Eqs. (17) and (22) below. We begin with the interatomic potential v(r) between two $H^\downarrow$, $D^\downarrow$ or $T^\downarrow$ atoms. This has been calculated accurately by Kolos and Wolniewicz[3] for separations $0.5 \leq r \leq 6.5$ Å. The long range van der Waals attraction has been evaluated by Bell[24] and by Hirschfelder and Meath.[25] We use Silvera's fit[10] to v(r) as quoted by Friend and Etters,[26]

$$\begin{aligned} V(r) &= \exp[0.09678 - 1.10173r - 0.3945r^2] \\ &\quad - [\theta(r-r_c) - \theta(r_c-r)\exp[-(r_c/r-1)^2]] \\ &\qquad \times \left[\frac{6.5}{r^6} + \frac{124}{r^8} + \frac{3285}{r^{10}}\right] \end{aligned}$$

where θ is the Heaviside step function and $r_c = 10.0378$. The r and v(r) are in atomic units.

We first evaluate the Galitskii-Feynman (GF) T-matrix from v(r) by solving the Bethe-Salpeter equation in the particle-particle channel. This equation is depicted in Fig. 3. This is a many-body GF T-matrix which takes account of the Fermi sea. It is solved iteratively using self-consistent (complex) energy denominators in the particle Green functions. These energies are calculated from the GF T-matrix within the Hartree-Fock (HF) approximation (Fig. 4). These HF energies and GF T-matrix give a self consistent GFHF approximation having continuous and complex single particle energies.

III. GROUND STATE ENERGY

The GSE of $D^{\downarrow}$ in the GFHF approximation is

$$E = \frac{3}{5}\varepsilon_F^{\,o} + \tfrac{1}{2}\sum_{\substack{1,2\\ \sigma_1}} n_s \Gamma_{\sigma_1}^{\,s}(12)n(1)n(2) \; . \tag{15}$$

Here $\varepsilon_F^{\,o}$ is the Fermi energy and $n_s = 1, 2, 3$ for $D_1^{\downarrow}$, $D_2^{\downarrow}$, $D_3^{\downarrow}$. The spin symmetric Γ^s is the usual HF interaction

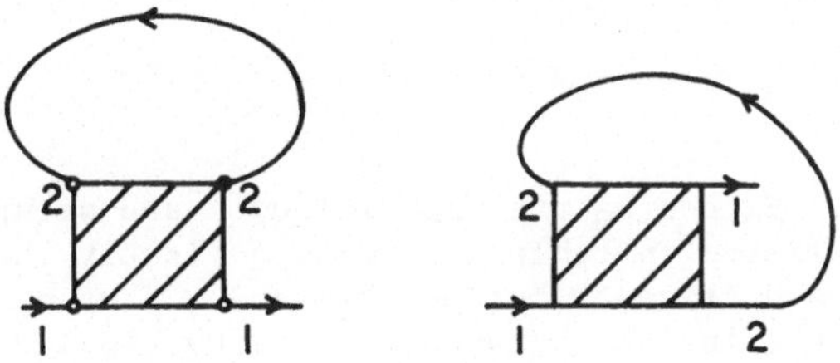

Fig. 3. The particle-particle Bethe-Salpeter equation. [hatched box] = $\Gamma(12,34)$, [open box] = $v(1-3)$.

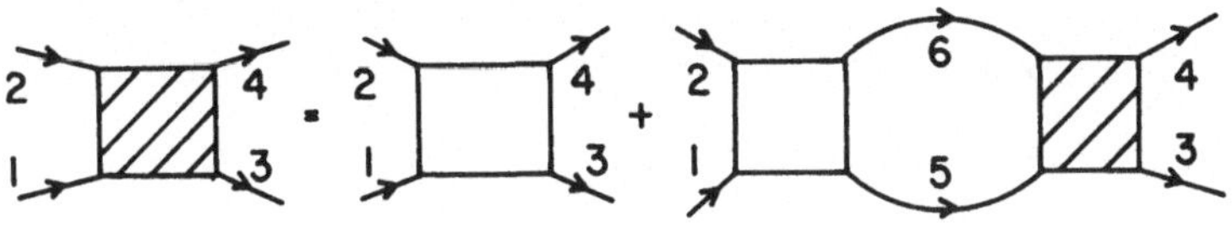

Fig. 4. The Hartree-Fock direct and exchange contributions to the single particle energies.

$$\Gamma^s = \frac{1}{n_s} \sum_{\sigma_2} [\Gamma(12,12) - \delta_{\sigma_1\sigma_2}\Gamma(12,21)]$$
$$= \frac{1}{n_s} \sum_{\sigma_2} \Gamma_{\sigma_1\sigma_2} . \qquad (16)$$

Specifically,

$$\begin{array}{ll} D_1^{\downarrow} & \Gamma^s = \Gamma_{\uparrow\uparrow} = 2a_o \\ D_2^{\downarrow} & \Gamma^s = \tfrac{1}{2}(\Gamma_{\uparrow\uparrow} + \Gamma_{\uparrow\downarrow}) = \tfrac{1}{2}(3a_o + a_e) \\ D_3^{\downarrow} & \Gamma^s = \frac{1}{3}(\Gamma_{\uparrow\uparrow} + \Gamma_{\uparrow\rightarrow} + \Gamma_{\uparrow\downarrow}) = \frac{1}{3}(4a_o + 2a_e) \end{array} \qquad (17)$$

where

$$a_o(12) = \sum_{L\ \mathrm{odd}} (2L+1)\Gamma_L(12)$$
$$a_e(12) = \sum_{L\ \mathrm{even}} (2L+1)\Gamma_L(12). \qquad (18)$$

Thus the GSE of $D_1^{\downarrow}$, $D_2^{\downarrow}$ and $D_3^{\downarrow}$ differ by (a) the value of $3/5\ \varepsilon_F{}^o \propto n_s{}^{-2/3}$ and (b) the angular momentum components, Γ_L, of the GF T-matrix entering Γ^s. Note that in each case the odd L components dominate.

In Fig. 4 we show E for $D_1^{\downarrow}$, $D_2^{\downarrow}$, and $D_3^{\downarrow}$. The energy may be fitted by

$$E = E_o + Ax^2 + Bx^3 \qquad (19)$$

where $x = (n-n_o)/n_o$. E_o and n_o are the minimum E and saturation density. E_o, n_o, A and B are listed in table 1. We note firstly that $E_1^{\downarrow} > E_2^{\downarrow} > E_3^{\downarrow}$ as expected. The difference between $E1\downarrow$ and $E2\downarrow$ at n_o is $\sim$ 0.54 K. However, the difference in <KE> between $D_1^{\downarrow}$ and $D_2^{\downarrow}$ is 1.0 K. Thus the attractive PE is actually stronger in $D_1^{\downarrow}$ than in $D_2^{\downarrow}$. From (17) this suggests that the attractive binding comes chiefly from the a_o (odd L) and the PE of $D_2^{\downarrow}$ and $D_3^{\downarrow}$ is weakened as a_e is switched in. Also Pauli repulsion is greatest in $D_1^{\downarrow}$ and least in $D_3^{\downarrow}$. The saturation volumes in each case are V $\sim$ 150 cm^3/mole. This is $\sim$ 5 times that of liquid ^3He.

Table 1. Ground State Energy parameters for Eq. (19).

	$10^3 n_o$ (Å^{-3})	V_o (cm^3/mole)	E_o (K)	A (K)	B (K)
$D_1^{\downarrow}$	3.91	154	0.295	1.09	0.927
$D_2^{\downarrow}$	3.96	152	-0.237	1.26	0.721
$D_3^{\downarrow}$	4.25	142	-0.786	4.7	6.1

The ratio of the $D^\downarrow$ to ^{3}He r_n values of $v(r)$ is ~ 3.2.

Secondly, the GFHF $E_1^\downarrow$ agrees well with the CBF values of Krotscheck et al.[13] This suggests that the GFHF approximation describes (nuclear) spin polarized systems well. This was found to be true for ^{3}He$^\uparrow$. Essentially, with all nuclear spins polarized there can be no spin fluctuations. These are not included in the GFHF theory. Also, with all spins parallel, Pauli exclusion correlations operate between all spins. These important repulsions, which are included in the GFHF, tend to reduce the importance of other correlations (such as collective density excitations) not included in the GFHF.

The GFHF values of $E_2^\downarrow$, and $E_3^\downarrow$ especially, are not very reliable at $n \sim 4 \times 10^{-3}$ Å^{-1}. The $E_3^\downarrow$ value lies well below the MC value of Panoff et al.[14] The GFHF ground state energy of normal liquid ^{3}He was also substantially below the observed value. This again suggests that higher order terms, not included in the GFHF, are important in unpolarized (nuclear) systems such as normal ^{3}He and $D_3^\downarrow$. Three-body correlations and momentum correlations appear to be important in $D_3^\downarrow$. The GFHF theory employing a continuous single particle energy spectrum tends to predict a low E for unpolarized fluids.[22] It is also a low density theory and from Fig. 5 is probably reasonably accurate for $D_3^\downarrow$ at $n \lesssim 1 \times 10^{-3}$ Å^{-1}.

The pressure and compressibility $\kappa^{-1} = v\, \partial^2 E / \partial v^2$ may be evaluated from (19). We find $D^\downarrow$ can be readily compressed by modest pressures of 1–2 atm. and $(n\kappa)^{-1}$ is ~ 2–3 K compared to ~ 12 K for normal ^{3}He at saturation.

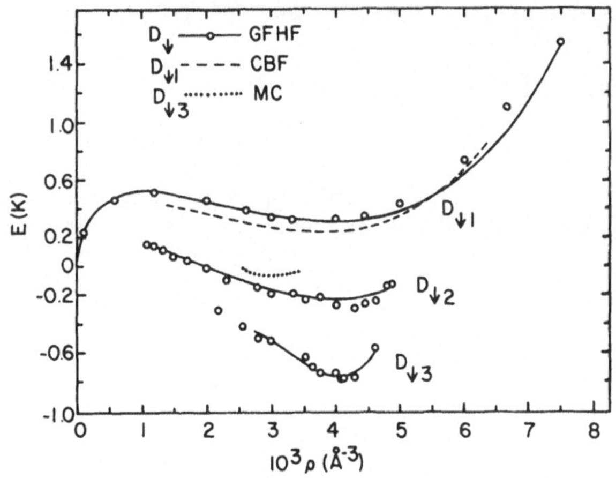

Fig. 5. The GFHF ground state energy for $D_1^\downarrow$, $D_2^\downarrow$ and $D_3^\downarrow$. CBF are the correlated basic function values of Krotscheck et al (Ref. 13) and MC are Monte Carlo values of Panoff et al (Ref. 14).

IV. SINGLE PARTICLE ENERGIES

The self energy of a single $D^{\downarrow}$ quasiparticle due to its interaction with other $D^{\downarrow}$ quasiparticles in the GFHF limit is

$$\sum_{\sigma_1}(1) = -\frac{i}{V}\int \frac{d^4p_2}{(2\pi)^4}\, n_s \Gamma^s(12) G^{HF}(2) \ . \qquad (20)$$

In (20) the sum over the second particle spin as already been performed as indicated in (16) and (17). Here G^{HF} is the HF single particle Green function evaluated using self consistent HF particle energies.

The real and imaginary parts of $\varepsilon(1)$ are shown in Fig. 6. There we see that $\mathrm{Im}\varepsilon(k)$ vanishes at $k = k_F$ but is large away from the Fermi surface. The Re $\varepsilon(k)$ values also display substantial attractive PE which is largely independent of k. This is required to give nearly zero total E since, for example, the KE of a $D_1^{\downarrow}$ atom at $n = 4\times10^{-3}$ Å^{-3} is $\simeq$ 2.8 K. For $D_1^{\downarrow}$ we see that the Hugenholtz-van Hove equality, Re $\varepsilon(k_F) = E$ at saturation is quite well satisfied. For $D_2^{\downarrow}$ and $D_3^{\downarrow}$ Re $\varepsilon(k_F)$ lies significantly below E suggesting that higher order terms are needed to get a completely consistent value of $\varepsilon(k)$ and E. This further displays that the GFHF describes (nuclear) polarized systems well, but unpolarized systems less well.

V. LANDAU PARAMETERS, M^* AND SOUND

A. Zero Order

At low temperature ($T \lesssim 0.1\ \varepsilon_F$) only quasiparticles on or near the Fermi surface are excited. The interaction between two quasiparticles having momentum k_1 and k_2 on the Fermi surface is represented by the dimensionless Landau parameters, F_L. To lowest order the F_L are given in terms of the GF T-matrix, $\Gamma(k_1,k_2,P)$, as

$$F_L^{s,a} = \left(\frac{dn}{d\varepsilon}\right)\frac{(2L+1)}{2}\int_{-1}^{1} d(\cos\theta) P_L(\cos\theta)\Gamma^{s,a}(\theta) \ . \qquad (21)$$

Since $|k_1| = |k_2| = k_F$, Γ depends only upon the angle θ between k_1 and k_2. The Γ^s are given by (17) and the corresponding spin-antisymmetric Γ^a are:

$$\begin{aligned} &D_1^{\downarrow} && \Gamma^a = 0 \\ &D_2^{\downarrow} && \Gamma^a = \tfrac{1}{2}(a_o - a_e) \\ &D_3^{\downarrow} && \Gamma^a = \tfrac{1}{3}(a_o - a_e) . \end{aligned} \qquad (22)$$

The F_L given by (21) are lowest order because the GF T-matrix here does not include spin fluctuations or momentum correlations which are known to contribute significantly to the F_L in liquid ^{3}He. The F_L from (21) for $D_1^{\downarrow}$, $D_2^{\downarrow}$ and $D_3^{\downarrow}$ at three volumes ($V \approx 150$ cm^3/mole is suatura-tion) are listed in Table 2.

The Landau parameters get much smaller as we go from D3↓ to D1↓. F_0^s drops by a factor of 10 and m^* is reduced from 1.5 to 0.9. This is partly due to a reduction in the density of states, i.e. $(dn/d\varepsilon) = 3.6\times10^{-3}$ (KÅ^3)$^{-1}$ in $D_3^{\downarrow}$ and 1.05×10^{-3} (KÅ^3)$^{-1}$ in $D_1^{\downarrow}$ at saturation. This

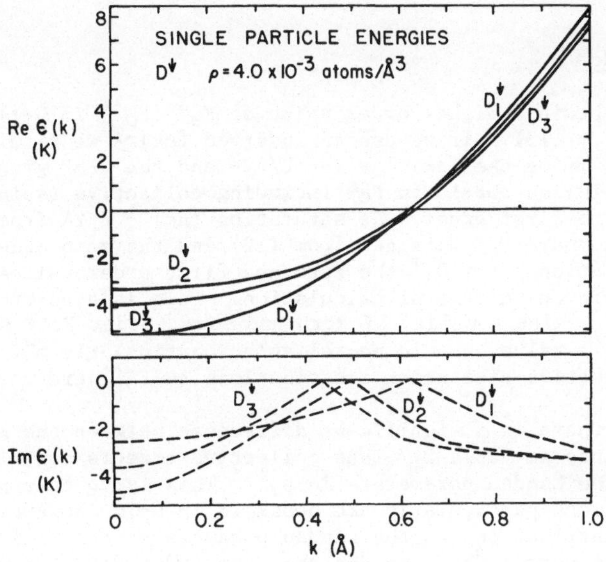

Fig. 6. The single particle energies in $D^{\downarrow}$.

Table 2. Landau Parameters from Eq. (21).

DENSITY (10^{-3}Å^{-3})		4.63		3.95		4.63	
VOLUME (cm^3/mole)		130		150		170	
	L	$F_L^{\uparrow\uparrow}$		$F_L^{\uparrow\uparrow}$		$F_L^{\uparrow\uparrow}$	
$D_1^{\downarrow}$	0	-0.36		-0.45		-0.50	
	1	-0.42		-0.37		-0.28	
	2	0.18		0.38		0.39	
	m^*	0.86		0.87		0.90	
	L	F_L^s	F_L^a	F_L^s	F_L^a	F_L^s	F_L^a
$D_2^{\downarrow}$	0	-2.7	-1.1	-2.7	-1.1	-2.9	-1.1
	1	0.84	-0.56	0.91	-0.36	1.0	-0.2
	2	1.0	-0.92	0.98	0.96	0.92	1.0
	m^*	1.28		1.30		1.34	
	L	F_L^s	F_L^a	F_L^s	F_L^a	F_L^s	F_L^a
$D_3^{\downarrow}$	0			-4.2	-1.2	-4.2	-1.4
	1			1.5	0.6	1.5	0.6
	2			0.6	0.9	0.8	1.2
	m^*			1.5		1.5	

reduction suggests that quasiparticles in $D_1{\downarrow}$ respond much like free, non-interacting Fermions. We found a similar dramatic reduction in the Landau parameters in going from normal ^{3}He to ^{3}He$\downarrow$, as did Bedell and Quader.[27]

B. First Order

We may obtain a higher order value of F_o^s ($F_o^{\uparrow\uparrow}$ in $D_1^{\downarrow}$) from the compressibility (12). If we use an observed $(n\kappa)^{-1}$ we obtain the "exact" or full F_o^s. If we use $(n\kappa)^{-1} = V^2\partial^2E/\partial V^2$ and the GFHF ground state energy E in (19) we obtain an F_o^s including collective (spin and momentum) correlations to first order. At saturation $(n\kappa)^{-1} = 2A$ from (19). In table 3, we compare F_o^s obtained from (12) and the zero order value from (21) at saturation. For $D_1^{\downarrow}$ the zero and first order values of $F_o^{\uparrow\uparrow}$ are the same within our errors of calculation. This is also true for all $n \leq 5\text{x}10^{-3}$ Å^{-3}. The equality of zero and first order $F_o^{\uparrow\uparrow}$ suggests the other zero order $F_L^{\uparrow\uparrow}$ values should be reliable, particularly m^*. This tells us that the collective excitation contributions to $F_o^{\uparrow\uparrow}$ are very small in $D_1^{\downarrow}$.

For $D_2^{\downarrow}$ there is a significant difference between the zero and first order F_o^s values in table 2. Thus collective effects contribute significantly to the Landau parameters in $D_2^{\downarrow}$. This is to be expected since spin fluctuations particularly are possible in $D_2^{\downarrow}$. However, the collective contributions to the Landau parameters are still much smaller in $D_2^{\downarrow}$ than in normal ^{3}He. In ^{3}He the zero order $F_o^s \simeq -8$ while the observed value is $F_o^s = 10.7$ suggesting very strong collective contributions, as is well known.

For $D_3^{\downarrow}$ the GFHF E value in (19) and table 1 is too unreliable to extract a meaningful $(n\kappa)^{-1}$ value. We have therfore fitted the form (19) to the Monte Carlo values of E computed by Panoff et al.[14] This gives

$$E = -0.061 + 2.59x^2 - 7.77x^3 \text{ K}$$

with $n_o = 2.9\text{x}10^{-3}$ Å^{-3}. The corresponding $(n\kappa)^{-1}$ and F_o^s is listed in table 2. There we see that collective contributions to F_o^s are larger in $D_3^{\downarrow}$ than in $D_2^{\downarrow}$. (The m^* is the zero order value obtained from F_1^s.) Again, the difference between the first and zero order F_o^s values suggests the zero order Landau parameters for $D_3^{\downarrow}$ in table 2 serve only as an estimate.

Table 3. Properties of $D^{\downarrow}$ at saturation density.

	$10^3 n$ (Å^{-1})	V (cm^3/mole)	ϵ_F (K)	$(n\kappa)^{-1}$ (K)	$(F_o^s)_{FIRST}$	$(F_o^s)_{ZERO}$	$\frac{m^*}{M}$	v_F (m/sec)	C_1 (m/sec)	C_o (m/sec)
$D_1^{\downarrow}$	3.91	154	5.60	2.2	-0.41	-0.45	0.9	225	95	-
$D_2^{\downarrow}$	3.96	152	2.23	2.5	0.69	-2.7	1.3	120	102	134
$D_3^{\downarrow *}$	2.9	207	1.2	5.2	5.5	-4.2	1.5	83	249	262

$^*(n\kappa)^{-1}$ and saturation V from Panoff et al (Ref. 14).

C. Sound

In table 3 are listed values of the Fermi velocity, v_F, the first sound velocity, C_1, and the zero sound velocity C_o. We see that $v_F = \hbar k_F/m^*$ increases significantly as $D\downarrow$ is nuclear polarized. This simply reflects the increase in k_F and the decrease in m^* with polarization (as we go from $D3\downarrow$ to $D1\downarrow$). Similar, at constant (saturation) density, C_o decreases with polarization until in $D1\downarrow$ it will no longer propagate. The C_o is obtained from

$$C_o = s v_F$$

where

$$\frac{s}{2} \ln\left(\frac{s+1}{s-1}\right) - 1 = (F_o^s + F_1^s s^2/m^*)^{-1}. \tag{23}$$

Thus C_o is proportional to v_F. However, it cannot propagate unless $s \geq 1$. This requires that the right hand side of (23) be > 0. For $n \gtrsim 6\times10^{-3}$ Å^{-3} in $D1$, the F_L^s become positive and large enough that zero sound again propagates and with a high velocity, $C_o \gtrsim 370$ m/sec.

Thus, we predict that at saturation zero sound propagates in $D2\downarrow$ and $D3\downarrow$ but not in $D1\downarrow$. However, in compressed $D1$, zero sound propagates with a very high velocity, ~ 30% greater than C_1. Thus the zero sound velocity is a sensitive function of both nuclear polarization and density in $D\downarrow$.

CONCLUSION

The agreement of the GFHF ground state energy with other calculations and the internal consistency of the single particle energies and Landau parameters show that the GFHF theory describes doubly polarized $D1\downarrow$ well. $D1\downarrow$ is predicted to behave like a gas of weakly interacting Fermions even at high density, i.e. like a Pauli paramagnet. It's $m^* \approx 1$ and the Landau parameters are small. The interactions between particles and collective effects increase significantly in $D2\downarrow$ and $D3\downarrow$. Zero sound is predicted to propagate in $D2\downarrow$ and $D3\downarrow$ but not in $D1\downarrow$ for $n \lesssim 4\times10^{-3}$ Å^{-3}.

REFERENCES

1. R. Sprik, J. T. M. Walraven, I. F. Silvera, Phys. Rev. Lett. 51, 479 (1983). H. F. Hess, D. A. Bell, G. P. Kochanski, R. W. Cline, D. Kleppner and T. J. Greytak, Phys. Rev. Lett. 51, 483 (1983).
2. I. F. Silvera and J. T. M. Walraven, Phys. Rev. Lett. 45, 1268 (1980).
3. W. Kolos and L. Wolniewicz, Chem. Phys. Lett. 24, 457 (1974).
4. M. D. Girardeau, Int. J. Quant. Chem. 17, 25 (1980).
5. R. M. More, Phys. Rev. Lett. 51, 396 (1983).
6. R. L. Danilowicz, J. V. Dugan and R. D. Etters, J. Chem. Phys. 65, 498 (1976) and earlier references.
7. W. C. Stwalley and L. H. Nosanow, Phys. Rev. Lett. 36, 910 (1976); W. C. Stwalley, Phys. Rev. Lett. 37, 1628 (1976).
8. L. H. Nosanow, L. J. Parish and F. J. Pinski, Phys. Rev. B11, 191 (1975).
9. M. D. Miller, L. H. Nosanow and L. J. Parish, Phys. Rev. Lett. 35, 581 (1975), Phys. Rev. B13, 214 (1976).
10. I. F. Silvera, Rev. Mod. Phys. 52, 393 (1980).

11. M. D. Miller and L. H. Nosanow, Phys. Rev. B15, 4376 (1977).
12. L. H. Nosanow, J. Low Temp. Phys. 23, 605 (1976); ibid 26, 613 (1977); J. Physique Colloq. C7-1 (1980).
13. E. Krotscheck, R. A. Smith, J. W. Clark and R. M. Panoff, Phys. Rev. B24, 6383 (1981); J. W. Clark, E. Krotscheck and R. M. Panoff, J. Physique Colloq. C7-197 (1980).
14. R. M. Panoff, J. W. Clark, M. A. Lee, K. E. Schmidt, M. H. Kalos and G. V. Chester, Phys. Rev. Lett. 48, 1675 (1982).
15. T. K. Lim, Phys. Rev. B25, 2057 (1982).
16. A. J. Leggett, J. Physique Colloq. C7-19 (1980).
17. A. G. K. Modawi, Thesis, University of Sussex, U.K. (1981).
18. K. S. Bedell and K. Quader, Phys. Rev. B31, 1627 (1984).
19. S. J. Buckle, J. Phys. C., 17, L633 (1984).
20. G. Baym and C. J. Pethick in The Physics of Liquid and Solid Helium, edited K. H. Bennemann and J. B. Ketterson (Wiley-Interscience, New York, 1978) Part II.
21. R. Gilmore, Lie Groups, Lie Algebras and Some of their Applications. (Wiley-Interscience, 1974, N.Y.) p. 124.
22. H. R. Glyde and S. I. Hernadi, Phys. Rev. B28, 141 (1983).
23. H. R. Glyde and S. I. Hernadi, Phys. Rev. B29, 3873 (1984).
24. R. J. Bell, Proc. Phys. Soc. (Lond.) B7, 594 (1966).
25. J. O. Hirschfelder and W. J. Meath, Adv. Chem. Phys. 12, 3 (1967).
26. D. G. Friend and R. D. Etters, J. Low Temp. Phys. 39, 409 (1980).
27. K. S. Bedell and K. F. Quader, Phys. Lett. 96A, 91 (1983).

A MOMENTUM DEPENDENT INDUCED INTERACTION MODEL APPLIED TO LIQUID ^{3}He

T. L. Ainsworth†

Department of Physics, University of Arizona

Tucson, AZ 85721 USA

The Babu-Brown[1] induced interaction model has recently been enjoying considerable success. It has been applied to liquid ^{3}He,[2,3] spin polarized ^{3}He,[4] ^{3}He-^{4}He mixtures,[5] paramagnetic metals,[2] and heavy Fermion systems.[6] The approach of this model is to treat the particle-hole reducible interaction explicitly by summing all one-particle-one-hole reducible diagrams to all orders. Thus this method is best at describing deformations and excitations about the Fermi surface.

The present work describes a generalization of the Babu-Brown induced interaction model. The initial reason for including a momentum dependence in the induced interaction was to maintain the antisymmetry of the quasiparticle interaction and thereby preserve the forward scattering sum rule (Pauli principle). Therefore in addition to the RPA particle-hole interaction (t-channel) the exchange particle-hole channel (u-channel, induced interaction) was included. Figure 1 shows diagrammatically the equations which self-consistently sum the one-particle-one-hole reducible diagrams.

The first equation is the sum of two terms; the first represents all diagrams that are particle-hole irreducible and the second those that are particle-hole reducible in the u-channel. The addition of these terms yields the set of t-channel irreducible diagrams. The second equation, the RPA equation, builds in the t-channel reducible diagrams. The symbol with a vertical (horizontal) line inside of a circle represents the set of diagrams that can not be cut vertically (horizontally), thus they are irreducible in

† Present address Physics Department, State University of New York at Stony Brook, Stony Brook, NY 11794 USA

the t-channel (u-channel). The open circle represents the full scattering amplitude, which is the sum of all possible diagrams. The function on the lefthand side of the first equation is the Landau F. The full momentum and frequency dependent quasi-particle interaction reduces to the Landau F in the ordered limit $q \to 0$ and $\omega \to 0$. In precisely this limit the second term in the second equation vanishes. The second term in the first equation depends upon the exchange momentum transfer, $q' = p_1 - p_4$, and therefore does not vanish in this limit. Therefore one identifies the sum of all t-channel irreducible diagrams with the Landau F. These equations are then solved self-consistently within some approximation. Due to the approximations made, the Babu-Brown formulation was only antisymmetric in the limit of zero momentum transfer and then only when the equations were truncated at $\ell = 0$. This shortcoming can easily be rectified by allowing for an explicit momentum dependence of the Landau parameters and the scattering amplitude.[7]

There are several advantages gained by generalizing the induced interaction model. The equations to be solved are explicitly crossing-symmetric and thus retain the same antisymmetry as the diagrams which they sum. Both

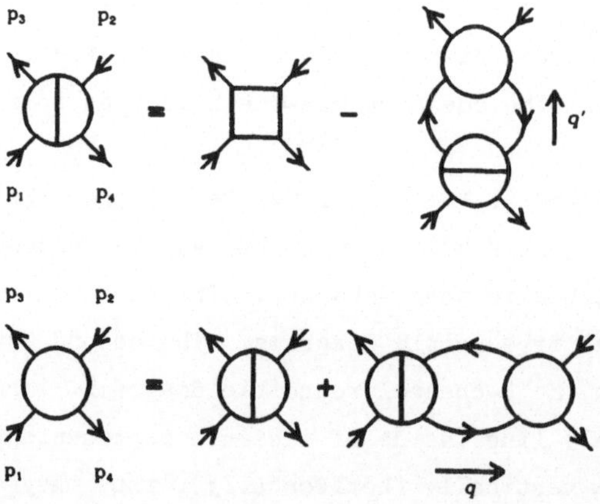

Fig. 1. Diagrammatic equations which self-consistently sum the particle-hole reducible diagrams, see text for further explanation.

the induced interaction channel and the RPA channel (u-channel and t-channel, respectively) are approximated in precisely the same manner and iterated to a self-consistent solution. This modification of the Babu-Brown model might be considered merely an aesthetic one except for one feature, one obtains a self-consistent antisymmetric momentum dependent scattering amplitude. Thus transport coefficients and the quasi-particle lifetime can be calculated without any additional modeling. Specifically, the S-P approximation or other means of forcing antisymmetry on the scattering amplitude becomes unnecessary. Within this model one calculates Landau parameters, which can be related to experimentally determined quantities (sound speed, spin susceptibility, specific heat) and the scattering amplitude, which when averaged over the Fermi surface yields transport coefficients (viscosity, thermal conductivity, spin diffusion).

The Landau parameters and the scattering amplitude are calculated on the Fermi surface, thus each external leg carries momentum k_F. This restriction implies that the scattering amplitude and Landau parameters depend only on two independent variables. While any two independent variables suffice, it is convenient to choose variables in which the equations at least partially decouple. The standard Landau expansion in Legendre polynomials of argument $\cos(\Theta_{12})$ does not decouple the momentum dependent equations and thus must be replaced. The angle in which the equations decouple, Θ_L, is the angle between $(p_1+p_3)/2$ and $(p_2+p_4)/2$. Figure 2 defines the various angles and momenta mentioned. The Legendre expansion in terms of $\cos(\Theta_L)$ reduces to the standard expansion in the limit of $q\to 0$. At $q=0$, Θ_L is identical to Θ_{12}. The simplification gained is signifigant, and the connection to the familar Landau expansion is retained. The momentum dependence of the Landau F's and A's are expanded in Legendre polynomials of argument $\cos(\Theta_{13})$. The momentum transfer, q, is related to Θ_{13} by $q=2k_F\sin(\Theta_{13}/2)$. The expansion of the momentum dependence is an extension of the Babu-Brown model. This choice of variables leads to a substantial decoupling of the non-linear coupled equations and to a clean description of the momentum dependence of the Landau parameters and scattering amplitude. The F's are given by the set of coefficients, $F_{\ell m}$, of the double Legendre expansion in terms of $\cos(\Theta_L)$ and $\cos(\Theta_{13})$;

$$F(\Theta_L,\Theta_{13}) = \sum_{\ell m} F_{\ell m} P_\ell(\cos(\Theta_L))\, P_m(\cos(\Theta_{13})). \tag{1}$$

A similar expression is used for the expansion of the A's in terms of $A_{\ell m}$. While this double expansion is not the most efficient in terms of the number of coefficients to be calculated, its simplicity and ease of application more than compensate for this inconvenience. In the limit $q=2k_F$, the angle Θ_L becomes undefined. (It is the angle between two vectors of zero length.) Therefore in this limit the scattering amplitude and the Landau parameters cannot depend on Θ_L, hence only the $\ell=0$ moments (F_0^s, F_0^a, A_0^s and A_0^a) can be nonzero at $q=2k_F$. All higher moments must identically vanish.[8] This requirement, a consequence of restricting the momenta of the external legs to the Fermi surface, provides a useful check on numerical solutions of the equations.

The other parts of this model are the form of the direct interaction, that which drives the equations, and the particle-hole propagators which link the pieces together. The direct interaction contains information about all particle-hole irreducible diagrams. In principle, any calculation of a particle-hole irreducible interaction can be used as the direct interaction without encountering any double counting problems in the t-channel or u-chan-

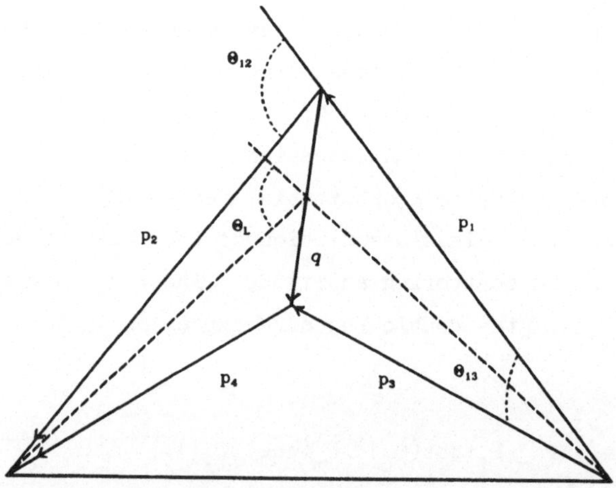

Fig. 2. Θ_{12} and Θ_{13} are the angles between p_1 and p_2, and p_1 and p_3, respectively. Θ_L is the third angle depicted, see text for definition. All momenta, p_i, have magnitude k_F.

nel. In practice, one should be a bit more careful. If, for instance, the G-matrix is used as the direct interaction then many important diagrams are omitted. Specifically, all particle-particle ladder diagrams with any particle-hole reducible rungs are ignored. This problem can be corrected by renormalizing the bare interaction with the particle-hole reducible diagrams, recalculating the G-matrix using the renormalized interaction, then recalculating the renormalization, and so on. Instead of this procedure a parameterized direct interaction is used.[3] The parameterized direct interaction approximates the sum of all particle-hole irreducible diagrams. For liquid ^3He, the local parts of this approximation reproduce the gross features of local direct interactions extracted from microscopic calculations.[9] The parameterization is in terms of three coefficients, a_s, a_t and b_t. The a_s parameter is the quasi-particle s-wave scattering length and is expected to have a strong density dependence. The others are p-wave in nature and should not be strongly density dependent. Using the forward scattering sum rule, an exact expression for a_s is derived;

$$\frac{4}{\pi}\frac{m^*}{m}a_s k_F = \sum_\ell \left(F^s_\ell - F^a_\ell \left\{ 1 + 2\,\frac{F^a_\ell/(2\ell+1)}{1+F^a_\ell/(2\ell+1)} \right\} \right) \tag{2}$$

For liquid ^3He the infinite sum is exhausted after the first two terms, therefore the experimental Landau parameters strictly constrain the value of a_s. Its density dependence is determined primarily by the density dependence of F^s_0, F^s_1 and k_F. Of the three direct interaction parameters, only one, a_s, is adjusted to fit the Landau parameters as a function of pressure. Both a_t and b_t are adjusted at one pressure and then held fixed for all remaining pressures.

The particle-hole Green's functions which connect the various pieces are the zero frequency Lindhard functions. These are simply angle averages of the available phase space for a particle-hole pair. Because the phase space is for quasi-particles the effective mass rather than the bare mass is used in the energy denominators. In principle, the momentum dependence of the effective mass can be determined through the frequency sum rule. However, this model has no frequency dependence and thus frequency sum rule arguments become dubious. A self-consistent treatment of an explicit frequency dependence in the Lindhard functions, the Landau parameters and the scattering amplitude completely destroys the decoupling of the non-linear equations and generally reaps havoc. Therefore a constant, momentum independent value of the effective mass is used throughout. This choice weights the couplings at large momentum transfer, $q \simeq 2k_F$, more heavily than one would expect on physical grounds, however the total weight for $q \simeq 2k_F$ is relatively small thus the errors made should be also. The Lindhard functions couple different moments of F and A in the $P_\ell(\cos(\Theta_L))$ expansions. Thus in general F_ℓ can couple to A_m yielding a contribution to A_n. In the q=0 limit $\ell=m=n$, as expected, however at finite q other more complicated couplings occur.[7,10]

At this point the pieces of the model have to be collected and put together. The Landau parameters and the scattering amplitude have been written in terms of two sets of coefficients, $F_{\ell m}$ and $A_{\ell m}$. The direct interaction and particle-hole propagators have been defined. Solution of the equations remains. a_s, a_t and b_t are adjusted, at one pressure, so that three experimentally determined Landau parameters are fit. Then at all other pressures only a_s is adjusted so as to fit, say, F_0^s. The value of a_s is tightly constrained by the sum rule relation, equation 2. Therefore there is very little room to make adjustments to improve the fit to experiment. Once the $A_{\ell m}$'s have been computed transport coefficients can be calculated. Essentially weighted averages of the square of the scattering amplitude are required.[11] The viscosity is of particular interest because the weighting is large near $q \simeq 1.4k_F$. Thus it should provide a good test of the momentum dependence of the scattering amplitude.

The model has several built-in numerical tests. The double Legendre expansions have to be truncated. The severity of the truncation can be estimated by watching the magnitudes of the expansion coefficients, $F_{\ell m}$ and $A_{\ell m}$, and the change in these coefficients when additional terms are kept in the Legendre expansions. Both of these estimates indicate that the truncation is not severely affecting the calculation. Another test relates to the redundancy of the double expansion. Not all of the coefficients are necessary. Specifically, at $q=2k_F$ the F_ℓ's and A_ℓ's must be identically zero for $\ell>0$. The difference of these quantities from zero provides an overall check on the numerical solutions. In only one case does this test suggest any possible problem. The A_1^a component of the scattering amplitude differs slightly from zero at $2k_F$. Additionally, the q dependence of this component shows some oscillatory structure. This structure matches that of the first Legendre polynomial dropped by the truncation of the $\cos(\Theta_{13})$ expansion. Inclusion of an additional term would correct this deviation, however the size of the correction is too small to change any results. Another test of the numerical solution is the forward scattering sum rule. The direct interaction is, by construction, antisymmetric. Diagrammatically the sum of the particle-hole reducible channels are antisymmetric. Therefore the numerical results for the scattering amplitude had better be. The amplitude for scattering two like-spin quasi-particles in the forward direction is zero, by the Pauli principle. In the forward limit the scattering amplitude for like-spin quasi-particles is numerically zero. These tests show no serious problems either in the truncation of the Legendre expansions or in the overall numerical solutions of the coupled non-linear integral equations.

The results of both the Landau parameters and transport coefficients are given in figure 3. Six experimental quantities are calculated at each pressure. The calculated values of F_0^s, F_1^s ($m^*/m = 1+F_1^s/3$) and F_0^a agree well with the experimantal values.[12-14] Both the present and the Babu-Brown formulations of the induced interaction model give comparable results for the Landau parameters.[2,3] The graphs of the transport coefficients show both the present calculation and a calculation using the S-P approximation to antisymmetrize the Babu-Brown scattering amplitude.[15] The largest difference between the two calculations is for the viscosity. The S-P approximation simply is not reasonable when $q \approx 1.4k_F$. Therefore one concludes that the present model does very well calculating half a dozen experimental results using essentially one free parameter.

One result of these calculations is an antisymmetrized scattering amplitude, therefore any quantity which depends upon the scattering of quasi-particles near the Fermi surface should be calculable. Specifically, the superfluid transition temperature, T_c, can be calculated, however the accuracy of the result is questionable. Using the $A_{\ell m}$'s in the Patton and Zaringhalam[16] equations for T_c points to a problem. The relevant quantities are the differences between roughly equal terms. The double Legendre expansion used for the F and A has this drawback; a quantity of interest, T_c, depends exponentially on a small number which is itself the difference of two large numbers. Thus an accurate determination of T_c is difficult within this model. The strong and weak coupling corrections to T_c may be more reasonably calculated since they are weighted averages of the scattering amplitude over the Fermi surface. Additionally, they can be compared to the experimentally observed specific heat jumps across the phase transitions. At present this has not been done.

Bedell and Meltzer[17] have used the momentum dependent Landau parameters in a model for the spin relaxation time, T_1, of liquid ^{3}He. They consider the linear response of the induced interaction to a weak dipole-dipole spin non-conserving interaction. This momentum dependent induced interaction is taken as the point about which the linear response is calculated. Their calculated T_1 is in good agreement with the experimental values. The momen-

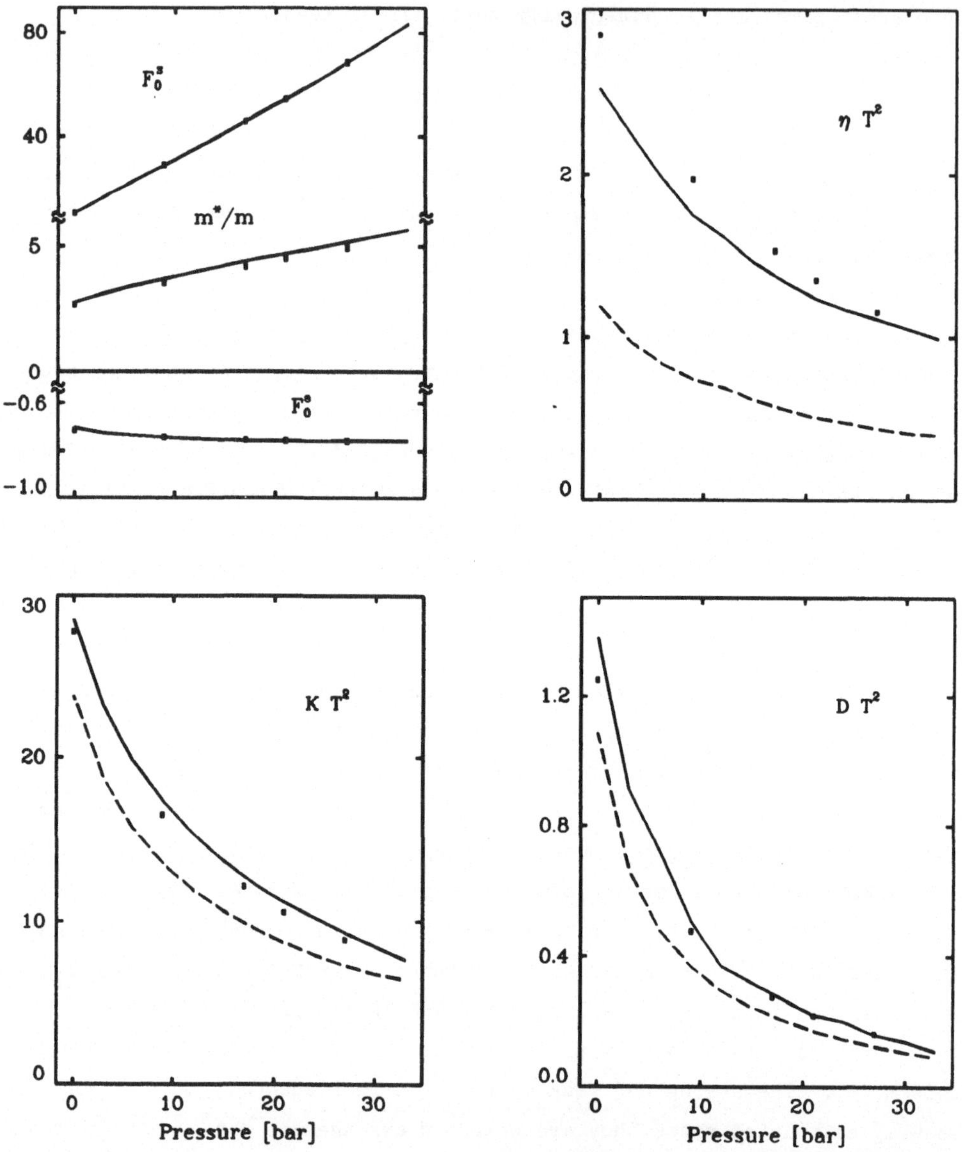

Fig. 3. The experimental values for three Landau parameters[12-14], the viscosity[18] (ηT^2), the thermal conductivity[19] (KT^2) and the spin diffusion coefficient[20] (DT^2) are denoted by solid curves. The calculated values, at five pressures, are denoted by the boxes. The S-P approximation results for the transport coefficients are shown with broken curves.[15] The units are; dimensionless for the Landau parameters, poise-$(mK)^2$ for viscosity, erg/sec-cm for thermal conductivity and cm^2-$(mK)^2$/sec for the spin diffusion.

tum dependence of the Landau parameters is critical to their model and thus provides another test of the present work.

An alternate formulation of the momentum dependent induced interaction has been presented by Pfitzner and Wölfle.[21] Diagrammatically the two are identical. These two approaches differ primarily in the treatment of the particle-hole propagators. Pfitzner and Wölfle employ a function, $\alpha(q)$, to cut down the strength of the particle-hole propagator for large momentum transfer. They also use a momentum dependent effective mass in the energy denominator of the propagator. These are the two differences in the physical content of the models. The model input, the direct interaction, also differ. They use a local direct interaction extracted from microscopic calculations. The parameterized form used in the present work contains both local and non-local components. The overall structure of the local components of the two direct interactions is similar. The non-local component is needed in the present model to obtain a good fit to the effective mass. The $\alpha(q)$ and the momentum dependence of the effective mass probably play similar roles in their model. Different sets of basis functions are used to expand the scattering amplitude and the Landau parameters, however this is of no physical consequence. Some calculations might be performed more easily in one basis or another but the song remains the same.

A conclusion to be drawn from this work is that both particle-hole reducible channels are important for calculating the quasi-particle interaction near the Fermi surface and by including both on an equal basis the resultant scattering amplitude is inherently antisymmetric. The generalization of the Babu-Brown induced interaction model to a momentum dependent model was necessary so that identical approximations could be made in both the u-channel and the t-channel summations. Two obvious improvements would be to self-consistently include a microscopic direct interaction and to relax the approximations made in evaluating the particle-hole reducible diagrams. Specfically, to allow the external legs to go off the Fermi surface. These improvements, though obvious, may not be so easy to accomplish.

ACKNOWLEDGEMENT

I thank K. S. Bedell, G. E. Brown, E. Krotscheck, N.-H. Kwong and L. Yi for many useful and interesting discussions. This work was supported in part by US National Science Foundation grant no. PHY-8100141 and US DOE contact DE-AC02-76ER13001.

REFERENCES

1. S. Babu and G. E. Brown, Ann. Phys. 78, p. 1.
2. T. L. Ainsworth, K. S. Bedell, G. E. Brown and K. F. Quader, J. Low Temp. Phys. 50, p. 317.
3. K. S. Bedell and T. L. Ainsworth, Phys. Lett. 102A, p. 49.
4. K. F. Quader and K. S. Bedell, J. Low Temp. Phys. 58, p. 89.
5. L. Yi, K. S. Bedell and T. L. Ainsworth, Bull. Am. Phys. Soc. 28, p. 655.
6. K. S. Bedell and K. F. Quader, Phys. Rev. B32, in press.
7. T. L. Ainsworth, K. S. Bedell and L. Yi, Bull. Am. Phys. Soc. 29, p. 697.
8. M. Pfitzner and P. Wölfle, J. Low Temp. Phys. 51, p. 535.
9. E. Krotscheck, private communication.
10. T. L. Ainsworth and K. S. Bedell, preprint, to be published.
11. C. J. Pethick, in "Lectures in Theoretical Physics, XI^B", K. T. Mahanthappa and W. E. Brittin, eds. (Gordon and Breach, New York, 1969).
12. J. C. Wheatley, Rev. Mod. Phys. 47, p. 415.
13. D. S. Greywall and P. A. Busch, Phys. Rev. Lett. 49, p. 146.
14. D. S. Greywall, Phys. Rev. B27, p. 2747.
15. L. Yi, private communication.
16. B. R. Patton and A. Zaringhalam, Phys. Lett. 55A, p. 95.
17. K. S. Bedell and D. E. Meltzer, Phys. Lett. 106A, p. 312.
18. J. M. Parpia, D. J. Sandiford, J. E. Berthold and J. D. Reppy, Phys. Rev. Lett. 40, p. 565.
19. D. S. Greywall, Phys. Rev. B29, p. 4933.
20. D. F. Brewer, D. S. Betts, A. Sachrajda and W. S. Truscott, Physica 108B, p. 1059.
21. M. Pfitzner and P. Wölfle, preprint, to be published.

HARD CORE SQUARE WELL QUANTUM MATTER

M. Fortes*, M. de Llano**, and J. del Río***

*Instituto de Física, Universidad Nacional Autónoma de México, 01000 Mexico, D.F.
**Physics Dept. North Dakota State University Fargo, N.D. 58105
***Physics Dept., Southern Illinois University Carbondal, IL 62901

INTRODUCTION

Quantum field theory[1] and Rayleigh-Schrodinger perturbation theory applied about an ideal gas have produced for the ground state energy per particle of a many-body system well-known low density expansions with three coefficients thus far determined for bosons[2] and four for fermions[3]. The particular approach we have been implementing attempts[4] to inquire how far one can go in constructing an accurate equation of state for several many-body Schrodinger systems, based only on these hard facts, and using extrapolation schemes like Padé[5] and generalizations.

For example, for identical bosons of mass m, with pair-interaction S-wave scattering length a and a number density $\rho=N/V$, one has for the ground state energy per particle

$$\frac{E}{N} \simeq \frac{2\pi\hbar^2}{m}\rho a\left[1 + C_1\sqrt{\rho a^3} + C_2\,\rho a^3 \ln \rho a^3 + C_3\,\rho a^3 + O\{(\rho a^3)^{3/2}\ln \rho a^3\}\right], \tag{1}$$

where $C_1 = 128/15\pi$, $C_2=8(\frac{4\pi}{3}-\sqrt{3})$, and C_3 is unknown but potential-shape-dependent. Expansions such as (1) were early abandoned[2] perhaps for two fundamental reasons: i) for negative scattering length (viz., most two-body potentials of interest in condensed matter science) imaginary terms appear on the rhs of (1) which are not easy to deal with and ii) many-body systems like the helium liquids are definitely not low-density ones as (1) assumes. Difficulty (i) is bypassed very naturally by expanding (analytically or numerically)

$$a = a_0 + a_1\lambda + a_2\lambda^2 + \ldots, \tag{2}$$

where λ is some appropriate measure of the attraction of the two-body potential. Substutition into (1) then leads to

$$\frac{E}{N} = \sum_{i=0}^{\infty} \epsilon_i(x)\,\lambda^i, \quad x \equiv \sqrt{\rho a_0^3}, \tag{3}$$

which is evidently real no matter how large λ (and thus how negative a). The "rearrangement" (3) is nothing more than a perturbation scheme, but now about the gas of attractionless particles, i.e., the scheme recommended by van der Waals more than a hundred years ago for classical fluids. Difficulty (ii) can perhaps be surpassed with the extapolation procedures mentioned above.

SUCCESSES OF THE NEW METHOD

a) Boson Hard Spheres

For this system there are available Green Function Monte Carlo (GFMC) results for 256 particles, i.e., we very probably know the "exact" answer, although only at four intermediate-density datapoints. A generalized Pade representation of (1), with a replaced by the hard sphere diameter c, which gives an excellent global fit to these data points is[4], with $x = \sqrt{\rho c^3}$,

$$\frac{E}{N} \hat{=} \frac{2\pi\hbar^2}{mc^2} \frac{x^2}{\left[1 - \dfrac{1}{1 - \frac{2C_2}{C_1} x \left(\ln x + \frac{[C_3 - \frac{3}{4}C_1^2]}{2C_2} \right)}\right]^2}, \tag{4}$$

if C_3 is taken to have the value 26.2. The symbol $\hat{=}$ stands for "represented by". The form[4] upon expansion about x = o reproduces (1); hence the name of Padé is associated with this extrapolant. More importantly, the form (4) was chosen so as to possess, at some finite x, the second-order uncertainty principle divergence physically expected at the so-called Bernal or random-close-packing density which should be strictly less than the regular-close-packing one given by e.g., the packing density of a face-centered-cubic arrangement which presumably is the closest possible packing of an infinite array of identical rigid spheres.

b) Quantum Bernal Densities

The well-known experimental[5] Bernal density for classical hard spheres is $0.86\frac{\sqrt{2}}{c^3}$. Quantum mechanically, however, one would expect the Bernal density to be somewhat smaller since the high energy (but non-relativistic, of course) total cross section for the scattering[6] of two hard spheres each of diameter c is not πc^2, as expected classically, but rather[7]

$$\sigma \xrightarrow[kc \gg 1]{} \pi(\sqrt{2}c)^2 \left\{ 1 + \frac{0.99615}{(kc)^{2/3}} + \cdots \right\} \tag{5}$$

where k is the wave number of the relative motion. A quantum estimate for the Bernal density is therefore provided by the passage

$$0.86\frac{\sqrt{2}}{c^3} \longrightarrow 0.86\frac{\sqrt{2}}{(\sqrt{2})^3 c^3} \simeq 0.304\rho_0. \tag{6}$$

This estimate does not include statistics. Boson statistics would tend to make the result (6) somewhat larger because this effect manisfests itself in terms of an equivalent attraction between ideal, but quantum, particles or, alternatively, smaller diameters. On the other hand, fermions equivalently have larger diameters because of the Pauli repulsion effect of the exclusion principle and would thus be expected to show a Bernal density somewhat smaller than (6). In fact, both of these physical expectations are borne out by the Padé analyses[4] that have been applied to quantum hard sphere systems. Based upon the corresponding low-density expansions one arrives[4], for two-species fermion hard sphere systems, at the extrapolant, with $x = k_F c$, given by

$$\frac{E}{N} \hat{=} \frac{3}{5}\frac{\hbar^2 k_F^2}{2m}\left[\frac{1+q_1 x}{1+p_1 x+p_2 x^2+p_3 x^3}\right]^2, \qquad (7)$$

where the p's and q's are determined only by the low-density coefficients in the initial low-density series. For four-species ones, there emerges the form

$$\frac{E}{N} \hat{=} \frac{3}{5}\frac{\hbar^2 k_F^2}{2m}\left[1+\frac{p_1 x+p_2 x^2+p_3 x^4 \ln x}{1+q_1 x}\right]^{-2}, \qquad (8)$$

with an identical comment as before regarding the p's and q's. In both cases, of course, $\rho \equiv \nu k_F^3/6\pi^2$ where ν is the number of species. A further physical expectation, namely that the Bernal densities of the 2- or 4-species systems should be very similar (since these ν values differ neglibibly from each other compared with the boson case where ν may be taken to be infinity) is also corroborated. Specifically, (7) and (8) have (second-order) divergences at Bernal densities

$$0.174\,\rho_0 \quad \text{and} \quad 0.173\,\rho_0, \quad (\rho_0 \equiv \sqrt{2}/c^3) \qquad (9)$$

respectively. These are to be compared with the boson value from eq. (4) of $0.35\,\rho_0$.

c) Hard Core Square Well Neutrons

Adequate treatment of attractive interactions is clearly of crucial importance in any condensed matter theory. Within the present method this means faithfully representing, in say eq. (3), the different order corrections $\epsilon_1(x), \epsilon_2(x), \epsilon_3(x), \ldots$ to sufficiently large values of density, i.e., of x. For fermions (in fact, neutrons) interacting via a hard core surrounded by an attractive square well (HCSW) potential this was accomplished[8] by searching first for the best Padé approximant, based on the low-density ladder approximation energy series[3], that would fit the exactly known (to within 0.1%) $\epsilon_1(x)$ through $\epsilon_4(x)$ corrections associated with the essentially exact calculation of the ladder energy as obtainable by numerical integration of the Bethe-Goldstone equation. Greater flexibility was achieved by introducing[8] the so-called fractional Padé approximants, and varying the exponent to optimize the fit. Figure 1 illustrates such a fit, in second order. The circles are the exact values; the dashed curve is the fractional Padé based on the three-term, low-density series (the leading term being factored out). These fractional-exponented Pade approximants were then employed in

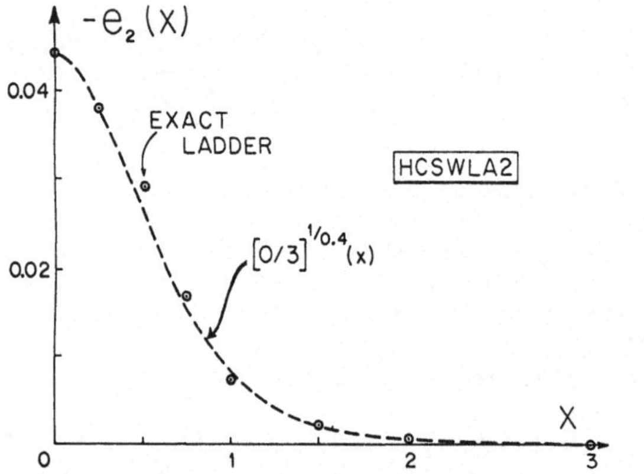

Fig. 1. Example of fit between fractional Padé approximant based on low-density ladder energy series and exact results (circled dots).

the complete series energy, i.e., the low-density expansion which correctly reproduces four coefficients (vs. only two for the ladder case). A final additional constraint was imposed on the complete energy, to be discussed more fully in the following section, and which in essence makes $\epsilon_1(x)$ a constant and make $\epsilon_2(x)$, $\epsilon_3(x)$, $\epsilon_4(x)$ zero---at the Bernal density $x=x_B$ = 1.93915 for 2-species fermions. These constraints convert our one-point (about x=0) Padé approximant, say $[0/3](x)$ to a two-point (now about $x=x_B$ also) one designated by a double slash, $[2//2]$. Figure 2 illustrates this for second order. Finally, Figure 3 compares the different Padé approximant results, in λ now, for the energy per particle vs. density (in units of ρ_{sat}) for both the "ladder" and "complete" cases as applied to the particular HCSW adopted in this study. Here $\lambda \equiv \frac{mV_0}{\hbar^2}(R-c)^2$, where V_0 is the attractive well depth, R its range and c the hard sphere diameter. We observe that fast convergence is accomplished in both cases as one proceeds from 1st through 4th order where the result varies little from second and third orders. Hence, stability is quickly achieved in this application of the van der Waals scheme. We also remark that the stabilized ladder results coincide very closely to the exact ladder values to densities just below where the result has an energy minimum.

Of course, real neutron matter is not expected to be self-bound so that our HCSW model is not quite adapted to reflect the right interaction. A better model for a neutron pair-potential is the HCSW potential of Baker, Hind & Kahane[9] (BHK) which has zero attraction in odd relative angular momentum states as well as hard core diameter, attractive range and depth parameters such that several low-energy scattering and bound-state empirical properties are reasonably well reproduced for the nucleon-nucleon system. For this more realistic model we obtained the results on Figure 4 for the complete theory. Again rapid convergence in the λ-series approximants is observed, third and fourth order being essentially stabilized. The vertical bar represents the spread in values for the energy per particle of neutron matter as extracted from the semi-empirical mass formulae found in the literature[10], for lack of any better indication of what the true ground state energy might be at any density for neutron matter. A peculiar, as yet unexplained, splitting down of the approximant is noted. Such a phenomenon was also observed in bosons and may have a deeper significance. Recent FHNC variational calculations by Clark & Flynn[11], essentially reproduced in at least one density value by K. Schmidt[12] who did a J-MC calculation, lie about twice as high as our converged result at $\rho=\rho_{sat}$. This serious discrepancy can probably be traced[12], at least in part, to the use of a spin-independent trial function.

Before dealing with bosons we shall summarize the basic physical ingredients of the method.

SUMMARY OF PHYSICAL IDEAS

There are three physical considerations which go into the what we are calling the van der Walls perturbation theory of quantum fluids: 1) first of all the observation, recently emerged from computer simulations in both classical and quantum many-body systems, that the pair distribution function for the hard sphere gas is qualitatively the same as for the liquid with repulsions plus attractions (as e.g., with Lennard-Jones interactions). This motivates, and to a large extent justifies, expansions of the kind eq. (1), and thus vindicates a priori the whole scheme. 2) The scheme can be made to generate second-order energy divergences at (random-close-packing) density values which differ in physically convincing ways from the well-established classical value. 3) In the expression for the ground state energy

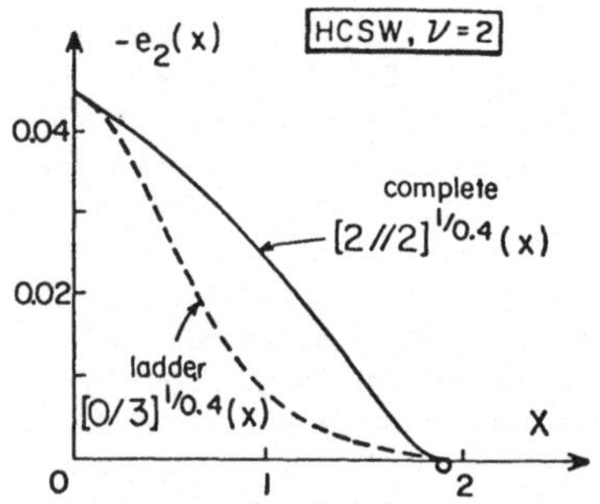

Figure 2. Fractional Padé approximant of Fig. 1 compared with the two-point one based on the complete energy series.

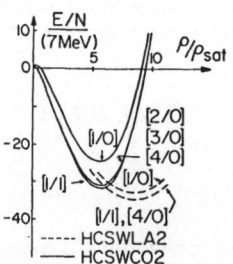

Figure 3. Energy per particle vs. density (in units of nuclear saturation density) as predicted by ladder (dashed) and complete (full) series.

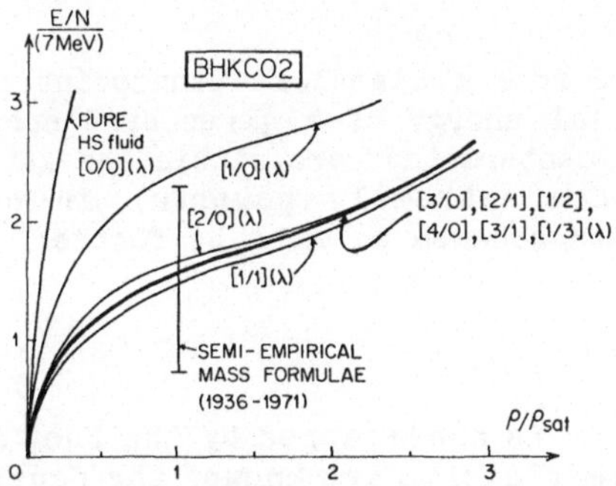

Fig. 4. Energy per particle of neutron matter interacting via the BHK potential in various orders of the van der Waals perturbation theory. Our converged result is the thick curve.

$$E(\rho) = E_0(\rho) + \lambda E_1(\rho) + \lambda^2 E_2(\rho) + \cdots \tag{10}$$

we may surmise that, at the Bernal density $\rho = \rho_B$, the total energy is linear in $\lambda \equiv \frac{mV_0}{\hbar^2}(R-c)^2$, namely

$$E(\rho_B) \simeq \infty + \lambda E_1(\rho_B) \quad + \text{neglible}, \tag{11}$$

where, if the attractive well depth is V_0, we should have for the full potential energy just

$$\lambda E_1(\rho_B)/N \simeq -\frac{1}{2} V_0 \left(\rho_B \frac{4\pi}{3} R^3 - 1\right), \tag{12}$$

while

$$E_i(\rho_B) \simeq 0 \quad (i=2,3,\ldots). \tag{13}$$

Eq. (12) is reasonable from a classical standpoint since the rhs is just the (negative) potential energy of a given hard sphere due to the number of neighboring sphere centers that are within the attractive well range R. This surmise has been extremely powerful in constructing rapidly convergent perturbation schemes as will be further illustrated below for the case of bosons.

APPLICATION TO BOSONS

The boson problem is handicapped by the fact that only three low-density expansion coefficients are known, the fourth one, C_3 in eq. (1), never having been calculated by the field theorists. Our determination of C_3 for the rigid sphere case (cf. below eq. (4)) is not of much help for non-pure-hard-core potential since it is known[2] that C_3 is potential shape-dependent. In fact, we shall deduce that its value for, say, a particular HCSW potential, is not 26.2 as for hard spheres but -0.061, a difference which can grossly affect the final equation of state curve.

To estimate C_3 recall that the corresponding term for 2-species fermions, i.e., the $k_F^6 \propto \rho^2$ term, is[3]

$$\frac{E}{N} \simeq \cdots + \left[\frac{1}{2} C_7 \frac{r_0}{a} + C_8 \frac{A_0''(0)}{a^3} + C_9\right] \rho^2 a^4 + \cdots \tag{14}$$

where the C's are pure numbers, given in ref. [4], while a and r_0 are the S-wave scattering length and effective range, and $A_0''(0)$ a shape-dependent, second-moment S-wave quantity defined in ref. [3] For the $HCSW_4$ potential these quantities have been explicitly written down in ref.[4] in terms of the hard sphere diameter c, the range parameter $\alpha \equiv (R-c)/c$ and the coupling constant $\lambda \equiv mV_0 c^2 \alpha^2/\hbar^2$ with V_0 the well depth and R its range. For example, one has the (better known) formula

$$a = c\left[1 + \alpha\left(1 - \frac{\tan\sqrt{\lambda}}{\sqrt{\lambda}}\right)\right], \tag{15}$$

as well as similar though more tedious expressions for r_o and $A_o''(o)$. The fermion ρ^2 coefficient (14) is very probably not in the correct form for <u>bosons</u> since the ρ^2 coefficient in (14) will <u>vanish</u> in the special case of a HCSW with a=0, given that from ref.[4], for $\alpha=1$,

$$r_o a^3 = \left(\frac{4}{3}+\frac{1}{\lambda}\right)a^2 \xrightarrow[a\to 0]{} 0$$
$$A_o''(o)a = \frac{1}{3}\left[15+\left(1-\frac{6}{\lambda}\right)\sqrt{1+4\lambda}\right]a \xrightarrow[a\to 0]{} 0 \qquad (16)$$

in that case. On the other hand, Moszkowski[13] has arrived at a non-zero ρ^2 term from an optimized-Jastrow energy calculation and, indeed, has extracted the form

$$E_{J\text{-}MC}/N \underset{\rho\to 0}{\simeq} 12.8\,\frac{\hbar^2}{mc^2}\rho^2+\cdots \geqslant E/N \qquad (17)$$

from Jastrow-Monte Carlo calculations performed by Panoff[14] on the boson HCSW system with a=o and $\alpha=1$. Assuming that the upper bound (17) is at least highly suggestive of a non-zero ρ^2 term for the a=o HCSW we proceed to propose as an <u>ansatz</u> for the boson C_3 the form

$$C_3 = A\left(\frac{c}{a}\right)^4 + B\left(\frac{r_o c}{a^2}\right) + C\left(\frac{A_o''(0)\,c}{a^4}\right) \qquad (18)$$

which is a slight generalization of (14) but which is such that C_3a^4 does <u>not</u> vanish as $a\to o$. Here A, B and C are unknown pure numbers, to be determined below.

We have studied the many boson ground state energy problem for the Burkhardt[15] HCSW which is phase-shift-equivalent to the He-He Lennard-Jones potential. Burkhardt obtains

$$V_o=1.43167\text{K},\qquad c=1.685\text{Å},\qquad R=5.5\text{Å}, \qquad (19)$$

which in turn gives $\alpha=2.264095$ and $\lambda=2.281417$. Also, from ref.[4] one gets for this potential

$$a=-36.2938\text{Å},\qquad r_o=1.64633\text{Å}\qquad \text{and}\qquad A_o''(0)=1{,}245.7433\text{Å}^3. \qquad (20)$$

We now formally expand

$$a = c\left[1+\sum_{i=1}^{\infty}\alpha_i\lambda^i\right] \qquad (21)$$

$$r_o = r_{oo}\left[1+\sum_{i=1}^{\infty}\beta_i\lambda^i\right] \qquad (22)$$

$$A_o''(0) = A_{oo}''(0)\left[1+\sum_{i=1}^{\infty}\gamma_i\lambda^i\right] \qquad (23)$$

and insert into eq. (1) with C_3 given by our ansatz eq. (18). There results the perturbation expansion

$$\frac{E}{N} = 2\pi \frac{\hbar^2}{mc^2} \sum_{i=0}^{\infty} e_i(x)\lambda^i , \qquad x \equiv \sqrt{\rho c^3}. \tag{24}$$

Here, for example,

$$e_0(x) \equiv x^2 \Big[1 + C_1 x + C_2 x^2 \ln x^2 + + \Big(A + B\frac{r_{oo}}{c} + C\frac{A''_{oo}(0)}{c^3}\Big) x^2 + \dots \Big], \tag{25}$$

and

$$e_2(x) \equiv \alpha_2 x^2 \big[1 + C_{12} x + C_{22} x^2 \ln x^2 + C_{32} x^2 + \dots \big] \tag{26}$$

$$C_{12} \equiv \Big(\frac{15}{8}\frac{\alpha_1^2}{\alpha_2} + \frac{5}{2}\Big) C_1 ; \qquad C_{22} \equiv \Big(6\frac{\alpha_1^2}{\alpha_2} + 4\Big) C_2 \tag{27}$$

$$C_{32} \equiv \Big(\frac{21}{2}\frac{\alpha_1^2}{\alpha_2} + 3\Big) C_2 + \Big(\frac{2}{3}\frac{\alpha_1^2}{\alpha_2} + \frac{4}{3} + \frac{4}{3}\frac{\alpha_1\beta_1}{\alpha_2} + \frac{2}{3}\frac{\beta_2}{\alpha_2}\Big) B - \frac{1}{3}\frac{\gamma_2}{\alpha_2} C. \tag{28}$$

Also, we have

$$e_3(x) \equiv \alpha_3 x^2 \big[1 + C_{13} x + C_{23} x^2 \ln x^2 + C_{33} x^2 + + C_{43} x^3 \ln x^2 + \dots \big] \tag{29}$$

with C_{13}, C_{23}, C_{33} being given similarly to eqs. (26) to (28), but with C_{43} unknown and to be discussed below. We further note that for HCSW potential[4] $\alpha_1/\alpha_2 \simeq \alpha_2/\alpha_3 \simeq \alpha_3/\alpha_4 \simeq 2.5$ so that in principle one expects fast convergence at least a low densities.

To determine the pure constants A, B and C we need three equations. The first one is naturally provided by the generalized Padé approximant (4) to the series (25) since for a pure hard sphere a=c, $r_{oo} = \frac{2}{3}c$ and $A''_{oo}(0) = -\frac{1}{3}c^3$. Thus

$$A + \frac{2}{3}B - \frac{1}{3}C = 26.2 \tag{30}$$

is the first equation. It essentially follows from physical idea #2 of the previous section. The other two equations follow from physical idea #3 stated in that section, applied in turn to first and second order, namely

$$\epsilon_1(x_B) = -\Big[\frac{4\pi}{3} x_B^2 (\alpha^3 + 3\alpha^2 + 3\alpha + 1) - 1\Big] / 4\pi\alpha^2 , \tag{31}$$

$$\epsilon_2(x_B) = 0, \tag{32}$$

where x_B is the Bernal density and $\epsilon_1(x)$ and $\epsilon_2(x)$ are appropriate generalized Padé approximants to the series $e_1(x)$ and $e_2(x)$, respectively. These approximants were found[16] to be

$$e_i(x) \triangleq \alpha_i x^2 \left[1 + \frac{C_{1i}\, x}{1 - 2\frac{C_{2i}}{C_{1i}} x \left(\ln x + \frac{1}{2}\frac{C_{3i}}{C_{2i}} \right)} \right] \equiv \epsilon_i \tag{33}$$

Thus, the needed 2nd and 3rd equations come from (31) and (32), or

$$3.739 = 0.002B + 0.232C \tag{34}$$

$$-35.301 = 1.151B + 0.546C. \tag{35}$$

Solving (30), (34) and (35) we get

A = 57.439, B = -38.555 and C = 16.606

or, using (18) and (20),

$$C_3 = -0.06084 \text{ (Burkhardt HCSW)} \tag{36}$$

as stated before, which is to be compared with the pure-hard-core value of 26.2 deduced from a <u>global</u> fit to GFMC data in ref.[4]

To incorporate <u>third</u> order van der Waals perturbation corrections one must deal with $\epsilon_3(x)$. However, an adequate approximant $\epsilon_3(x)$ to $e_3(X)$ of (29) (<u>without</u> the C_{43} term) should, if consistent, give

$$\epsilon_3(x_B)/x_B^2\alpha_3 \simeq 0. \tag{37}$$

Instead, it gave 0.675, a value which is sizeable compared with unity, the value of the lhs of (37) at x=0. Bad convergence in the λ-series perturbation scheme then ensued. We thus proceed to incorporate the unknown C_{43} term of eq. (29) and write as our approximant

$$\epsilon_3(x) = \alpha_3 x^2 \left[1 + \frac{C_{13}x + \left(C_{33} - \left[\frac{C_{13}C_{43} - C_{33}C_{23}}{C_{23}} \right] \right) x^2}{1 - \left\{ \frac{C_{13}C_{43} - C_{33}C_{23}}{C_{13}C_{23}} x + \frac{C_{23}}{C_{13}} x \ln x^2 \right\}} \right] \tag{38}$$

The constraint $\epsilon_3(x_B) = 0$ then allows us to determine C_{43} as 7,938.4. This procedure markedly improved the convergence, which we now discuss.

Our energy per particle is finally represented by

$$E/N \triangleq 2\pi \frac{\hbar^2}{mc^2} \left[\epsilon_0(x) + \sum_{i=1}^{3} \epsilon_i(x) \lambda^i + \cdots \right] \triangleq [L/M](\lambda), \quad 0 \leq L + M \leq 3. \tag{39}$$

The results are graphed in Figure 5 for liquid helium four, where $[0/0](\lambda)$

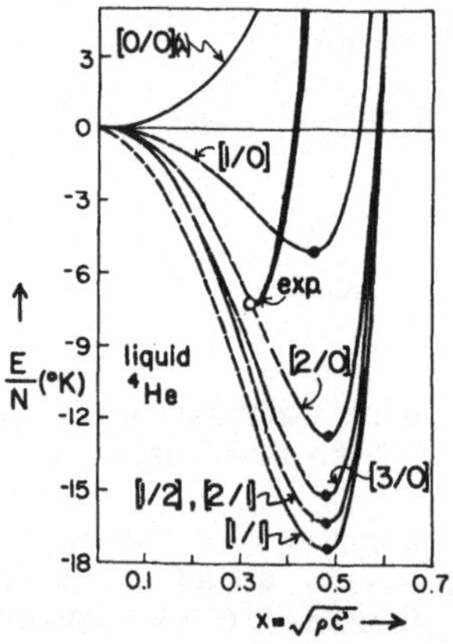

Fig. 5. Energy per particle of liquid helium four interacting via a HCSW potential which is phase-equivalent to the He-He Lennard-Jones 6-12 potential.

refers to the pure hard sphere fluid. We note quick convergence in going from first-order through third-order i.e., $[1/0](\lambda)$ through $[3/0](\lambda)$, and excellent convergence in the saturation density. It will probably be necessary to go somewhat beyond the third order calculated until now to achieve a staisfactorily converged final prediction for the ground state energy. We remark that Burkhardt's application[15] to liquid helium <u>three</u> of his HCSW potential also produced gross over-binding as compared with the corresponding empirical values. The puzzling split-off of the $[1/1](\lambda)$ approximant is again observed. Finally we note that GFMC calculations with the LJ potential, which is phase-equivalent to the HCSW one employed here, place the ground state curve just <u>above</u> the empirical one in Figure 5.

CONCLUSIONS

The low-density-expansion-based van der walls perturbation scheme, i.e., about not the ideal gas but about the fluid of repulsive cores, has thus far made several specific predictions, some of which as yet remain untested for lack of, e.g., GFMC calculations. The latter would be highly desireable, even in the apparently problem-free case of bosons, for HCSW bosons and spin-polarized atomic hydrogen. For fermions, GFMC data are badly needed for hard sphere fluids (2- and 4-species) and even the simple, purely-repulsive Bethe homework problem of v_0 neutrons where only variational Monte Carlo calculations are available. Spin-dependent variational trial function calculations, in both FHNC and Monte Carlo schemes, would be desireable for BHK neutron matter, not to speak of course of GFMC itself. Finally, nuclear matter (even with simple schematic interactions like the BHK potential), liquid helium three with both LJ and Aziz interactions, the different fermion spin-polarized systems, etc., still remain open frontiers.

ACKNOWLEDGEMENT The authors thank Susana Ramírez for assistance.

REFERENCES

1. A. L. Fetter and J. D. Walecka, "Quantum Theory of Many-Particle Systems," McGraw-Hill, New York (1971).

2. N. M. Hugenholtz and D. Pines, <u>Phys</u>. <u>Rev</u>. 116:489 (1959)

3. G. A. Baker, Jr., <u>Rev</u>. <u>Mod</u>. <u>Phys</u>. 43:479 (1971)

4. G. A. Baker, Jr., M. de Llano and J. Pineda, <u>Phys</u>. <u>Rev</u>. B24:6304(1981); G. A. Baker, Jr., L.P. Benofy, M. Fortes, M. de Llano, S. Peltier and A. Plastino, <u>Phys</u>. <u>Rev</u>. A26:3575 (1982); G.A. Baker, Jr., G. Gutiérrez and M. de Llano, <u>Ann</u>. <u>Phys</u>. 153:283 (1984).

5. G. D. Scott and D. M. Kilgour, J. Phys. D2:863 (1969).

6. A. Messiah, "Quantum Mechanics", North-Holland, Amsterdam (1962) vol. I, p. 393.

7. S. I. Rubinow and T. T. Wu, J. <u>Appl</u>. <u>Phys</u>. 27:1032 (1959).

8. V. C. Aguilera-Navarro, G. A. Baker, Jr., L.P. Benofy, M. Fortes and M. de Llano, Constructive Methods for the Ground State Energy of Neutron Matter", to be submitted to <u>Phys</u>. <u>Rev</u>. C.

9. G. A. Baker, Jr., M. F. Hind and J. Kahane, Phys. Rev. C2:841 (1970).

10. A.E.S. Green, T. Sawada and D. S. Saxon, "The Nuclear Independent Particle Model," Academic, New York (1968) p. 52; G. Baym, H. A. Bethe and C. J. Pethick, Nucl. Phys. A175:225 (1971)

11. J. W. Clark and M. Flynn, priv. comm.

12. K. Schmidt, priv. comm. (through G.A. Baker, Jr.)

13. S.A. Moszkowski, priv. comm.

14. R. M. Panoff, priv. comm.

15. T. W. Burkhardt, Ann. Phys. (N.Y.) 47:516 (1968).

16. J. del Rio, Master's Degree thesis (Carbondale, IL, 1985, unpubl.)

VARIATIONAL DENSITY MATRIX THEORY I

C. E. Campbell

School of Physics and Astronomy
University of Minnesota
116 Church St. S. E.
Minneapolis, Minnesota 55455, U.S.A.

K. E. Kürten, G. Senger, and M. L. Ristig

Institut für Theoretische Physik
Universität zu Köln
D-5000 Köln 41, W. Germany

A generalization to finite temperatures of the Jastrow-Euler-Lagrange variational theory of the ground state of quantum fluids may be achieved by replacing the wave function by the coordinate space representation of the N-body density matrix, where N is the number of particles in the system.[1,2] In that case the most appropriate minimum principle is the Gibbs-Delbrück-Moliere extremum principle for the Helmholtz free energy F:

$$F_0 \leq F[W] = \mathrm{Tr}(HW) + \beta^{-1}\,\mathrm{Tr}(W\,\ell n\,W)$$

where W is the trial density matrix defined on the Hilbert space of the Hamiltonian H of the system. Thus, it must be non-negative, hermitian, and normalized under the trace operation. Equality is achieved if and only if $W = W_0$, where W_0 is the equilibrium density matrix of the system:

$$W_0 = [\exp - \beta H]/Z_0$$

where Z_0 is the partition function

$$Z_0 = \mathrm{Tr}\,\exp -\beta H = e^{-\beta F_0}\,.$$

Schematically, the Bloch equation for the density matrix may be replaced by the Euler-Lagrange equation

$$\frac{\delta F[W]}{\delta W} = 0$$

where W satisfies the properties described above.

As a first step toward a tractable variational solution of this equation for a quantum fluid, it is convenient to write the coordinate representation of W in the form

$$W(\underline{R},\underline{R}') = \phi(\underline{R})^* \; Q(\underline{R};\underline{R}') \; \phi(\underline{R}') \tag{1}$$

where $\underline{R}$ represents the 3N continuous representation "row" indices $\vec{r}_1 \ldots \vec{r}_N$, and $\underline{R}'$ the corresponding "column" indices. This expression for W is unique if one requires that ϕ contain all factors in W which depend on $\underline{R}'$ alone. Thus Q is an intrinsic connection between the primed and unprimed indices, and therefore represents the incoherent part of the density matrix. If Q is just a constant, then W is a single-state density matrix corresponding to the wave function $\phi(R)$. More generally, one must exercise some care in the choice of Q so that W is a statistical operator.

The simplest system of interest is liquid ^4He. The Bose statistics then require that $\phi(R)$ be symmetric under particle interchange, and that $Q(\underline{R};\underline{R}')$ be individually symmetric in the primed coordinates and the unprimed coordinates.

Examination of the path integral formulation of a boson system shows that $W(\underline{R},\underline{R}')$ is real and non-negative for all choices of $\underline{R}$ and $\underline{R}'$. (This should not be confused with the fact that W must be a non-negative operator, i.e., that its eigenvalues must be non-negative.) Consequently, both ϕ and Q may be taken to be non-negative, real functions of the coordinates. One way to assure that this feature is preserved for trial density matrices is to choose ϕ and Q to be the exponentiated sum of real many-body functions, similar to the Feenberg function form of the boson ground state. In the present work we have taken the simplest non-trivial choice for these two functions, whereby ϕ has the Jastrow form and the incoherence factor Q has a similar two-body structure:

$$\phi(\underline{R}) = \exp\tfrac{1}{2} \sum_{i<j}^{N} u(r_{ij})/\sqrt{I} \tag{2}$$

and

$$Q(\underline{R};\underline{R}') = \exp \sum_{i,j}^{N} \gamma(|\vec{r}_i - \vec{r}_j{}'|) \quad . \tag{3}$$

(I is the normalization integral for this density matrix.) This form for Q is already familiar from the cluster analysis of the low temperature structure of the density matrix by Reatto and Chester, or, equivalently, from the summation of non-interacting phonon contributions to the density matrix; both of these are low temperature, long wavelength theories.

It can be shown that $W(\underline{R},\underline{R}')$ is a statistical operator if the fourier transform of $\gamma(r)$ is positive.

The two Euler-Lagrange equations for the "Jastrow" density matrix are written schematically as

$$\frac{\delta F[u,\gamma]}{\delta u(r)} = 0 = \frac{\delta F[u,\gamma]}{\delta \gamma(r)} \; .$$

In order to carry out this functional differentiations it is necessary to derive a manageable expression for the dependence of the functional F upon u and γ. This is a simple matter for the internal energy U, but poses difficulties for the entropy S, where

$$F = U - TS \quad .$$

In particular,

$$\frac{U}{N} = \frac{\rho}{2} \int g(r) \left[v(r) - \frac{\hbar^2}{4m} \nabla^2 \{u(r) + 2\gamma(r)\} \right] d^3r + \sum_k \frac{\hbar^2 k^2}{2m} \gamma(k)$$

where ρ is the average number density, v is the bare two-body potential, and g(r) is the radial distribution function for the N-body probability density $W(\underline{R},\underline{R})$.

The entropy is a more difficult quantity to calculate because the product W ℓn W is a matrix product; a simple expression for ℓn W is not immediately evident.

A method which leads to a practical approximation scheme is to use the identity

$$\frac{\partial}{\partial\alpha} \; \mathrm{Tr}\; W^{\alpha+1} \Big|_{\alpha=0} = \mathrm{Tr}\; W \;\ell n W \tag{4}$$

When W has the form of equations (1-3), it is a straightforward matter to see that Tr W^n is mathematically equivalent to the partition function for a classical n component mixture:

$$\mathrm{Tr}\; W^n = \exp - \beta_c [F_c(n) - n\, F_c(1)]$$

where $F_c(n,\beta_c)$ is the Helmholtz free energy of the classical mixture with interactions simply related to u(r) and γ(r), at a fictitious temperature β_c^{-1}.

The procedure we follow is to calculate F_c at integer values of n and then continue to non-integer values so that Eq. (4) can be evaluated.

The simplest tractable approximation is the separability approximation in paired-phonon space, which provides the leading (gaussian) contributions in γ(k). The result of this approximation is a Bose liquid expression for the entropy:

$$\frac{TS}{N} \simeq \frac{TS_0}{N} = \frac{1}{\beta N} \sum_k [(1+n_k)\;\; \ell n\,(1+n_k) - n_k \;\; \ell n\, n_k]$$

where the positive real function n_k is defined by

$$n_k[1+n_k] = \gamma(k)\; S(k)$$

where S(k) is the liquid structure function for the density matrix W.

With this approximation, the internal energy U is conveniently rewritten

$$\frac{U}{N} = \frac{\rho}{2} \int v^*(r) g(r) d^3r - \frac{1}{2(2\pi)^3\rho} \int \nu^*(k) S(k) d^3k$$

where

$$v^*(r) = v(r) - \frac{\hbar^2}{4m} \;\nabla^2\; (u(r) + 2\gamma(r))$$

and

$$\nu^*(k) = \frac{1}{2}\,\frac{\hbar^2 k^2}{2m}\; [1-(1+2n_k)^2]/S(k)^2$$

Note that at T = 0, where γ(k) = 0, we have $n_k = 0$ and thus $\nu^*(k) = 0$.

It is convenient to shift notation somewhat by replacing u(r) + 2γ(r) by u(r), so that $v^*(r) = v(r) - \hbar^2/4m\; \nabla^2 u(r)$. Then the first Euler-Lagrange equation,

$$\frac{\delta F}{\delta\gamma(r)} \Big|_u = 0$$

can be written in the suggestive form

$$\varepsilon(k) = \frac{\hbar^2 k^2}{2mS(k)} \quad \coth \frac{\beta\varepsilon(k)}{2} \;,$$

which is a transcendental equation for the auxiliary function ε(k) which has been <u>defined</u> in the obvious way

$$n_k = \frac{1}{\exp \beta\varepsilon(k) - 1}$$

This remarkably simple result may be taken as the definition of the statistical elementary excitation energy, which in turn is seen to be a natural finite temperature generalization of the Bijl-Feynman spectrum.

The second Euler-Lagrange equation,

$$\left.\frac{\delta F}{\delta u(r)}\right|_{\gamma} = 0$$

must be solved simultaneously with the above equation. By introducing the generating function

$$W(R;R;\alpha) \equiv I(\alpha)^{-1} \prod_{i>j} \exp \tfrac{1}{2}\, u\,(r_{ij};\alpha)$$

where

$$u(r;\alpha) \equiv u(r) + \alpha[v^{*}(r) + \nu^{*}(r)] \ ,$$

one may obtain this second Euler-Lagrange equation in the familiar paired-phonon form:

$$\dot{S}(k) = \frac{\hbar^2 k^2}{4m}\,[1 - S(k)]$$

where

$$\dot{S}(k) \equiv d\, S(k;\alpha)/d\alpha \left.\vphantom{\frac{1}{1}}\right|_{\alpha=0} ,$$

or, equivalently,

$$\dot{g}(r) = \hbar^2 \nabla^2\, g(r)/4m \quad .$$

For the sake of a stability analysis it is convenient to rewrite this equation in the equivalent Bogoliubov form:

$$\varepsilon(k) = \{ \frac{\hbar^2 k^2}{2m}\, [\, \frac{\hbar^2 k^2}{2m} + 2\nu(k)]\, \}^{\frac{1}{2}}$$

where we have introduced yet another effective potential defined by

$$2\nu(k) = [\, \frac{\hbar^2 k^2}{2m}\, S(k) + 2\, \dot{S}(k)]/S^2(k) - \frac{\hbar^2 k^2}{2m} - 2\nu^{*}(k) \ .$$

Clearly, then, the system is unstable unless

$$\frac{\hbar^2 k^2}{2m} + 2\, \nu(k) \ < 0 \ .$$

We find that this instability sets in first at long wavelengths, where the condition becomes

$$mc^2 \equiv \nu(0) > 0$$

Note that c enters $\varepsilon(k)$ as the isothermal sound velocity:

$$\varepsilon(k) \underset{k\to 0}{\to} \hbar kc$$

With this interpretation, the locus defined as $\nu(0)$ in the thermodynamic phase plane is just the isothermal spinodal line.

We have solved the above set of Euler-Lagrange equations by using the HNC/0 approximation as the relationship between u(r) and g(r) (or S(k)).[2] This approximation can be easily improved (or replaced by a Monte Carlo calculation), although the additional work is not merited until we can improve upon the separability approximation.

The spinodal line defined by $mc^2 = 0$ encloses a region in the $T - \rho$ plane where there are no uniform density solutions to the Euler-Lagrange equations. The maximum temperature of this curve and its corresponding density defines our estimate of the liquid gas critical point: $T_c = 4.3K$ and $\rho_c = 0.009$ Å^{-3}, compared to the experimental values of $T_c = 5.2$ K and $\rho_c = 0.010$ Å^{-3} .

The elementary excitation spectrum and the liquid structure both behave according to experiment at long wavelengths.

The deficiencies of our theory appear in part at intermediate and short wavelengths. For example, the roton region of the excitation curve is no better than the Bijl-Feynman T = 0 curve, where it is off by approximately a factor of two and has the wrong density dependence and temperature dependence. This is not surprising since the trial density matrix omits backflow and excitation interactions.

Moreover, there is no evidence for the disappearance of the Bose condensate in the bulk system (in contrast to two dimensions, where it is already known that thermally populated phonons destroy the condensate). Thus, the liquid-gas critical point is between a superfluid liquid and a superfluid gas. Similarly one can easily show that our simple trial density matrix cannot achieve the correct high temperature limit, where the incoherence factor is a permanent of gaussians of thermal de Broglie length scale.

Within these limitations the separability approximation is still more restrictive, limiting the validity of the numerical results to low temperatures. This part of the approximation must be improved in order to achieve a sensible liquid-gas metastable region between the coexistence curve and the spinodal curve.

Improvements on this approximation are discussed in the next paper. Elaborations on the trial density matrix are discussed in Ref. (2).

This Research has been supported in part by the National Science Foundation under Grants DMR-77-18329 and DMR-79-26447, by the Department of Energy through Contract DE-AC-02-76ER-03077, by the Deutsche Forschungsgemeinschaft under Grant Ri-267 and the Alexander von Humboldt Stiftung.

References

1. C. E. Campbell, K. E. Kürten, M. L. Ristig and G. Senger, Phys. Rev. B30, 3728 (1984).

2. G. Senger, M. L. Ristig, K. E. Kürten, and C. E. Campbell, to be published.

VARIATIONAL DENSITY MATRIX THEORY II

M.L.Ristig, G.Senger and K.E.Kürten

Institut für Theoretische Physik, Universität zu Köln
D-5000 Köln 41, West-Germany

C.E.Campbell

School of Physics and Astronomy, University of Minnesota
116 Church Street S.E., Minneapolis, Minnesota 55 455

An Euler-Lagrange variational theory of boson fluids in thermal equilibrium may be developed from the Gibbs-Delbrück-Moliere extremum principle for the Helmholtz free energy employing a suitably chosen set of trial density matrices (1 - 4) . A brief outline of the main features of such a theory and its simplest realization based on the separability assumption has been given in part I of our contribution to this workshop (5) .

Part II continues this study and develops further the density matrix approach by focussing particularly on our present efforts and achievements in overcoming the limitations set by the separability assumption. This approximation provides us with a first contribution to function $\gamma(r)$ and its Fourier inverse $\gamma(k)$ which generates the incoherence factor $Q(\underline{R},\underline{R}')$ of the trial density matrix (4,5) . As described in part I we find in separability approximation a Bose fluid expression for the entropy per particle,

$$\frac{TS}{N} \simeq \frac{TS_o}{N}$$

$$= \frac{1}{\beta N} \frac{1}{(2\pi)^3\rho} \int d\underline{k} \left\{ [1 + n(k)] \ln [1 + n(k)] - n(k) \ln n(k) \right\}$$

where the positive function $n(k)$ is defined by

$$2n(k)[1 + n(k)] \equiv m(k) = 2\gamma(k) S(k)$$

involving the liquid structure function $S(k)$ of the Bose fluid. The latter equation may be written, equivalently,

$$1 + 1/m(k) = \{1 - X(k) + 2\gamma(k)\} / 2\gamma(k)$$

reducing in the 'uniform limit' to

$$1 + 1/m\ (k) = \left\{1 - u(k) + 2\ \gamma\ (k)\right\} / 2\ \gamma\ (k).$$

The particularly simple form of these relations is, of course, a consequence of the separability assumption. From now on we discard this assumption trying to derive better approximations for the crucial quantities $F_c\ (\alpha)$, $\alpha = 0, 1, 2, \ldots$. These functionals have been introduced in part I representing the traces of powers of the density matrix W, TrW^{α}. They are the essential ingredients for generating an explicit form for the entropy functional. In particular, at $\alpha = 2$ we find

$$F_c(2) = - \frac{1}{(2\pi)^3\rho} \int X_{11}\ d\underline{k} - \frac{1}{2(2\pi)^3\rho} \int \ln \begin{vmatrix} 1-X_{11} & X_{12} \\ X_{12} & 1-X_{11} \end{vmatrix} d\underline{k}$$

$$+ \rho \int \left\{X_{11} + X_{12} - \frac{1}{2}\ (g_{11} - 1)^2 - \frac{1}{2}\ (g_{12} - 1)^2 \right\} d\underline{r}.$$

This expression may be formally interpreted as the free energy (in units of k_BT) of a binary mixture of classical fluids(6). The various integrals involve the radial distribution functions $g_{11}(r)$ and $g_{12}(r)$, their corresponding direct portions $X_{11}(r)$ and $X_{12}\ (r)$ and the associated structure functions $S_{11}(k)$ and $S_{12}(k)$ and the Fourier inverse $X_{11}(k)$ and $X_{12}(k)$ for like and unlike particles, respectively. The direct correlation functions X_{11} and X_{12} are generated - via the familiar hypernet relations - by the bare functions $u - 2\ \gamma$ and $2\ \gamma$, respectively,

$$X_{11}(r) = \exp\left\{u - 2\ \gamma + N_{11}\right\} - N_{11} - 1,\quad X_{12}\ (r) = \exp\left\{2\ \gamma + N_{12}\right\} - N_{12} - 1.$$

These expressions suggest that we may improve relation $1 + 1/m(k) = \left\{1 - u + 2\ \gamma\right\} / 2\ \gamma$ (which derives from the separability assumption) by replacing functions $u - 2\ \gamma$ and $2\ \gamma$ by the corresponding dressed quantities X_{11} and X_{12}. The result is

$$1 + 1/m\ (k) = \left\{1 - X_{11}(k)\right\} / X_{12}(k).$$

We may now proceed as in Ref. (5) expressing the free energy of the Bose fluid, $F = U - TS$, as a functional of the structure function $S(k)$ and the positive distribution function $m\ (k) = 2\ n\ (k) \left[1 + n\ (k)\right]$. At the present stage of development we evaluate functional F only up to order $o\ (m^2)$ in the internal energy U and $o\ (Tm)$ in the entropy term TS, inclusively. On this level of approximation the Helmholtz free energy per particle reads

$$\frac{F}{N} = \frac{F_1}{N} - \frac{1}{(2\pi)^3\rho}\int U_o \rho^* d\underline{k}$$

$$+ \frac{1}{2(2\pi)^6\rho^2}\int \nu^* S^2 G(\underline{k}, \underline{k}')\, S'^2 \rho'^* d\underline{k}\, d\underline{k}'$$

where F_1 is the expression for the free energy in separability approximation as derived in part I. The first additional term contains a single particle effective potential

$$4\, U_o(k) = W_o + M_o^* \varepsilon_o\, (k)$$

($\varepsilon_o \equiv \hbar^2 k^2/2m$) with the self-consistent parameters

$$W_o = \frac{1}{(2\pi)^3\, \rho}\int \varepsilon_o S^2 \rho^* d\underline{k},$$

$$M_o^* = \frac{1}{(2\pi)^3\rho}\int S^2 \rho^* d\underline{k},$$

and an effective density of (excited) states

$$\rho^*(k) = \frac{m}{S(1 + 2m)}\,.$$

The second integral describes interaction effects between the elementary excitations. The interaction is mediated by a propagator G which is determined by the solution of a Bethe-Salpeter equation

$$G = G_o + G_o KG.$$

The operators G_o and K are defined in terms of their matrix elements in $\underline{k}$-space,

$$G_o(\underline{k}, \underline{k}') = 1 - S(\,|\underline{k} - \underline{k}'|),$$

$$K\,(\underline{k}, \underline{k}') = \delta(\underline{k} - \underline{k}')\left[1 - S^2(k)\right].$$

Note that on the level of approximation adopted these elements are independent of function m(k).

Having at our disposal the explicit expression for the Helmholtz free energy functional we may proceed in standard fashion to arrive at a coupled set of Euler-Lagrange equations which determine the optimal structure function S(k) and the optimal thermal distribution function n(k) at given temperature and density. Varying the energy F with respect to quantity n(k) yields the spectral dispersion relation

$$\varepsilon(k) = M(k) \frac{\varepsilon_o(k)}{S(k)} \coth \frac{\beta}{2} \varepsilon(k)$$

where we employ the statistical excitation energies $\varepsilon(k)$ defined (as in part I) by $n(k) = [\exp\beta\varepsilon(k) - 1]^{-1}$.

In contrast to our result in separability approximation (5) this improved result involves an effective mass factor M(k) which contributes important corrections to the excitation energies at elevated temperatures. Varying the energy functional F with respect to the distribution function g(r) we arrive at the second Euler-Lagrange equation which may be written in the familiar paired phonon form involving a suitably generalized structure function (4,5). Defining an appropriate potential $\nu(k)$ this paired phonon equation may be cast into the more practical form

$$S(k) = \left\{ 1 + 2\nu(k)/\varepsilon_o(k) \right\}^{-\frac{1}{2}} \coth \frac{\beta}{2} \varepsilon(k).$$

The potential $\nu(k)$ as well as the mass factor M(k) are known explicitly in terms of quantities S(k) and m(k) or $\varepsilon(k)$. Since these functional expressions are somewhat lengthy we omit their presentation in this rather brief contribution referring to a forthcoming more detailed publication.

We have applied the presently developed formalism for a numerical study of the optimal solutions S(k) and $\varepsilon(k)$ and of the associated thermodynamic quantities such as the free energy and the entropy to explore theoretically the phase diagram of liquid ^{4}He. Here, we list briefly the most prominent features of these numerical results:

(i) In contrast to our findings based on the separability assumption (4,5) the optimized excitation energies $\varepsilon(k)$ decrease appreciably in the roton region as the temperature increases (Figure 1). This behavior is in accord with results of semi-phenomenological roton liquid theories (7,8) and in qualitative agreement with experimental results (9).

(ii) A phase instability occurs at a critical temperature $T_c \simeq 3.1$ K - 3.2 K depending on the actual density. There are no physical solutions of the Euler-Lagrange equations at temperatures $T>T_c$.

(iii) At the critical temperature T_c the thermodynamic quantities such as energy F and entropy S are singular. In particular, they each approach a finite value at the critical point but the associated specific heat at constant volume is infinite and a cusp appears in the isothermal compressibility.

(iv) The Euler-Lagrange equations permit a second branch of thermodynamically unstable solutions at temperatures $T<T_c$.

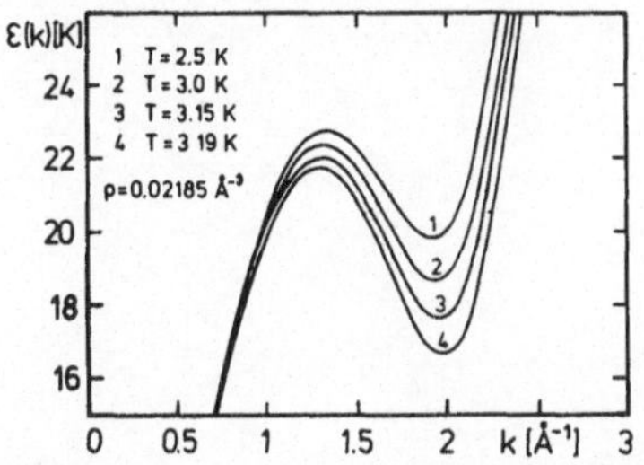

Fig. 1 Theoretical elementary excitation spectrum of liquid ^{4}He at experimental saturation density and at various temperatures $T \leq T_c \simeq 3.19$ K. The calculation is based on the present realization of the density matrix formalism.

A detailed report on our numerical results and the statements (i) - (iv) together with a thorough discussion of the status of the present realization of the density matrix approach will be given elsewhere. Some general quantitative as well as qualitative improvements are outlined and discussed in Ref. (4).

REFERENCES

1. A. Huber, in Methods and Problems of Theoretical Physics, 37 (edited by J.E.Bowcock, North-Holland, Amsterdam, 1970).

2. C.E. Campbell, K.E. Kürten, M.L. Ristig and G. Senger, Phys.Rev. B 30 (1984) 3728.

3. C.E. Campbell, M.L. Ristig, K.E. Kürten and G. Senger, Proceedings of the 17th International Conference on Low Temperature Physics, 1209 and 1211 (edited by U. Eckern. A. Schmid, W. Weber and H. Wühl, 1984 Elsevier).

4. G. Senger, M.L. Ristig, K.E. Kürten and C.E. Campbell, preprint.

5. C.E. Campbell, M.L. Ristig, K.E. Kürten and G. Senger, proceedings, this workshop.

6. K. Hiroike, Prog. Theor. Phys. 24 (1960) 317.
7. K. Bedell, D. Pines and I. Fomin, J.Low Temp. Phys. 48 (1982) 417.
8. K. Bedell, D. Pines and A. Zawadowski, Phys. Rev. B 29 (1984) 102.
9. F. Mezei, Phys. Rev. Lett. 44 (1980) 1601.

THERMAL RESPONSE OF HOT NUCLEI

J. P. Vary and G. Bozzolo

Department of Physics
Iowa State University
Ames, Iowa 50011

H. G. Miller and R. M. Quick

Theoretical Physics Division
National Research Institute for Mathematical Sciences
CSIR, Pretoria 0001, Republic of South Africa

Given an effective Hamiltonian H_{eff} in a chosen model space for A interacting nucleons, the finite temperature Hartree-Fock approximation (FTHF) requires minimizing the thermodynamic potential

$$\Omega = \langle H_{eff} \rangle_T - TS - \mu A \tag{1}$$

with respect to the single-particle (s.p.) FTHF orbitals $|\nu\rangle$ and the thermal occupation probabilities f_ν satisfying the constraints

$$\sum_{neutrons} f_\nu = N \qquad \sum_{protons} f_\nu = Z \quad . \tag{2}$$

For Eq. 1 $\langle H_{eff} \rangle_T$ represents the ensemble average at absolute temperature T of the effective Hamiltonian, μ is the chemical potential and A = Z + N, is the number of nucleons. The entropy S is given by

$$S = -\sum_\nu \{(f_\nu)\ell n(f_\nu) + (1-f_\nu)\ell n(1-f_\nu)\} \tag{3}$$

The s.p. thermal occupation probabilities f_ν are given by a Fermion distribution

$$f_\nu = \left[1 + \exp\left[\frac{\varepsilon_\nu - \mu}{T}\right]\right]^{-1} \tag{4}$$

where ε_ν are the FTHF s.p. energies.

We have evaluated H_{eff} in a sequence of model spaces, abbreviated as two-space (0s, 0p and 1s-0d shells), the three-space (two-space plus 1p-0f shell), the four-space (three-space plus 2s-1d-0g shell) and the five-space (four-space plus 2p-1f-0h shell). This sequence of model spaces permits us to estimate convergence properties of the FTHF results. For all these calculations all particles are active and only isospin symmetry is insured by requiring that the neutron and proton orbitals are degenerate.

An additional phenomenological adjustment of H_{eff} is made in order to achieve agreement with measured ground state properties before proceeding with the FTHF calculations. Two overall factors, λ_1 and λ_2 for the kinetic energy and effective interaction terms, respectively, are introduced and later adjusted, simultaneously with $\hbar\omega$, the oscillator spacing, to achieve the desired rms radius and binding energy for a given nucleus for each choice of model space. The theoretical rationale for the approximate values of these parameters and expected dependence on model space, together with the resulting numerical values are thoroughly discussed in Refs. 1 and 2.

Spherical Solutions

In our first set of calculations[1,2] we make a further approximation by restricting the FTHF orbitals to have good angular momentum, good total angular momentum and its projection. Only the radial wave function of the s.p. state is varied, and we concentrate on ^{16}O and ^{40}Ca which are expected to have spherically symmetric ground states.

In this approximation, we solve for the temperature dependence of global and s.p. properties. The model space dependence of the calculated thermal properties leads us to conclude that for each size of the model space there exists a value T_s such that the results are reliable for $T < T_s$. For light nuclei, $T_s \sim 7$ MeV for the five-space, ~ 5.5 MeV for the four-space and ~ 4 MeV for the three-space. All results given below correspond to the five-space, for $0 \leq T \leq 7$ MeV.

The excitation energy E* admits a very simple parametrization

$$E^*(T) = \sigma T^2 \tag{5}$$

with $\sigma = 0.185A$ for ^{16}O and $\sigma = 0.104A$ for ^{40}Ca. These results for ^{40}Ca follow the $\sigma = 0.1A$ results of Sauer et al.[3] obtained for systems with $A > 40$ using phenomenological Hamiltonians. The larger thermal sensitivity of ^{16}O is clearly seen through the coefficient σ in this parametrization. Also, we expect greater thermal sensitivity in ^{16}O because of its larger surface to volume ratio.

A similar convenient parametrization is obtained for the rms mass radii R(T),

$$R(T) = R_o(1+bT^2) \tag{6}$$

with $R_o = 2.74$ (3.50) fm and $b = 5.1 \times 10^{-3}(3.2 \times 10^{-3})$ MeV^{-2} for ^{16}O (^{40}Ca), respectively. This rms radial expansion is also considerably greater than that reported by Sauer et al. In addition, we see a greater thermal response for ^{16}O.

A convincing demonstration of the dependence of nuclear thermal properties on surface, volume and shell properties combined will require the more careful parametrization of these results in terms of a thermal liquid drop model plus shell corrections.[4]

The temperature dependence of s.p. properties can be summarized in Fig. (1), where the lowest 14 proton FTHF s.p. energies of ^{40}Ca at integral values of T from 0 to 7 MeV are presented, together with a plot of the fermion occupation function

$$f(\varepsilon) = \left[1 + \exp[\tfrac{\varepsilon-\mu}{T}]\right]^{-1}$$

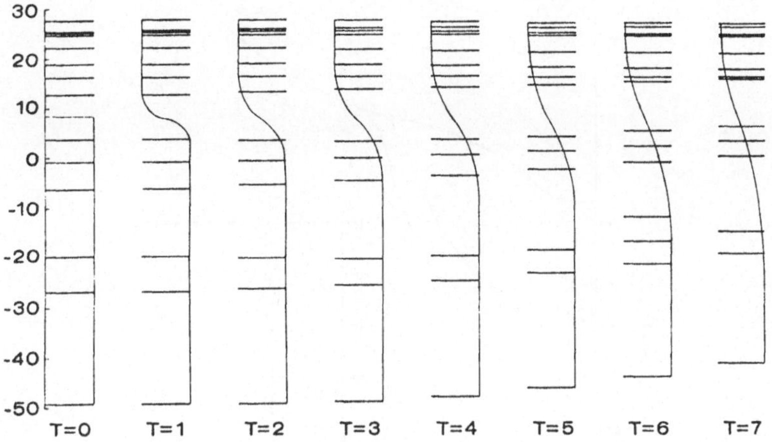

FIGURE 1.

which uses the self-consistently determined value of μ at each T. Although the most deeply bound orbitals are not significantly depopulated, their s.p. energies change considerably. This sensitivity of the s.p. energies contrasts sharply with the results of Sauer et al.[3] who reported the s.p. energies virtually unchanged for $T < 5$ MeV. The most striking feature of our results is that the spin-orbit splitting dissolves fast enough that the gap between the highest orbit of one shell and the lowest orbit of the next is actually preserved out to $T \sim 7$ MeV. Of course, the splitting between shell centroids is decreasing with T as expected.

Deformed Solutions

At zero temperature, it is relatively easy to find a number of spherical and deformed solutions of the Hartree Fock (HF) equations for some nuclei. Since the HF equations are non-linear, these solutions do not, in general, form an orthonormal set. However, they are usually linearly independent and appear to form a good basis for configuration mixing calculations[5-7], yielding good approximates to the energies of the low-lying eigenstates of the many particle Hamiltonian operator.

All together, relatively few solutions of the finite temperature Hartree Fock (FTHF) equations have been reported and most solutions are for the lowest state. The lowest deformed FTHF solutions exhibit a transition to a spherical shape as temperature is increased[8,9]. These reported shape transitions are continuous in excitation energy but are expected to be discontinuous in measures of the deformation. A measure of the deformation such as the electric or mass quadrupole moment is a logical analogue of the order parameter. Therefore previous studies have concluded that nuclear shape transitions are like that of a first order phase transition. We have reported[10] calculations in ^{24}Mg in which we find no such discontinuity in the quadrupole moments of the two

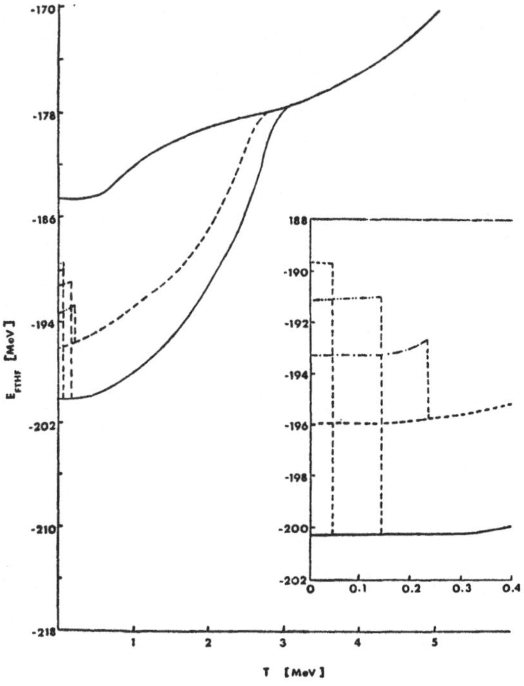

FIGURE 2.

lowest lying solutions to within the numerical accuracy of our calculations. This indicates the character of these transitions is second order. This type of behaviour has not been previously reported for fully self-consistent microscopic calculations. Furthermore we examine the temperature dependence of six distinct HF solutions within ^{24}Mg. In all we obtain five phase transitions: three deformed to deformed occurring with $T < 0.3$ MeV and two deformed to spherical occurring at $T \approx 2.8$ MeV. We find that the three deformed to deformed transitions are clearly first order. The calculations in ^{24}Mg were performed in an oscillator model space consisting of the 0s, 0p, 0s-1d shells (the "two-space"). Since the matrix elements of H_{eff} were initially calculated for a system with $A = 16$ particles with $\hbar\omega = 14$ MeV they were scaled here in a manner described in Refs. 1 and 2 to accommodate A dependence. The matrix elements of the scaled effective Hamiltonian for ^{24}Mg are given by

$$\langle H_{eff}\rangle = \frac{\omega'}{\omega}\left[\frac{A}{A'}\langle T_{rel}\rangle + \langle V_{eff}\rangle\right] \tag{7}$$

with $A' = 24$ and $\hbar\omega' = 13.05$ MeV. Here we denote the two body relative kinetic energy operator by T_{rel} and the realistic effective nucleon-nucleon interaction by V_{eff}. We neglect the coulomb interaction in the ^{24}Mg calculations.

As discussed in Refs. 1 and 2 one needs a final adjustment of H_{eff} to bring the ground state properties of ^{24}Mg at $T = 0$ into close agreement with the experimental values. We multiply T_{rel} by 0.99 and V_{eff} by 1.04 to obtain ^{24}Mg ground state properties.

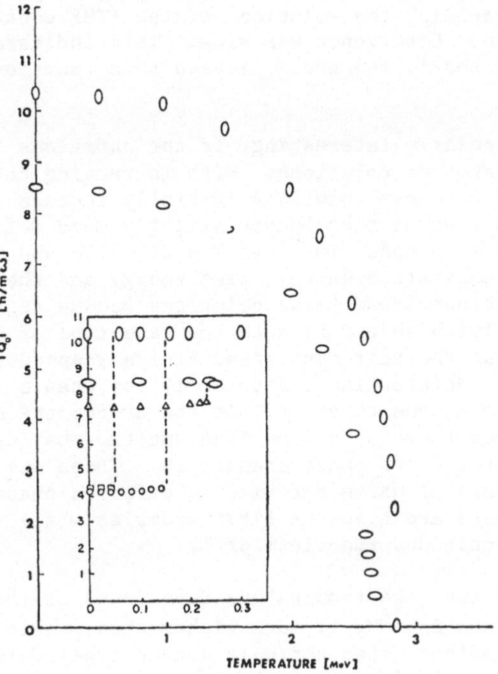

FIGURE 3.

The variational energies, the rms radii, the electric quadrupole moments and chemical potentials of the five lowest lying solutions of the FTHF equations are given in Ref. 10. All five solutions are found to be axially symmetric. As expected, at $T = 0$ where the FTHF reduces to the HF approximation, the ground state is prolate and the first excited state is oblate. Furthermore, the quadrupole moments of higher lying HF solutions are smaller in absolute magnitude than those of the two lower lying HF solutions.

The temperature dependence of the excitation energy of the lowest-lying FTHF solutions are given in Fig. 2. Somewhat surprisingly the spherical solution displays structure at the lower values of T. This structure appears to persist even if the size of the model space is increased and may possibly be due to spurious excitations of the center of mass[9].

In the case of the lowest lying prolate and oblate solutions, they become less deformed with increasing temperature. At $T \approx 2.80$ MeV and $T \approx 2.85$ MeV respectively they become degenerate with the spherically symmetric solution of the FTHF equation. The quadrupole moment, (see Fig. 3) which we take as an order parameter of the system, exhibits no discernable discontinuity, to within our numerical accuracy, at the transition temperatures for both of these solutions. Thus we observe second order prolate to spherical and oblate to spherical phase transitions at different critical temperatures. This situation differs somewhat from the suggestion of Levit and Alhassid[8] in a model calculation that the deformed to spherical phase transition is first order becoming second order in the limit when the oblate and prolate solutions are degenerate at the transition temperature. In the narrow range of temperatures where the magnitudes of the quadrupole moments are falling most rapidly, the solutions of the FTHF equations were more difficult to obtain. Covergence was slow. This indicates that the free energy surface had nearly reached a plateau as a function of the deformation.

What is particularly interesting, is the anomalous behaviour of the higher lying deformed solutions. With increasing temperature, the quadrupole moments of these solutions initially increase in absolute magnitude. In other words they become slightly more deformed rather than less deformed in shape. Then, at $T \approx .23$, .16 and .06 MeV respectively, the excitation energy, free energy and the quadrupole moments are discontinuous and these solutions become degenerate with one of the two lowest lying solutions with the same type of deformation. Thus it appears that the self-consistent fields respond to increasing T by increasing their deformation sufficiently for stable solutions of the FHTF equations to continue to exist. At the aforementioned critical temperatures however these solutions then undergo what definitely appears to be a first order phase transition. These are the first examples we are aware of where deformed to deformed phase transitions are predicted. These are also the first examples where the excitation energies are discontinuous functions of T.

We have found that the temperature dependence of the low-lying deformed HF solutions in ^{24}Mg is much richer than previously reported. Our calculations indicate that not only second order deformed to spherical "phase transitions" occur but that first order deformed to deformed "phase transitions" also take place.

We conclude with some speculative remarks on the physical significance of these states. Since they occurred as distinct variational solutions of the FTHF equations they are interpreted as the one stable (S) and five metastable (MS) states of ^{24}Mg at low temperature. Their relative population, for example, in the interior of a star depends on the difference of their free energy via $P_{MS}/P_S = \exp((F_S-F_{MS})/T)$ and affects the equation of state since it permits energy storage in metastable states at low temperature.

The more physically interesting situations occur when $(\frac{F_{MS}-F_S}{T} < 1)$ as occurs here for the first MS plate near $T = 2.50$ MeV. As one approaches the respective critical temperatures one obtains relative probabilities approaching 1.0 for population of a MS state to the stable state. In other nuclei different MS states may have F_{MS} closer to F_S at lower temperatures. It is important to check this for nuclei of astrophysical interest in star interiors.

Time Dependence of Thermal Response

Recently, Sagawa and Bertsch[11] have introduced time dependent constraints in the FTHF approximation. This allows the examination of the time evolution of thermally excited nuclei in the mean field limit. The constrained Hamiltonian

$$H'' = H_{eff} - \lambda_1 r^2 - \frac{\lambda_2}{2}(\vec{r}\cdot\vec{p} + \vec{p}\cdot\vec{r}) \tag{8}$$

replaces H_{eff} in Eq. (1). Then λ_1 and λ_2 are chosen self-consistently as functions of time to satisfy the equations of motion

$$\frac{d\langle r^2\rangle}{dt} = \langle\ [H_{eff}, r^2]\ \rangle = 2\ \lambda_2\langle\ r^2\ \rangle \tag{9}$$

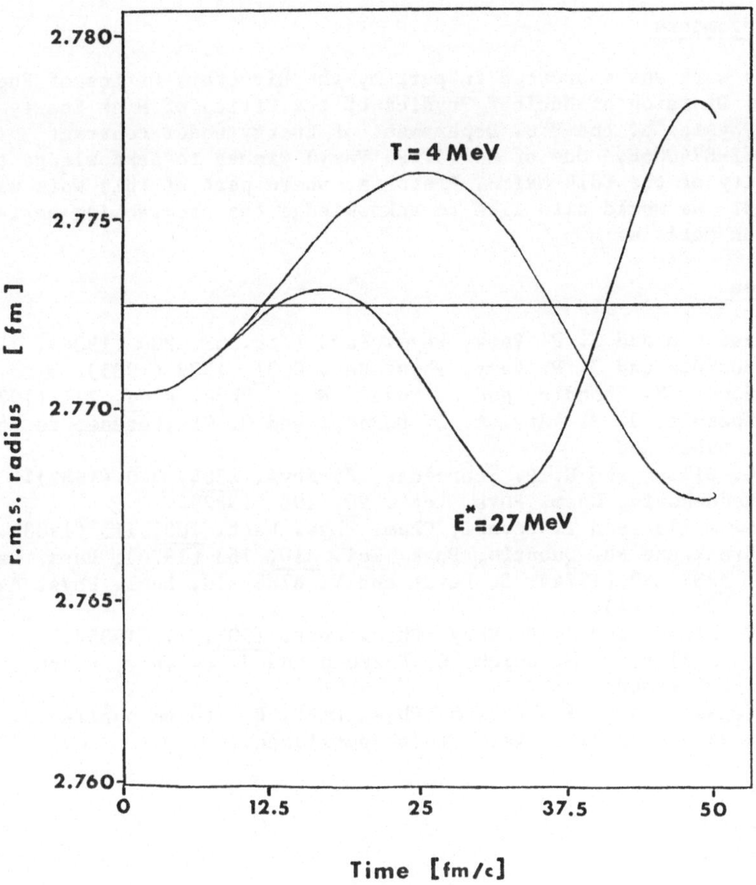

FIGURE 4.

$$\frac{d\langle \frac{1}{2}(\vec{r}\cdot\vec{p} + \vec{p}\cdot\vec{r})\rangle}{dt} = \langle\, [H_{eff}, \tfrac{1}{2}(\vec{r}\cdot\vec{p} + \vec{p}\cdot\vec{r})]\,\rangle = -\, 2\lambda_1\langle r^2\rangle \qquad (10)$$

Here, we consider ^{16}O in the two-space to illustrate how this approach is implemented. Initially we choose $\lambda_2 = 0$ and a small value of λ_1 such that the nucleus is compressed from its equilibrium shape. We also select an initial temperature, T = 4 MeV. Then, we self-consistently solve the equations of motion and follow the nuclear expansion either isothermally or adiabatically. The results are displayed in Fig. 4. At the present stage of these calculations we do not consider the effect of nucleon emission which would be expected to be important at these temperatures and excitation energies. Consequently, one should not put great physical significance on what happens after the first maximum in the rms radius is achieved.

These results represent an initial attempt to simulate the time dependent thermal expansion of a heated compressed nucleus. A more ambitious project would include rotations, deformations and coupling to the gas phase in order to provide a realistic simulation of events produced in heavy ion collisions. Further work in these directions are in progress.[12]

Acknowledgements

This work was supported in part by the Director, Office of Energy Research, Division of Nuclear Physics of the Office of High Energy and Nuclear Physics of the U.S. Department of Energy under contract DE-AC02-82-ER40068. One of us (J. P. Vary) wishes to acknowledge the hospitality of the CSIR-NRIMS, Pretoria, where part of this work was completed. We would also like to acknowledge the programming assistance of Mr. Jan Matlala.

References

1. G. Bozzolo and J. P. Vary, Phys. Rev. Lett. 53, 903 (1984).
2. G. Bozzolo and J. P. Vary, Phys. Rev. C 31, 1909 (1985).
3. G. Sauer, H. Chandra, and U. Mosel, Nucl. Phys. A264, 221 (1976).
4. G. Bozzolo, J. P. Vary, A. L. DePaoli and O. Civitarese, to be published.
5. H. G. Miller and H. P. Schroeder, Z. Phys. A304, 340 (1982).
6. J. Hendekovic, Chem. Phys. Lett. 90, 198 (1982).
7. H. G. Miller and T. Geveci, Chem. Phys. Lett. 100, 115 (1983).
8. M. Brack and Ph. Quentin, Phys. Sci. A10, 163 (1974); Phys. Lett. 52B, 159 (1974); S. Levit and Y. Alhassid, Nucl. Phys. A413, 439 (1984).
9. H. G. Miller and J. P. Vary, Phys. Lett. 150B, 11 (1985).
10. H. G. Miller, R. M. Quick, G. Bozzolo and J. P. Vary, to be published.
11. H. Sagawa and G. F. Bertsch, Phys. Lett. B., to be published.
12. G. Bozzolo and J. P. Vary, to be published.

DYNAMICAL BEHAVIOUR OF STRONGLY CORRELATED COULOMB PLASMAS: A NONLINEAR-RESPONSE APPROACH

K.I. Golden*, F. Green, and D. Neilson

School of Physics
University of New South Wales
Kensington 2033, Australia

In this paper we survey the nonlinear-response approach to strongly correlated plasmas developed by Golden, Kalman, and co-workers. Its successes to date include the dynamical description of one-component plasmas in two and three dimensions and, most recently, the dynamics of binary ionic mixtures. In each case the resulting theory provides a microscopic picture of the collective-mode behaviour - and of its underlying physics - which is accurate and remarkably consistent over the entire range of coupling strength. The nonlinear-response approach secures this by conforming to the basic conservation laws at the level of microscopic dynamics.

INTRODUCTION

The nonlinear response-function approach of K. Golden, G. Kalman, and their collaborators[1-3] addresses itself to the structure of plasma kinetic equations. Its distinguishing features are:

(i) It is a means of preserving the symmetries and the structural relations among the correlation and response functions; their preservation is essential in order to obtain a conserving dynamic theory. At the same time, the approach produces a tractable approximation to the dynamic response functions.
(ii) The method does not need to make uncontrolled functional approximations when the hierarchy of kinetic equations is closed, because the time dependence of the dynamic correlation functions is maintained implicitly.
(iii) The method stresses the language of response functions as the most natural way of discussing time-dependent plasma behaviour, notably the collective modes.

The present nonlinear-response formalism has been worked out in the context of classical Coulomb systems. After some additional development of its main elements, the same philosophy is expected to generate powerful results in the quantum-plasma regime. This paper outlines the logic of the existing classical formalism and surveys its achievement in accurately describing plasma dynamics over an unprecedented range of parameters.

---

*Permanent address: Department of Electrical and Computer Engineering
Northeastern University, Boston MA 02115

For general reviews of strongly coupled Coulomb problems see, for example, Kalman and Carini,[4] and Ichimaru.[5] Full details of the nonlinear-response approach can be found in Golden and Kalman,[1] Golden and Lu,[2] and Golden, Green, and Neilson.[3]

FORMALISM

We consider homogeneous classical plasmas consisting of one or more ionic species in a neutralizing background of degenerate electrons, whose Fermi degeneracy energy E_F is much greater than the thermal energy k_BT. Realizations of such models exist in some stellar cores;[6] further important sources of data include molecular-dynamics[7] and Monte Carlo[8-10] simulations, and hypernetted-chain calculations.[11,12]

For brevity we discuss the one-component plasma. Define the coupling strength Γ as the ratio of the mean ion-ion electrostatic potential to the thermal energy

$$\Gamma = \frac{(Ze)^2}{a} / k_B T \tag{1}$$

where Ze is the ionic charge and a is the interionic separation. The theory will cover, within a single microscopic prescription, the range of Γ over the entire fluid regime.

As usual we seek a linear-response theory with respect to a weakly coupled *external* perturbing potential $\hat{\phi}(\underset{\sim}{x},t)$. The nonlinear-response feature of the approach will describe the *internal* dynamic response, which is conditioned by the strong ion-ion interactions.

In the presence of $\hat{\phi}(\underset{\sim}{x},t)$ the nonequilibrium one-particle distribution function $F(\underset{\sim}{x},\underset{\sim}{v};t)$ in phase space obeys the kinetic equation

$$\left(\frac{\partial}{\partial t} + \underset{\sim}{v}_1 \cdot \frac{\partial}{\partial \underset{\sim}{x}_1} + \frac{1}{m}\left(-\frac{\partial\hat{\phi}}{\partial \underset{\sim}{x}_1}\right)\cdot\frac{\partial}{\partial \underset{\sim}{v}_1}\right) F(1;t) =$$

$$-\frac{1}{m}\frac{\partial}{\partial \underset{\sim}{v}_1}\cdot\int d^3x_2\int d^3v_2\left(-\frac{\partial}{\partial \underset{\sim}{x}_1} V(\underset{\sim}{x}_1 - \underset{\sim}{x}_2)\right) G(12;t)\ , \tag{2}$$

where m is the ionic mass, $V(\underset{\sim}{x}) = (Ze)^2/|\underset{\sim}{x}|$ is the interionic potential, and $G(\underset{\sim}{x}_1,\underset{\sim}{v}_1;\underset{\sim}{x}_2,\underset{\sim}{v}_2;t)$ is the two-particle distribution function (we condense phase-space labels in Eq.(2) in an obvious notation). The distributions F and G determine the exact one- and two-particle nonequilibrium correlation functions respectively through

$$\int d^3v_1\ F(1;t) \equiv \langle n(\underset{\sim}{x}_1)\rangle\ (t) \tag{3a}$$

$$\int d^3v_1\int d^3v_2\ G(12;t) \equiv \langle n(\underset{\sim}{x}_1)n(\underset{\sim}{x}_2)\rangle\ (t) - \delta(\underset{\sim}{x}_1 - \underset{\sim}{x}_2)\langle n(\underset{\sim}{x}_1)\rangle(t). \tag{3b}$$

We now want to express Eq.(2) purely in terms of these correlation functions, and ultimately in terms of dynamic response functions whose properties will determine the collective modes. But the velocity integration on the right side of Eq.(2) precludes any direct representation in correlation-function language, and we are thus led to the pivotal element of the theory - the Velocity Average Approximation (VAA), in which G is replaced with its average over velocity space:

$$G(\underset{\sim}{x}_1, \underset{\sim}{v}_1 ; \underset{\sim}{x}_2, \underset{\sim}{v}_2 ; t) \overset{\text{VAA}}{\equiv} \frac{1}{2} f(\underset{\sim}{x}_1, \underset{\sim}{v}_1 ; t) \int d^3 v_1' \, G(\underset{\sim}{x}_1, \underset{\sim}{v}_1' ; \underset{\sim}{x}_2, \underset{\sim}{v}_2 ; t)$$

$$+ \frac{1}{2} f(\underset{\sim}{x}_2, \underset{\sim}{v}_2 ; t) \int d^3 v_2' \, G(\underset{\sim}{x}_1, \underset{\sim}{v}_1 ; \underset{\sim}{x}_2, \underset{\sim}{v}_2' ; t) \quad , \tag{4}$$

where $f(1;t) = F(1;t)/\langle n(\underset{\sim}{x}_1)\rangle(t)$. Note that, as an identity, Eq.(4) imposes a self-consistency constraint on the VAA form of G.

On applying this Ansatz to Eq.(2), one readily obtains the fundamental VAA kinetic equation

$$\left(\frac{\partial}{\partial t} + \underset{\sim}{v}_1 \cdot \frac{\partial}{\partial \underset{\sim}{x}_1} - \frac{1}{m} \frac{\partial \hat{\phi}}{\partial \underset{\sim}{x}_1} \cdot \frac{\partial}{\partial \underset{\sim}{v}_1} \right) F(1;t) =$$

$$- \frac{1}{m} \frac{\partial}{\partial \underset{\sim}{v}_1} f(1;t) \int d^3 x_2 \left(- \frac{\partial}{\partial \underset{\sim}{x}_1} V(\underset{\sim}{x}_1 - \underset{\sim}{x}_2) \right) \langle n(\underset{\sim}{x}_1) n(\underset{\sim}{x}_2) \rangle (t) \, . \tag{5}$$

Eq.(5) has the important property of being exact for any static external perturbation, since the velocity dependence of each distribution function is then exactly separable.[1] The VAA's *exact static limit* provides a strong constraint on its dynamic behaviour; for example, it guarantees satisfaction of the compressibility sum rule for the dynamic structure factor at long wavelengths.[13] It also validates the practical step of using independent simulation data[7-12] for the static structure factor as an initial input to the theory.

In contrast with traditional approaches to kinetic equations, no approximation is made for the form of G as a functional of F; the VAA is not a truncation of the hierarchical structure of Eq.(2). Rather, the strategy is to go over to the nonequilibrium correlation function $\langle n(1)n(2)\rangle$, a much more tractable and controllable object than G. It is largely for this reason that the nonlinear-responce approach displays excellent sum-rule behavior.[1-3,13]

The linearized VAA equation becomes, in Fourier form,

$$F^{(1)}(\underset{\sim}{k}, \underset{\sim}{v}; \omega) = - \left(\frac{1}{m} \frac{\partial F^{(0)}(v)/\partial \underset{\sim}{v}}{\omega - \underset{\sim}{k} \cdot \underset{\sim}{v}} \right) \cdot [\underset{\sim}{k} \hat{\phi}(\underset{\sim}{k}, \omega)$$

$$+ \frac{1}{N} \sum_{\underset{\sim}{q}} \underset{\sim}{q} \, V(q) \, \langle n(\underset{\sim}{q}) n(\underset{\sim}{k} - \underset{\sim}{q}) \rangle^{(1)}(\omega)] \, . \tag{6}$$

Here $F^{(0)}(v)$ is the equilibrium one-particle distribution and $F^{(1)}$ and $\langle nn\rangle^{(1)}$ are the parts of F and $\langle nn\rangle$ linear in $\hat{\phi}$. On integrating over velocity, one obtains the equation for the *macroscopic* linear density response $\hat{\chi}(\underset{\sim}{k},\omega)$:

$$\hat{\chi}(\underset{\sim}{k},\omega) = -\frac{\langle n(\underset{\sim}{k})\rangle^{(1)}(\omega)}{\hat{\phi}(\underset{\sim}{k},\omega)}$$

$$= \chi_0(\underset{\sim}{k},\omega)\,[1+V(k)\,\hat{\chi}(\underset{\sim}{k},\omega)-\hat{v}(\underset{\sim}{k},\omega)] \quad , \tag{7}$$

where χ_0 is the Vlasov (RPA) polarization function. The second term on the right side of Eq.(7) is the familiar Hartree/RPA correlation, while the last term $\hat{v}$ contains the collisional effects carried by $\langle nn\rangle^{(1)}$. To make an explicit identification of the collisional effects, it is convenient to rewrite Eq.(7) as an equation for the *microscopic* (or, screened) response $\chi(\underset{\sim}{k},\omega)$:

$$\chi(\underset{\sim}{k},\omega) = \hat{\chi}(\underset{\sim}{k},\omega)/(1+V(k)\,\hat{\chi}(\underset{\sim}{k},\omega))$$

$$= \chi_0(\underset{\sim}{k},\omega)\,[1-v(\underset{\sim}{k},\omega)] \;; \tag{8}$$

we have introduced $v=\hat{v}/(1+V\hat{\chi})$, thereby abstracting from $\hat{v}(\underset{\sim}{k},\omega)$ the collective plasmon mode. All the purely non-RPA dynamic correlations are now explicitly retained in $v(\underset{\sim}{k},\omega)$.

The form of the correlation piece $v(\underset{\sim}{k},\omega)$ is obtained through the second major step in the Golden-Kalman analysis: the dynamical nonlinear fluctuation-dissipation theorem.[14,15] In essence, the coefficient $\langle nn\rangle^{(1)}/\hat{\phi}$ is a three-particle *equilibrium* correlation, containing lower-order clusters of the form $\langle n\rangle^{(0)}\langle nn\rangle^{(0)}$ plus a collisional $\langle nnn\rangle^{(0)}$ correlation term. By means of the *linear* fluctuation-dissipation theorem the $\langle n\rangle^{(0)}\langle nn\rangle^{(0)}$ correlations readily transform into the induced Hartree response $V\hat{\chi}$ (this can then be factored out, as in Eq.(8)). Analogously, the collisional correlation $\langle nnn\rangle^{(0)}$ transforms into the quadratic-response polarization term $\hat{v}$ through the *nonlinear* fluctuation-dissipation theorem. The associated term v, which is conceptually related to the irreducible diagrams of cluster analysis, has the form

$$v(\underset{\sim}{k},\omega) = -V(k)\left(\frac{ik_BT}{\pi N}\right)\sum_{\underset{\sim}{q}}\frac{\underset{\sim}{k}\cdot\underset{\sim}{q}}{q^2}\int_{-\infty}^{\infty}\frac{d\mu}{\mu-i0^+}\times$$

$$\left[\frac{{}_2\chi(\underset{\sim}{k}-\underset{\sim}{q},\,\omega-\mu;\,\underset{\sim}{q},\mu)}{\varepsilon(\underset{\sim}{k}-\underset{\sim}{q},\,\omega-\mu)\varepsilon(\underset{\sim}{q},\mu)}+\frac{{}_2\chi(\underset{\sim}{k}-\underset{\sim}{q},\mu;\,\underset{\sim}{q},\,\omega-\mu)}{\varepsilon(\underset{\sim}{k}-\underset{\sim}{q},\,\mu)\varepsilon(\underset{\sim}{q},\omega-\mu)}\right]\;; \tag{9}$$

ε is the dielectric function, $\varepsilon = 1-V\chi = (1+V\hat{\chi})^{-1}$, and ${}_2\chi$ is the microscopic quadratic polarization as defined by ${}_2\chi\varepsilon = \delta^2\langle n\rangle/\delta\phi\delta\phi$, where $\phi=\hat{\phi}/\varepsilon$ is the total, dynamically screened, perturbation.

In obtaining a dynamical equation, Eq.(8), for the plasma's screened response χ, we have so far introduced a velocity-averaging procedure for two-particle distribution functions, and we have expressed the resultant two-particle dynamic correlations entirely in the language of dynamic response functions. It now remains to effect closure of Eq.(8) by approximating the VAA quadratic response ${}_2\chi$ in Eq.(9) as a functional of the linear response χ. This is most easily done by replacing ${}_2\chi$ with its long-wavelength RPA structure. A routine calculation[1] from Eq.(5) leads to a Vlasov-like expression for ${}_2\chi$ in terms of the linear Vlasov function χ_0; one then keeps the functional form of ${}_2\chi$ but replaces χ_0 with its full VAA counterpart χ. Equations (8) and (9) are then solved self-consistently for χ.

The RPA Ansatz for ${}_2\chi$ manifests the same gauge-invariant symmetry as the exact quadratic polarization.[15] Consequently it is particle-conserving at all values of the coupling Γ, although it starts out as a weak-coupling expression.[1]

Physically, substitution of the quadratic-RPA form for ${}_2\chi$ into (9) generates the leading dynamical correlations beyond linear RPA, as is clear from the cluster structure of the integrand in Eq.(9), which is now

$$ {}_2\chi_0(\underset{\sim}{p},\nu\,;\,\underset{\sim}{q},\mu)\left(\frac{1}{[1-V(p)\chi(\underset{\sim}{p},\nu)][1-V(q)\chi(\underset{\sim}{q},\mu)]}\right), \tag{10} $$

with ${}_2\chi_0$ being the pure RPA quadratic polarization.[1]

At this point we should remark that the expression (10) is considerably more complicated in the case of a binary ionic plasma; in particular, the introduction of an ion-species index leads to a matrix analogue of Eq.(8) for partial species-response functions,[3] whose trace then gives the overall system response. The nonlinearity of the matrix dynamical equation gives rise to qualitatively different physics for binary mixtures in comparison with the one-component plasma. This is discussed in the next section.

All applications of the Golden-Kalman approach to date have featured the dynamical-cluster form (10). The ensuing calculations of the collective-mode dispersions are notable for their accuracy at all coupling strengths. This speaks for the high degree of control over the approximations used, and especially for the soundness of the basic VAA assumption.

RESULTS

First we summarize the general properties of the nonlinear response-function approach in two important asymptotic regions, and then discuss actual implementations of Eqs.(8),(9), and (10) for the dynamic response.

Static Limit ($\omega = 0$).

As noted, the VAA per se is exact in the classical static limit. In both two and three dimensions it necessarily satisfies not only the common long-wavelength sum rules, but it also exactly preserves the entire hierarchy of relations among the static structure factors at all wavelengths.[1-3]

High-Frequency Limit ($\omega \gg k\sqrt{3k_BT/m}$)

It has been shown[1-3] that the VAA form of the polarization function $\chi(k,\omega)$ satisfies the third frequency-moment sum rule expression

$$ V(k)\chi(\underset{\sim}{k},\omega) \underset{\omega\to\infty}{\sim} \frac{\omega_0^2}{\omega^2} + \frac{\omega_0^4}{\omega^4}\left(\frac{3k_BT}{m\omega_0^2}\,k^2 + \sum_{\underset{\sim}{q}}\left(\frac{\underset{\sim}{k}\cdot\underset{\sim}{q}}{kq}\right)^2 [S(\underset{\sim}{k}-\underset{\sim}{q}) - S(\underset{\sim}{q})]\right), \tag{11} $$

where ω_0 is the appropriate plasma frequency for the system. The ω^{-4} term displays a thermal-kinetic-energy component, and a correlational component determined by the usual static structure factor S(q).

Recently there has been renewed interest in the role of the third-moment sum rule in characterizing multiparticle correlations beyond the RPA. In the classical domain the nonlinear-response approach has, for the first time, quantitatively tracked the evolution of the collective mode through the whole range of coupling strength Γ; for $\Gamma \to \infty$ one is led to the conclusion that the mode dispersion coincides with that for the Wigner lattice and is *entirely* determined by the third-moment correlational term in Eq.(11).[3,18]

The same third-moment sum rule is implicated in strong-coupling problems for quantum plasmas;[19] that a model as tractable as the VAA should correctly reproduce the physics underlying this sum rule is therefore of importance beyond the classical context.

We now examine the calculational results of the nonlinear-response approach.

One-Component Plasma

The correlation term $v(k,\omega)$ of Eq.(9) can be resolved into a static and a dynamic component.[1-3] The first piece is the correlational part of the third-moment coefficient in Eq.(11). This dominates at high frequency. In the limit of *weak coupling* ($\Gamma \ll 1$) and long wavelength ($\underset{\sim}{k} \cdot \underset{\sim}{v} \ll \omega$) the dynamic component determines plasmon damping completely, and augments the static contribution to the plasmon dispersion. Our Table 1 (Table 2 of Ref. 20) compares damping and dispersion coefficients in the VAA with exact results. While the VAA performs well, it misses the characteristic $k^2\Gamma^{3/2}\,\ell n\,1/\Gamma$ damping term. This is attributed to a loss of long-ranged collisional information following the velocity-averaging process.[1]

Table 1. Exact vs. VAA Plasmon Damping Coefficients at Long Wavelength for the One-Component Plasma. $\nu = \omega/\omega_0$, $\gamma = \sqrt{3}\,\Gamma^{3/2}$ Results in units of γk^2.

ν	Exact	VAA
$\to 0$	$0.271 + i\nu^{-1}(-0.087 + 0.301\,\ell n\gamma^{-1})$	0.149
1	$0.214 + i(0.030 + 0.301\,\ell n\gamma^{-1})$	$0.133 + i0.046$
2	$0.148 + i(0.119 + 0.150\,\ell n\gamma^{-1})$	$0.106 + i0.078$
$\to\infty$	$-0.457\nu^{-1} + i\nu^{-1}(0.78 + 0.301\,\ell n(2\gamma^{-1}\nu^{-1}))$	$0.41\nu^{-2} + i\nu^{-1}0.338$

At stronger coupling ($\Gamma > 1$), the University of Paris molecular-dynamics simulations[7] reveal a marked crossover from plasmon-like to optic-mode-like dispersion [see Fig. 1 this paper (Fig. 2., Ref. 16)]. Recent numerical calculations[17] of Eqs.(8) and (9) indicate that this crossover occurs at $\Gamma \approx 8.8$ in very good agreement with the molecular dynamics data. This is one of the highlights of the nonlinear-response-function approach. Moreover, the Wigner-lattice-like limiting dispersion and its association with the third frequency-moment sum rule consolidates the theory's success in this area.[18]

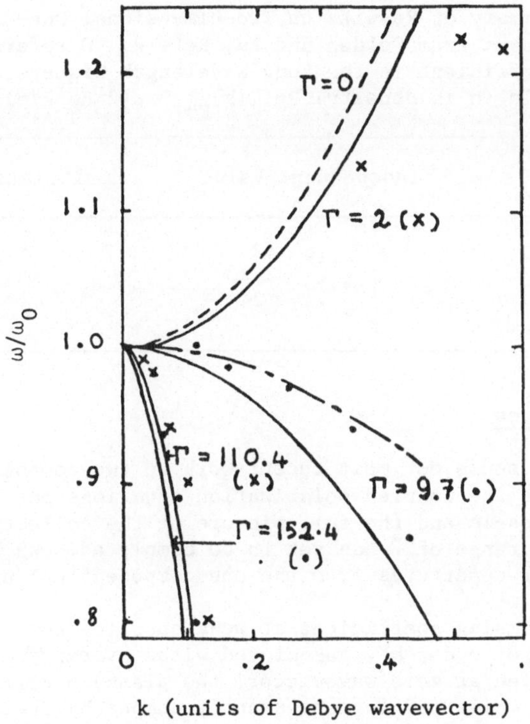

Fig. 1. Calculated VAA Dispersion at Various Values of Γ Compared with Data from Hansen et al.[7]. Dashed Line: Bohm-Gross Curve. Dot-dashed Line: Revised Calculation from Ref. 17.

The collective-mode behaviour in *two dimensions* follows a similar pattern to the three-dimensional case. However, the results here are not dependent on the VAA assumption, following instead a hydrodynamic formulation which in two dimensions is superior to the VAA.[2] The approach leads to even tighter numerical results and confirms that the dispersion can never be described by mean-field-theory or RPA approaches[21] no matter how small the coupling may be.[22]

As can be seen from Table 2, the calculated crossover value of Γ compares favourably with independent estimates.[23] The Wigner-lattice-like dispersion of the collective mode at extreme coupling almost exactly reproduces the Bonsall-Maradudin result[24] for the hexagonal Wigner lattice in two dimensions.

Table 2. Summary of Results on Two-Dimensional One-Component Plasma from Golden and Lu, Ref. 2. C refers to the Coefficient in the Long-Wavelength Dispersion Formula (n is density): Re $\omega(k)_{\Gamma\to\infty} = \sqrt{(2\pi ne^2k/m)}(1 - Ck/\sqrt{\pi n})$

Parameter	Independent Value	Calculated Value
$\Gamma_{crossover}$	$\geq$ 2.29 [22] 3.55 [2,23]	3.76
C	0.173 [24]	0.1728

Binary Ionic Mixtures

Finally, we discuss our most recent work on two-component plasmas.[3] The nonlinearity of the coupled polarization equations for two ionic species is more formidable and the full picture of the collective-mode dispersion over the full range of Γ has yet to be completed. We have, however, confirmed two major departures from the one-component calculation:

(a) the plasmon damping coefficient at weak coupling now contains an explicit $\Gamma^{3/2}\ell n 1/\Gamma$ term, of order k°, associated with *interdiffusion* between the two ion species; even at zero wavevector, the plasma oscillation acts partly to segregate the species and is consequently damped by their mixing together again. The plasmon linewidth at zero wavevector takes the form[3]

$$\mathrm{Im}\,\Omega_0\Big|_{VAA} \propto \omega_0 \left(\frac{e_A}{m_A} - \frac{e_B}{m_B}\right)^2 \Gamma^{3/2}\,\ell n\, 1/\Gamma \quad , \tag{12}$$

where ω_0 is the plasma frequency for the mixture and e_A/m_A and e_B/m_B are the charge-to-mass ratios for the two species. For a H^+-He^{2+} mixture at equal concentrations, the calculated plasma frequency exhibits the *coupling dependent shift*

$$\mathrm{Re}\,\Omega_0\Big|_{VAA} = \omega_0(1 + 0.053\Gamma^{3/2}) \quad , \tag{13}$$

in good agreement with Baus' prediction:[25] $\mathrm{Re}\,\Omega_0 = \omega_0(1 + 0.08\Gamma^{3/2})$.

(b) At strong coupling our model generates a *coupling-independent* shift in Ω_0 which agrees exactly with the estimate of Hansen et al.:[26] $\Omega_0 = \omega_0(1 + 0.0198)$. We find that the strong-coupling dispersion of the collective mode is once more determined totally by the correlational contribution to the third-moment sum rule.

Note that, from Eq.(12), the $\Gamma^{3/2}\ell n\, 1/\Gamma$ damping contribution is absent for so-called symmetric mixtures (He^{2+} - D^+ is an example), for which $e_A/m_A = e_B/m_B$. From this point of view the theory treats such mixtures effectively as one-component plasmas; some information is lost which, in the one-component case, is presumably related to self-diffusion effects as distinct from interdiffusion between species.

Our nonlinear-response approach to binary mixtures makes possible a comprehensive treatment of these systems over the whole fluid regime; it remains for us to study the detailed form of the collective-mode dispersion and its crossover behaviour in the intermediate range of the coupling strength.

SUMMARY

We have reviewed the essential structure of the Golden-Kalman approach to strongly coupled Coulomb plasmas. The hallmark of the nonlinear-response analysis is the replacement of the hierarchy of kinetic equations for particle distribution functions by a kinetic-equation hierarchy for *dynamic response functions* which directly determine the collective modes. This is achieved in three stages.

(i) The Velocity-Average Approximation, which converts two-particle distribution functions into two-particle nonequilibrium correlation functions.

(ii) Linearization of the new VAA kinetic equation for the one-particle nonequilibrium correlation function. This converts the collisional term in the equation from a nonequilibrium two-particle object to an equilibrium three-particle object.

(iii) The Dynamical Nonlinear Fluctuation-Dissipation Theorem, which converts three-particle equilibrium correlation functions into dynamical quadratic-response functions.

The resulting equation for the dynamic linear-response function can then be solved self-consistently within an appropriate model for the quadratic response function appearing in the collisional term. It is crucial that any such model should manifestly preserve the symmetries of the complete quadratic response function, in order to maintain particle conservation. This is the case for the simple RPA Ansatz (10) which forms the basis of all existing calculations within the nonlinear-response approach.

We have seen that the theory yields excellent results over the complete range of classical plasma situations: one-component plasmas in two and three dimensions, and binary ionic mixtures in three dimensions. The crossover from plasmon to longitudinal-mode dispersion is well predicted, and the dominant role of the third-moment sum rule at extremely strong coupling is clarified in detail. To our knowledge, this last feature is not adequately treated in any other theory covering as extensive a range of plasma parameters as does the nonlinear-response theory.

The completeness and cohesiveness of the method's classical results makes it strongly desirable to extend it to quantum plasmas. The most natural way to accomplish a transition to the quantum domain appears to be a reformulation starting from Wigner distribution functions. At present there is no quantum analogue of the dynamical nonlinear fluctuation-dissipation theorem; its derivation constitutes the foremost step in the programme.

ACKNOWLEDGEMENTS

The authors wish to recognize support from the following sources. For K.I.G.: U.S. National Science Foundation Grant No. ECS-83-15801; the Australian-American Educational Foundation; the Gordon Godfrey Bequest at the University of New South Wales. For D.N. and F.G.: Australian Research Grants Scheme. For K.I.G. and D.N.: U.S. National Science Foundation Grant No. INT-83-20531.

REFERENCES

1. K. I. Golden and G. Kalman, Phys. Rev. A19, 2112 (1979), and Ann. Phys. (N.Y.) 143, 160 (1982).
2. K. I. Golden and De-Xin Lu, Phys. Rev. A31, 1763 (1985).
3. K. I. Golden, F. Green, and D. Neilson, Phys. Rev. A31, 3529 (1985); Phys. Rev. A32, to appear.
4. *Strongly Coupled Plasmas*, edited by G. Kalman and P. Carini (Plenum, New York, 1978).
5. S. Ichimaru, Rev. Mod. Phys. 54, 1017 (1982).
6. V. Trimble, Rev. Mod. Phys. 54, 1183 (1982).
7. J.-P. Hansen, E. L. Pollock, and I. R. McDonald, Phys. Rev. Lett. 32, 227 (1974); J.-P. Hansen, I. R. McDonald, and E. L. Pollock, Phys. Rev. A11, 1025 (1976).
8. E. L. Pollock and J.-P. Hansen, Phys. Rev. A8, 3110 (1973).
9. H. Totsuji, Phys. Rev. A17, 399 (1978).
10. R. C. Gann, S. Chakravarty, and G.V. Chester, Phys. Rev. B20, 326 (1979).
11. J. F. Springer, M. A. Pokrant, and F. A. Stevens, J. Chem. Phys. 58, 4863 (1973).
12. F. Lado, Phys. Rev. B17, 2827 (1978).
13. K. I. Golden and G. Kalman, Phys. Rev. A17, 390 (1978).
14. K. I. Golden, G. Kalman, and M. Silevitch, J. Stat. Phys. 6, 87 (1972).
15. K. I. Golden and De-Xin Lu, J. Stat. Phys. 29, 281 (1982).
16. P. Carini, G. Kalman, and K. I. Golden, Phys. Lett. 78A, 450 (1980).
17. P. Carini and G. Kalman, Phys. Lett. 105A, 229 (1984).
18. K. I. Golden (unpublished).
19. N. Iwamoto, E. Krotscheck, and D. Pines, Phys. Rev. B29, 3936 (1984); N. Iwamoto, Phys. Rev. A30, 3289 (1984).
20. P. Carini, G. Kalman, and K. I. Golden, Phys. Rev. A26, 1686 (1982).
21. N. Studart and O. Hipolito, Phys. Rev. A22, 2860 (1980); P. M. Platzman and N. Tzoar, Phys. Rev. B13, 3197 (1976).
22. M. Baus, J. Stat. Phys. 19, 163 (1978).
23. H. Totsuji and N. Kakeya, Phys. Rev. A22, 1220 (1980).
24. L. Bonsall and A. A. Maradudin, Phys. Rev. B15, 1959 (1977).
25. M. Baus, Phys. Rev. Lett. 40, 793 (1978).
26. J.-P. Hansen, I.R. McDonald, and P. Vieillefosse, Phys. Rev. A20, 2590 (1979).

DENSITY FUNTIONAL THEORY AS AN ALTERNATIVE TO THE EXTENDED THOMAS-FERMI THEORY IN CONDENSED MATTER CALCULATIONS

Eduardo V. Ludeña

Centro de Química, Instituto Venezolano de Investigaciones Científicas, IVIC, Apartado 1827, Caracas 1010-A
Venezuela

Some new developments of density functional theory are discussed. In particular, it is argued that an alternative to the gradient expansion for the kinetic energy density is given by its decomposition into a local Weiszacker term plus a non-local correction. Several approximations to this non-local correction are discussed. It is shown that for these functionals, there occurs a virial partitioning of R^3 into subvolumes defined by zero flux surfaces.

I.- INTRODUCTION

Although in principle, the Schrodinger Equation provides the means to attain a complete description of a many Ferminon system, in practice, the computational difficulties which arise in its application to atomic, molecular, solid state and nuclear physics, have made it desirable to develop alternative formulations based on mathematical objects simpler than the wavefunction, such as the density ρ or the first order reduced density matrix γ(1-4).

The advantage of relying on $\rho(\vec{r})$ or on $\gamma(\vec{r},\vec{r}')$comes from the fact that regardless of the size of the system, the problem remains in three dimensions for ρ, or in six dimensions for γ, whereas in the case of the wavefunction it grows as 3n, where n is the number of particles.

The earliest theory which was based explicitly on the density ρ is the Thomas-Fermi theory (5,6). Ever since its inception it has played a very important role in several branches of physics both as a guide to physical insight and also as an expedite procedure for reaching approximate numerical results (7-10).

An important aspect of the Thomas-Fermi theory is that it is exact in the large Z (nuclear-charge) limit (10). This makes it an interesting theory on its own accord, regardless of its possible applications to more realistic systems.

Extensions of the Thomas-Fermi theory have been geared at introducing corrections in order to better describe the kinetic energy density and the exchange and Coulomb correlation effects (11-15). A general formulation of these corrections have been given in terms of an infinite gradient expansion for both the kinetic and the potential energy densities.

In addition to zero temperature applications for isolated systems, the Thomas-Fermi and other related theories have also been extensively applied in the context of a thermodynamical-statistical description of condensed matter. Although the extension of the Thomas-Fermi theory for the case when $T > 0$ was given a long time ago (16-19) and it has been widely applied to a number of situations, recently, there has been a renewed effort (20) to develop related theories valid at $T > 0$ which would at the same time, avoid some of its shortcomings.

In view of this renewed interest and of the possible practical applications to condensed matter theory, it becomes desirable to discuss some recent developments of density functional theory, which present an alternative to the usual formulation of extensions to the Thomas-Fermi theory in terms of a gradient expansion. Strictly speaking, this discussion refers to the case of zero temperature, but it could be easily extended to finite temperature.

In Section II-A we present a critique of the gradient expansion and review some important results for the case of Fermions interacting through Coulomb forces (21). In addition, in Section II-B we advance a different type of expansion for the kinetic energy density where the Weizsacker term becomes the leading local term and where the corrections are given by non-local terms which incorporate statistical correlation corrections among Fermions (22,23). In Section II-C, we discuss some recent alternative formulations for this non-local functional (24-27).

In Section III we point out a general property of Fermion systems whose kinetic energy density is given by a gradient expansion or by the functional described in Section II-C, namely, that there occurs a partition of R^3 into subvolumes where the virial theorem holds locally (28). This general property, which in molecular physics sets the basis for a description of a molecule as a collection of quasi-atoms (or virial fragments) (29) and which also permits a rigorous description of a solid as a collection of distorted atomic fragments bound by zero flux surfaces, may also be of use in the case of condensed matter theories in the sense of allowing the identification of a given subsystem enclosed in a subvolume or virial fragment, whose microscopic properties could then be employed in the usual statistical mechanical treatment leading to the calculation of macroscopic properties.

II.- A CRITIQUE OF THE GRADIENT EXPANSION

A. The Thomas-Fermi Term and its Corrections

One way of looking at a system of electrons under the influence of an external potential is to consider that as a first approximation it may be described as if it were a free-electron gas and then to correct it by adding terms in order to introduce the physical effects which are absent in this initial description. To each electron in a free electron gas one associates a plane wave defined in a cubic box

$$\phi_{\vec{k}}(\vec{r}) = \frac{1}{\sqrt{V}} \exp(i\vec{k}.\vec{r}) \tag{1}$$

and then transforms the discrete summation over all states in this box having a momentum lower or equal to the momentum of the Fermi level F, into an integral. Working on the details of this procedure, which is well known, and which is almost as old as quantum mechanics, one obtains the Thomas-Fermi term for the kinetic energy density. This term is given as a function of the charge density $\rho(\vec{r})$ by the equation

$$t_{TF}\ [\rho] = \frac{3}{10}\ (3\pi^2)^{2/3}\ \rho(\vec{r})^{5/3} \tag{2}$$

Strictly speaking, there is a logical jump in going from the correct expression for a free electron gas

$$t_{TF}\ [\rho] = \frac{3}{10}\ \rho(\vec{r})k^2{}_F \tag{3}$$

to Eq (1). The difficulty may be readily appreciated by noticing that this identification is not rigorously possible because in the definition of the momentum at the Fermi level

$$k_F = (3\pi^2\rho)^{1/3} \tag{4}$$

ρ is a constant. Consequently, since ρ in this case is not a function of $\vec{r}$, one may not assume that Eq (1), where ρ is indeed a function of $\vec{r}$, is valid for finite systems (30). The Thomas-Fermi term, however, has been customarily considered to be the leading term of an expansion, where the remaining terms introduce finite size effects. But as it was mentioned in the introduction, the Thomas-Fermi theory may be regarded as one of the corner-stones of physics since it is exact for the case when Z goes to infinity (10). Also, let us mention that this theory was fundamental for proving from first principles the stability of matter (31).

The first "correction" to the Thomas-Fermi term is the Weizsacker term $t_w\ [\rho]$ given by (11)

$$t_w\ [\rho] = \frac{1}{8}\ \frac{\nabla\rho(\vec{r})\ \nabla\rho(\vec{r})}{\rho(\vec{r})} = \frac{1}{2}\ (\nabla[\rho(\vec{r})]^{1/2})^2 \tag{5}$$

This correction arises when modified plane waves of the type

$$\phi_k = (\vec{r}) = \frac{1}{\sqrt{V}}\ (1+ar)\exp(i\vec{k}\cdot\vec{r}) \tag{6}$$

are used.

Higher order corrections have been obtained by Kirzhnits (13) and by Hodges (15). Their treatment lays the foundation for the gradient expansion of the kinetic energy density. The convergence of the gradient expansion has been discussed by Hohenberg and Kohn (32) who also have pointed out its intrinsic limitation for properly describing the radial oscillations of the electronic density in an atom which are responsible for the electronic shell structure. From considerations concerning invariance they have been able to explicitly give the energy density functional up to fourth order.

The general form of the gradient expansion for the kinetic density is given by (21)

$$t_G\,|\rho| = \sum_{k=o}^{oo} \sum_{\ell=o}^{k} t_{2K}^{2k-\ell}(\rho)(\nabla\rho\,\nabla\rho)^{k-\ell}(\nabla^2\rho)^{\ell} \tag{7}$$

In this expression, the subscript G on the kinetic energy density simply expresses the fact that $t_G|\rho|$ need not be the exact one, which we denote by $t_E|\rho|$. It has been shown by Tal and Bader (21) that in order for the virial theorem to be satisfied, the coefficients must have the form

$$t_{2k}^{2k-\ell}\,[\rho] = a_{k\ell}\,\rho^{(5+3\ell-8k)/3} \tag{8}$$

and thus, the gradient expansion becomes

$$t_G[\rho] = \sum_k \sum_\ell a_{k\ell}\,\rho^{(5-2k)/3}\left(\frac{\nabla\rho\,\nabla\rho}{\rho^2}\right)^{k-\ell}\left(\frac{\nabla^2\rho}{\rho}\right)^{\ell} \tag{9}$$

Explicitly writing out the first few terms, we have

$$t_G[\rho] = a_{oo}\,\ell^{5/3} + a_{10}\left(\frac{\nabla\rho\,\nabla\rho}{\rho}\right) + a_{11}\nabla^2\rho$$

$$+ \rho^{1/3}[a_{20}(\nabla\rho\,\nabla\rho/\rho^2)^2 + a_{21}(\nabla\rho\,\nabla\rho/\rho^2)(\nabla^2\rho/\rho)$$

$$+ a_{22}(\nabla^2\rho/\rho)^2] + \ldots \tag{10}$$

where one can easily recognize the Thomas-Fermi term, the Weizsacker term and the higher order quantum corrections (13-15).

For systems of Fermions interacting through Coulomb forces, when the asymptotic behavior at infinity is considered, it has also been shown by Tal and Bader (21) that the only terms which do not diverge are $t_{TF}[\rho]$ and $t_W|\rho|$. This is due to the fact that when r approaches infinity $\rho(r)$ has the form (33)

$$\rho(r) \underset{r\to\infty}{=} r^{\beta}\,e^{-\alpha r} \tag{11}$$

or more generally (34)

$$\rho(r) = (1+r)^{\beta} e^{-\alpha r}$$

$$r \to \infty \qquad (12)$$

It follows from these expressions that all terms in the gradient expansion for $k > 2$ will result in negative powers of ρ and will therefore diverge. Furthermore, when the variation of $t_G|\rho|$ with respect to ρ is carried out, the ensuing Euler equation will have, in view of the asymptotic behavior of $\rho(r)$, divergent terms for $k > 1$. Consequently, the only terms that satisfy this asymptotic behavior are those for $I=o$ and $k=1$, namely, the Thomas-Fermi and the Weizsacker terms, so that $t_G|\rho|$ simplifies to

$$t_G [\rho] = t_{TF} [\rho] + t_w [\rho] \qquad (13)$$

It may be easily shown, however, that $t_G|\rho| \neq t_E|\rho|$, where $t_E|\rho|$ is the exact kinetic energy density (21).

From the above considerations, it is clear that the gradient expansion for a system of Fermions interacting through Coulomb forces, is far from being adequate. The main reason for this inadequacy stems from the fact that all terms other than t_{TF} and t_w diverge when the asymptotic behavior of ρ at large r is taken into account and that these two terms alone do not lead to the correct representation of the kinetic energy density. This inadequacy was already forseen in the work of Hohenberg and Kohn (32) when it was noticed that the gradient expansion does not strictly converge for finite values of r.

We may conclude from this situation that it is not sufficient to assume a given gradient expansion and hope for it to converge to the correct result. This, for Coulomb systems is not the case and for any other applications of this expansion such as, for example, nuclear matter, it should be imperative to analyze the specific conditions which validate its use. A particularly critical aspect is the determination of the asymptotic behavior of the density for non-coulombic interactions . As an example of a work in this direction let us mention the conditions obtained by Merkur'ev (35) for short range interactions.

B. The Weizsacker Term and its Correlation Corrections

Let us consider a pure state described by the n-particle wavefunction Ψ (1...n) or, equivalently, by its nth-order density operator Γ^n,

$$\Gamma^n (1,...n; 1'...n') = |\Psi (1...n)\rangle \langle\Psi (1',...n')| \qquad (14)$$

The reduced first order density operator $\gamma(1,1')$ is obtained by integrating over coordinates 2...n on Γ^n:

$$\gamma(1,1') = n \int d2... \int dn\ \Gamma^n(1,2, ...n; 1',2, ...n) \qquad (15)$$

In an orbital representation

$$\gamma(1,1') = \sum_i \sum_j \Gamma_{ij}\, \phi_i^*(1)\, \phi_j(1') = \vec{\phi}^{*T}(1)\, \Gamma\vec{\phi}(1') \quad (16)$$

The density $\rho(1)$ is defined by

$$\rho(1) = \gamma(1,1) \quad (17)$$

Clearly, $\gamma(1,1')$ is a non-local operator, whereas $\rho(1)$ is a local one. Defining

$$\gamma(1,1') = \rho(1)^{1/2}\, \rho(1')^{1/2}\, G(1,1') \quad (18)$$

it follows that

$$G(1,1') = |1 + B(1,1')/\rho(1)\rho(1')|^{1/2} \quad (19)$$

where

$$B(1,1') = \vec{\phi}^{*T}(1)\, \Gamma d(1,1')\Gamma\, \vec{\phi}^{*}(1') \quad (20)$$

and where $d(1,1')$ is a matrix defined by

$$d(1,1') = \vec{\phi}(1')\, \vec{\phi}^{T}(1) - \vec{\phi}(1)\, \vec{\phi}^{T}(1') \quad (21)$$

Let us notice that when $1'=1$, $d(1,1) = 0$ and hence, $B(1,1) = 0$. This property reflects the fact that $B(1,1')/\rho(1)\rho(1')$ is closely related to the statistical correlation factor which describes how the presence of a particle (with a given spin) at position 1, affects the probability of finding another particle at position 1'. In view of Eq.(21), $G(1,1')$ describes, therefore, this type of correlation, or equivalently, this non-local effect.

The importance of writing $\gamma(1,1')$ in terms of local and non-local components becomes evident when we write the exact expression for the kinetic energy density

$$t_E|\gamma| = (1/2)\, \nabla_1 \nabla_{1'}\, \gamma(1,1')|_{1'\to 1} \quad (22)$$

(Notice that it ia function of γ and not of ρ). Substituting Eq.(18) into Eq.(22), we obtain

$$t_E|\gamma| = (1/8)\, \frac{\nabla\rho(1)\, \nabla\rho\,(1)}{\rho(1)} + (1/2)\, \rho(1)\, \nabla_1 \nabla_{1'}\, G(1,1')|_{1'\to 1} \quad (23)$$

We recognize the first term as the Weizsacker term and we see that it is the local component of the kinetic energy density. The second term contains the correlation function $G(1,1')$ and thus incorporates the non-local effects. This second term has also been obtained by Sears,Parr and Dinur (23), using information theory. It has been shown to correspond to the average of the one-particle density of Fisher's information associated with the conditional density. Also, in terms of a natural orbital representation, this second term may be exactly expressed as (21)

$$t_E - t_W = (1/8)\, \rho(1) \sum_i \sum_i (\lambda_i \lambda_j/\rho_i \rho_j)|\, \rho_j(1)\nabla\rho_i(1) - \rho_i(1)\nabla\rho_j(1)|^2 \quad (24)$$

where the λ_i's are the occupation numbers and

$$\rho_i(1) = \chi_i^*(1)\chi_i(1) \tag{25}$$

and where the functions χ_i are the natural spin orbitals of the system.

Equation (23) is exact, presents an alternative to the gradient expansion, and allows for the gradual incorporation of the physical correlation effects. For example, if one uses for G(1,1') the correlation function for a free electron gas

$$G(1,1') = 3\ (\sin k_F r - k_F\ r \cos k_F r)/(k_F r)^3 \tag{26}$$

where r is the modulus of $\vec{r}_1 - \vec{r}_2$, one obtains

$$t = t_W + (3/10)\ \rho(1)\ k_F^2 \tag{27}$$

The second term is the Thomas -Fermi term given by Eq.(3).

C. Other Approximations to the Non-local Correlation Function

In an orbital basis, the non-local correlation function G(1,2), for the case of a single Slater determinant representation of the wavefunction is given by (26):

$$G(1,2) = \left[\frac{\sum\limits_{i=1}\sum\limits_{j=1} \phi_i^*(1)\,\phi_j(2)\,\phi_i(2)\,\phi_j^*(1)}{\left(\sum\limits_{i=1}\phi_i^*(1)\,\phi_i(1)\right)\left(\sum\limits_{j=1}\phi_j^*(2)\,\phi_j(2)\right)}\right]^{1/2} \tag{28}$$

An explicit evaluation of G(1,2) can be carried out by using the orthonormal orbital set introduced by Harriman (36):

$$\phi_k(1) = |\rho_1(1)|^{1/2} \exp(i(k\ f(r) + g(r)) \tag{29}$$

where

$$f(r) = 2\pi\int_0^r dr'(r')^2\ \bar{\rho}_1(r') \tag{30}$$

$$\bar{\rho}_1(r) = \int_0^\pi d\theta\ \sin\theta \int_0^{2\pi} d\phi\, \rho_1(r,\theta,\phi) \tag{31}$$

and where g(r) is a phase factor to be determined by minimizing the kinetic energy(Ref. (24)). Explicitly, this phase factor is

$$g(r) = -\ ((n+1)/2)f(r) \tag{32}$$

so that the corrected Harriman's orbitals become

$$\phi_k(1) = \left[\rho_1(1)\right]^{1/2} \exp\left[i\left(k - \frac{n+1}{2}\right) f(r)\right] \tag{33}$$

Notice that in Eq (33), $\rho_1(1)$ is normalized to unity, so that the relation to the usual ρ (1) is

$$\rho_1(1) = \frac{1}{n} \rho(1) \tag{34}$$

Substituting Eq (33) into Eq (28), we obtain (cf. Eqs (23) to (27) of Ref. (26))

$$G(1,2) = \frac{1}{n} e^{-i\left(\frac{n+1}{2}\right) F(r_1,)} \sum_k e^{ik F(r_1,r_2)} \tag{35}$$

where $F(r_1,r_2) = f(r_2) - f(r_1)$. This yields the non-local correlation function

$$G(1,2) = \sin \frac{n}{2} F \,/\, n \sin \frac{F}{2} \tag{36}$$

and the kinetic energy term

$$t = tw + \frac{8}{3}\pi^4 \left(1 - \frac{1}{n^2}\right) r^4 \rho^3(1) \tag{37}$$

This equation corrects the one previously derived in Ref (26), where the phase factor was not included. Equations similar to Eq (37) have been derived by Percus (25) a few years ago, and more recently by Ghosh and Parr (27).

Zumbach and Maschke (24) have introduced a complete set of orbitals which are three dimensional generalizations of Harriman's orbitals. They are defined by

$$\Phi_{\vec{k}}(\vec{r}) = \left[\rho(r)/n\right]^{1/2} \exp i\left[(\vec{k}.\vec{R}(\vec{r}) + \theta\,)\right] \tag{38}$$

where $\vec{k}$ is a number vector (k_1,k_2,k_3) and $\vec{R}(\vec{r}) = R_1(\vec{r}),R_2(\vec{r}),R_3(\vec{r})$. These functions are obtained by partial integration over spatial coordinates.

The kinetic energy expression corresponding to Zumbach and Maschke's orbitals is

$$t = t_W + \frac{1}{n} \rho(1) \left[\sum_{\vec{k}} \left[\nabla(\vec{k}.\vec{R})\right]^2 - \frac{1}{n}\left[\sum_{\vec{k}} \nabla(\vec{k}.\vec{R})\right]^2\right] \tag{39}$$

It has been pointed out, however, that this expression does not lead to an improved kinetic energy functional as there appear singularities (27).

In view of these difficulties, another possible generalization of Harriman's orbitals has been discussed by Ghosh and Parr (27). This may be accomplished by introducing the angular part by means of spherical harmonics

$$\phi_{k,\ell,m}(r,\theta,\phi) = \rho(r)/n^{1/2} g_\ell(r) \exp i2\pi k f_\ell(r) \; Y_\ell^m(\theta,\phi) \tag{40}$$

The function $g_1(r)$ has been included so that the singularity in the kinetic energy corresponding to the term $\ell(\ell+1)/r^2$, be removed. The new kinetic energy density is given by

$$t = t_W + \sum_\ell (\lambda_\ell/n) \; (1/2)\rho \, \nabla g_\ell^{\,2} + (1/2)\nabla\rho . \, g_\ell \nabla g_\ell + \frac{\ell(\ell+1)}{2} \frac{g_\ell^{\,2}}{r^2} \rho + \sum_\ell \frac{32\,\pi^4}{n^3} \sum_{k,1} \lambda_k^\ell k^2 \; g_\ell^{\,6} r^4 \rho^3 \tag{41}$$

where λ_k^ℓ is the occupation number of the kth-state for a particular ℓ value. As reported by Ghosh and Parr (27), the results obtained using Eq.(41) lead toa significant improvement on previous formulations, but, however, they do not reproduce the shell structure in atoms. Thus, the way to properly describe the shell structure of atoms by density functional theory, still remains as an open question.

III.- VIRIAL PARTITIONING FOR APPROXIMATE FUNCTIONALS

We discuss here what we consider to be an important property of matter, namely, the partitioning of R^3 into subvolumes separated by zero flux surfaces. Within each one of these subvolumes, the virial theorem is satisfied. For a collection of atoms, forming a molecule or solid, these subvolumes correspond to regions of R^3, which represent distorted atoms (or quasi-atoms). Thus, a molecule or a solid, many be viewed as a collection of these quasi-atoms confined to virial fragments.

This property arises from the basic quantum mechanics and it involves no approximation (29). Therefore, it would be desirable, to develop approximate theories that satisfy this virial partitioning. In what follows, we show that the functional given by Ghosh and Parr, satisfies this condition for a particular, but reasonable choice of $g_1(r)$.

As it has shown by Bader (29), the zero flux condition which defines the surfaces $S(\Omega)$ of a subvolume Ω is

$$\vec{\nabla}\rho(r).\vec{n} = 0 \qquad r \in S(\Omega) \tag{42}$$

It is not surprising that in this virial partitioning the density $\rho(r)$ also appears as a basic quantity.

Assuming, in the context of the Hohenberg-Kohn theory the variational functional for the energy to be given by

$$H[\rho] = E[\rho] - \mu N[\rho] = \int (\upsilon(\vec{r}) - \mu\,\rho(\vec{r}))\, d\vec{r} \tag{43}$$

where μ is the lagrange multiplier which incorporates the normalization condition, we have shown elsewhere (28,37) that the variation δH leads in general to

$$\delta H[\rho,\Omega] = \oint_{s(\Omega)} \eta(r)(\nabla_\rho \upsilon.\vec{n})\, ds + \oint_{s(\Omega)} ds \; \sum_x \Big[\frac{\partial\upsilon}{\partial\rho_{xx}} \eta_x - \eta \frac{\partial}{\partial x}\left(\frac{\partial\upsilon}{\partial\rho_{xx}}\right)\Big] n_x \tag{44}$$

Since in the first surface integral, ∇_ρ is a gradient operator with respect to ρ, it is easily seen that because the correction terms to the Weizsacker term may be expressed as polynomials in ρ, then they do not contribute to the first surface integral. Clearly, in the case of the functional described by Eq (41), this occurs for the choice of

$$g_1(r) = \sum_k a_{k,1} f(r)^{b_1} \tag{45}$$

The natural boundary condition coming from the Weizsacker term in the first surface integral is then, precisely, the condition given by Eq.(42). In conclusion, for this approximate functional, the virial partitioning of R^3 into subvolumes, also holds.

IV. DISCUSSION

We have emphasized in the present work, some recent developments in density functional theory, which bypassing the gradient expansion open new possibilities for the introduction of the non-local terms in the kinetic energy density. At the same time, we have reviewed some of the shortcomings of the gradient expansion which render it inadequate for the treatment of Fermion systems interacting through Coulomb forces.

The new picture has the advantage of differentiating between local and non-local effects and presents, therefore, a very convenient framework for the gradual incorporation of correlation effects.

Finally, we have shown that these approximate functionals satisfy the virial partitioning condition, which was proven by Bader to be a fundamental property of quantum systems. We hope that by emphasizing these aspects, which have been developed mainly in the context of molecular physics, some important cross-fertilization may be achieved in domains such as condensed matter theory, where the incorporation of these new developments might offer interesting possibilities for research.

REFERENCES

1. J. Callaway and N.H.March, Solid State Phys. 38,135 (1983).
2. J. Keller and J.L. Gazquez, Eds.,"Density Functional Theory", Springer Verlag, Berlin (1983).
3. J.P. Dahl and J. Avery, Eds., "Local Density Approsimations in Quantum Chemistry and Solid State Physics". Plenum, New York (1984).
4. R. Dreizler and J. da Providencia, Eds., Proceedings of the NATO Advanced Study Institute Summer School on Density Functional Methods in Physics, Alcabideche, Portugal, 1983, Plenum, New York (1985).
5. L.H. Thomas, Proc. Cambridge Philos. Soc. 23,542 (1927).
6. E. Fermi, Rend. Accad. Naz. Lincei 6,602 (1927).
7. N.H. March, Adv. in Phys. 6,1 (1957).
8. P. Gombas, "Die Statistischen Theorie des Atomes und ihre Anwendungen", Springer Verlag, Berlin (1949).
9. I.M. Torrens "Interatomic Potentials", Academic, New York (1972).
10. E.H. Lieb, Revs. Modern Phys. 53,603 (1981).
11. C.F, von Weizsäcker, Z. Phys. 96,431 (1935).
12. P.A.M. Dirac, Proc. Cambridge Philos. Soc. 26,376 (1930).
13. D.A. Kirzhnits, Zh. Eksp. Teor. Fiz 32,115 (1975). Sov. Phys. JETP 4, 328 (1957).

14. A.S. Kompaneets and E.S. Pavlovskii, Zh. Eksp. Teor. Fiz 31,427 (1956).
15. C.H. Hodges, Can. J. Phys. 51,1428 (1973).
16. E.C. Stoner, Phil. Mag. 28,257 (1939).
17. T. Sakai, Proc. Phys. Math. Soc. Japan, 24,254 (1942).
18. R.P. Feynman, N. Metropolis and E. Teller, Phys. Rev. 75,1561 (1940).
19. N.D. Mermin, Phys. Rev. A 137,1441 (1965).
20. M. Brack, Phys. Rev. Lett. 53,119 (1984); 54,851 (1984).
21. Y. Tal and R.F.W. Bader, Int. J. Quantum Chem. Symp. 12,153 (1978); see also, L. Szasz and I. Berrios, Z. Naturforsch. 30a,1516 (1975).
22. E.V. Ludeña, J. Chem. Phys. 76,3157 (1982).
23. S.B. Sears, R.G. Parr and U. Dimur, Israel J. Chem. 19,165 (1980).
24. G. Zumbach and K. Maschke, Phys. Rev.A28,544 (1983); 29,1585 (1984).
25. J.K. Percus, Int. J. Quantum Chem. 13, 89 (1978).
26. E.V. Ludeña, J. Chem. Phys. 79,6174 (1983); see also: E.V. Ludeña, "Recent Progress in Many Body Theories", H. Kümmel and M.L. Ristig, Eds, Springer-Verlag, Berlin (1984), p 370-376.
27. S.K. Ghosh and R.G. Parr, J. Chem. Phys. 82,3307 (1985).
28. E.V. Ludeña and V. Mujica, Int. J. Quantum Chem. 21,927 (1982).
29. R.F.W. Bader, T.T. Nguyen-Dang and Y. Tal, Rep. Prog. Phys. 44,893 (1981) and references therein.
30. E.V. Ludeña, Int. J. Quantum Chem. 23,127 (1983).
31. E.H. Lieb, Revs. Modern Phys. 48,553 (1976).
32. P. Hohenberg and W. Kohn, Phys. Rev. B 136,864 (1964).
33. M. Hoffmann-Ostenhof and T. Hoffmann-Ostenhof, Phys. Rev. A 16,1782 (1977).
34. R. Ahlrichs, M. Hoffmann-Ostenhof, T.Hoffmann-Ostenhof and J.D. Morgan III, Phys. Rev. A 23, 2106 (1981).
35. S.P. Merkur'ev, Sov. J. Nucl. Phys. 19,222 (1974).
36. J.E. Harriman, Phys. Rev. A 24,680 (1981).
37. E.V. Ludeña, in J.P. Dahl and J. Avery, Eds. "Local Density Approximations in Quantum Chemistry and Solid State Physics" Plenum, New York (1984), p 287-301.

LOCAL APPROXIMATIONS IN THE APPLICATIONS OF DENSITY FUNCTIONAL THEORY

Jaime Keller, Carlos Amador and Carmen de Teresa

División de Estudios de Posgrado, Fac. de Química
Universidad Nacional Autónoma de México,
Ciudad Universitaria, Delegación Coyoacán
04510 México, D.F., México

INTRODUCTION

We present two examples of the use of local approximations in density functionals (DF) theory: 1) the estimation of the ionization potentials and spectroscopic terms for atoms and 2) the calculation of spin susceptibility for atoms in metals, and use them to review some current trends in the applications of DF. In the first case we stress that symmetry and statistics should be built into the basic formulation of the computational schemes in order to obtain realistic results. In the second example we show that DF allows the treatment of small regions of a material as a system in itself, provided the proper boundary conditions are considered for the density function and for the auxiliary Kohn-Sham wave functions. Some numerical results are given for several materials.

DENSITY FUNCTIONAL THEORY

We should remember that besides the theoretical importance of DF, the main objective of the method is to obtain the ground state energy and the electronic density of a many-body (electrons in our examples) system in a self consistent way. The discussion of real problems is made from the energy parameters of the calculations, some auxiliary wave functions obtained in the Kohn-Sham (KS) scheme and the theoretical relationship between these functions and parameters to the actual experimental properties.

For the purpose of the first part of our presentation, we should remember the basic theorems of Hohenberg & Kohn[1] and of Kohn-Sham[2] stating that an external potential v determines the density ρ and as a consequence all the properties. The starting point is the assumption that some functionals $E_k[\rho]$, $V_{ee}[\rho]$ and $E[\rho]$ exist and that there is a stationary principle.

$$\delta\{E[\rho] - \mu[N[\rho] - N]\} = 0 \qquad (1)$$

where E_k is the kinetic energy, V_{ee} the electron-electron interaction potential energy and E the total energy functionals and μ the chemical potential obeying an Euler equation

$$\mu = \delta E/\delta\rho = v + \delta F/\delta\rho \qquad (2)$$

with $F[\rho] = E_k[\rho] + V_{ee}[\rho]$, in such a way that in changing from one state of the system to another we have a change in the energy

$$\delta E = \mu \delta N + \int \rho(1) \delta v(1) dv_1 \qquad (3)$$

which identifies the chemical potential as

$$\mu = (\delta E/\delta N)_v \qquad (4)$$

The Kohn-Sham theory introduces a set of auxiliary "wave functions" ϕ_k

$$\rho = \sum_k^N |\phi_k|^2 n_k \qquad (5)$$

to construct the exact density, with occupation numbers n_k used as constrains to minimize the total energy, leading to the KS equations

$$\hat{F}_{KS}\phi_k = \varepsilon_k^{KS}\phi_k \qquad (6)$$

with the effective Hamiltonian (ϕ is the electron-electron coulomb potential)

$$\hat{H}_{KS} = -(1/2)\nabla^2 + v + \phi + (\delta E_{xc}/\delta\rho) \qquad (7)$$

"VERTICAL" VERSUS "HORIZONTAL" COMPUTATIONAL SCHEMES IN DENSITY FUNCTIONALS THEORY.

Since the original derivation of the minimum principle in the H-K theory and the computational procedure, which we will call "vertical" in energy, of Kohn and Sham, where the concept of chemical potential is formally introduced, the standard calculation procedure selects a set {k} of elementary excitations of given symmetries k (s,p,d,f,... for atoms, Σ,π,Δ,... for diatomic molecules, etc., usually called atomic or molecular orbitals or one-electron states) and their occupation n_k to find an approximate ground state (sometimes an excited state) of the system, using the equations (5)-(7) of the previous paragraph. In this way a shell structure of the system is introduced and a practical labelling of the configuration is obtained as the configuration corresponding to a set of occupation numbers $\{n_k\}$ similar to the usual atomic or molecular orbitals self consistent field (SCF) procedure. For example we say that ground state Be is $1s^2 2s^2$, and an excited Be is $1s^2 2s^1 2p^1$. Nevertheless a procedure corresponding to the configuration interaction scheme (CI) for correlation would be partially missing, although the fact that the energy functional itself is constructed from an all electron density allows, either local or non local, approximations for exchange and correlation and the use of perturbation theory approaches for multiplets and magnetic interactions.

An "horizontal" computational scheme has been recently proposed by Levy, Perdue and Sanhi (L-P-S)[3] using only one auxiliary function: the square root of the total electronic density and one Lagrange multiplier, the ionization potential IP = $-\mu$.

It could be thought that the L-P-S scheme is different from the K-S scheme, but it is not and in fact it is one, although most important, particular case, where the auxiliar energy eigenvalue ε_k corresponds to μ, within the limit of the accuracy of the energy functional used.

In Thomas-Fermi (TF) theory, another "horizontal" scheme, the total electronic density can be obtained, not from the TF integral equation, but from the optimization of a model of the electron density, recovering in this way the possibility of a shell structure, but loosing all the information obtained from the one-electron atomic or molecular orbitals or energy bands of the solids.

In the following we will present, through a very simple modification of the K-S theory, a scheme which is at the same time vertical and horizontal in energy, where the orbitals and their eigenvalues are obtained and also the energy density of elementary excitations or quasiparticles of the system, through a collection of ionization potentials for different configurations (as defined above). In any case, the total energy of the system is obtained for each configuration as a byproduct of the calculational procedure. Then we will obtain the usual orbitals or bands and at the same time the ionization potentials from which the spectra are determined exactly. In the K-S procedure the spectra are either approximated from the one particle energy eigenvalues or from some ad hoc "transition state" calculations[4].

The procedure we will follow in this paper is to use the K-S theory in two, complementary, steps. The first is the standard one where the auxiliary functions and the corresponding energy eigenvalues (bands) are obtained from the interpretation of the total density as a sum of (vertical) states with occupation numbers n_i used as variational parameters[5].

In this case, the energy eigenvalues ε_i are only an approximation to the ionization potentials μ_i because

$$\mu_i = \rho \left.\frac{\partial E}{\partial n_i}\right|_{\{n_k\},v} + \frac{1}{2}\,\frac{\partial^2 E}{\partial n_i^2} + \ldots = \varepsilon_i + \Delta\mu_i\ , \tag{8}$$

usually the relaxation given by the second term of the right-hand side is either neglected or at best approximated by a transition state procedure.

In the second step, the density is rewritten in the form

$$\rho(r) = \rho^{*\frac{1}{2}}(r,\{n_k'\})\,\rho^{\frac{1}{2}}(r,\{n_k'\})\,n_t \tag{9}$$

Keeping the $\{n_k'\}$ as a <u>fixed</u> set of occupations corresponding to a proposed configuration, hence to an assumed symmetry of the density function. And, at the same time, the parameter n_t is the variational parameter for a change in the total number of electrons. Then the minimization of the total energy with respect to this parameter will, through the K-S analysis, define the total chemical potential of the system which is exact to the accuracy of the density functional $E[\rho]$ used.

It may seem strange that the square root of the density can be written from the auxiliary functions in a simple way but the use of a mathematical trick, introducing hypercomplex numbers α_j , with the following properties:

$$\alpha_i\alpha_j = -\,\alpha_j\alpha_i\ ;\ \alpha_i^2 = -1 \tag{10}$$

allows the procedure of taking the square root as:

$$\rho = \sum_i \phi_i^2 = \Big(\sum_j \alpha_j^{*}\phi_j^{*}\Big)\Big(\sum_k \alpha_k\phi_k\Big) \tag{11}$$

and the K-S theory to be used in the way stressed by L-P-S, where now the effective hamiltonian will include the hypercomplex numbers in such a way that the kinetic energy will be a sum over states of the radial and angular parts.

Formally we should write the total electronic energy functional as a sum over components

$$E[\rho] = E_K[\rho] + E_{ne}[\rho] + E_{ee}[\rho] \tag{12}$$

in the Kohn-Sham form

$$E[\rho] = E_k^{KS}[\rho] + E_{ne}[\rho] + E_{ee,c}[\rho] + E_{xc}[\rho] \tag{13}$$

where the different contributions, through the introduction of a functional form for the density

$$\rho = N n \sum_i \frac{n_i}{N} \rho_i = N^{\frac{1}{2}} \sum_i \frac{n_i^{\frac{1}{2}}}{N^{\frac{1}{2}}} \psi_i^* \alpha_i^* \; N^{\frac{1}{2}} \sum_j \frac{n_j^{\frac{1}{2}}}{N^{\frac{1}{2}}} \psi_j \alpha_j \tag{14}$$

and the Kohn-Sham definition of the (independent particle) kinetic energy

$$E_{KS} \equiv \sum_i <\psi_i| - \tfrac{1}{2}\nabla^2 |\psi_i> n_i = N n \sum_i <\psi_i| - \tfrac{1}{2}\nabla^2 |\psi_i> \frac{n_i}{N} = N e_{KS} \,, \tag{15}$$

the electron nuclear potential energy

$$E_{en} = \sum_i n_i \int \psi_i^2 v_n d\tau = N n \sum_i <\psi_i| v_n |\psi_i> \frac{n_i}{N} = N e_{en} \,, \tag{16}$$

the electron-electron coulomb part of the electronic interaction

$$E_{ee,c} = \sum_i n_i <\psi_i| \frac{1}{2} v_{ee,v} |\psi_i> = N^2 n^2 \sum_i <\psi_i| \frac{1}{2} v_{ee,c} |\psi_i> \frac{n_i}{N} = N^2 e_{ee,c} \tag{17}$$

and, finally the main term of the exchange correlation energy defined as

$$E_{xc}^o = \sum_i n_i <\psi_i| \frac{4}{3} v_{xc}^o |\psi_i> = N^{4/3} n^{4/3} \sum_i <\psi_i| \frac{4}{3} v_{xc}^o |\psi_i> \frac{n_i}{N} = N^{4/3} e_{xc}^o \tag{18}$$

which, comparing with (1) and (2) contains all the terms in (1) arising from the difference between independent particle approximations and the actual exchange correlation terms, written in a local approximation symbolic form, proportional to $\rho^{4/3}$ as it should from dimensional analysis.

In formulae (14-18) we have introduced two parameters: N as the total number of electrons (per representative system) and an occupation number $0 < n < 1$ which will be used in the "horizontal" scheme. The usual Kohn-Sham scheme is obtained if the "vertical" orbital occupation numbers n_i are varied.

For the horizontal procedure we prefer to write the total energy

$$E_T = N\left(e_{KS} + e_{en} + N e_{ee} + N^{1/3} e_{xc}^o\right) = N e_T \tag{19}$$

in such a way that the chemical potential μ is obtained from a variation of a symmetry constrained ground state:

$$\left(\frac{\partial e_T}{\partial n} \right)_{v_n} = \mu \qquad (20)$$

where we will obtain an eigenvalue equation for μ

$$\hat{H}_n \psi_n = \mu \psi_n \qquad (21)$$

with the effective hamiltonian $\hat{H}_n$ given by

$$N\hat{H}_n = \sum_i - \frac{A_i}{2} \nabla_i^2 \alpha_i + \mathcal{C} v_n + N^2 v_{ee} + N^{4/3} \frac{4}{3} v_{xc}^o \qquad (22)$$

But for practical computational procedure we will obtain the auxiliary function ψ_n from the vertical calculation. Then we will use the expression

$$N\mu = \langle \psi_n | \hat{H}_n | \psi_n \rangle = \frac{5}{3} E_k^{TF} + \frac{4}{3} E_{en} \mathcal{C}' + 2E_{ee,c} + \frac{4}{3} E_{xc}^o + \Delta E' \qquad (23)$$

to compute the exact ionization energy of the many particle system. It is to be noted that in (23) all terms are proportional, through universal numerical constants, to the standard contributions to the total energy, which are always computed, and usually printed out in all self-consistent calculation procedures. Moreover the different terms can be given independent adjustments to make them internally consistent using the Virial theorem and similar universal relationships.

In the following we present two types of examples for these calculations. First an analytical auxiliary function ψ_n is constructed as a sum of Slater type orbitals and equations (21)-(23) solved by adjusting the different parameters of Slater orbitals to minimize the total energy (19). In the second type of calculations, a standard Kohn-Sham (relativistic) program for atomic calculations is used to obtain the auxiliary functions ϕ_i and the contributions to the total energy which are then used to compute the ionization potentials, for some set of occupation parameters n_i, using the r.h.s. of (23), where E_k^{TF} is the Thomas-Fermi kinetic energy and $\Delta E'$ the main higher order corrections. The constant A_i in (22) is obtained from dimentional analysis consideration of the variation of each orbital ψ_i parameters if the volume is locally expanded.

Table 1. Computed values of contributions to the total energy and the negative of the ionization potentials for atoms of atomic number Z.

Z	E_k^{TF}	$E_{ee,c}$	E_{xc}^o	E_{en}	μ	μ(exp)
2	+5.121	+2.574	-2.341	-13.711	-.805	-.904
3	+13.359	+8.202	-3.797	-34.578	-.162	-.198
4	+26.256	+14.407	-5.385	-67.467	-.259	-.343
5	+43.930	+23.258	-7.429	-113.957	-.413	-.305
6	+67.296	+35.704	-9.967	-176.459	-.515	-.414
12	+368.000	+191.789	-31.051	-958.314	-.267	-.281
11	+297.560	+160.105	-27.276	-779.625	-.177	-.189
13	+446.740	+225.836	-35.032	-1157.207	-.295	-.220
15	+631.080	+307.169	-43.896	-1624.780	-.572	-.404
16	+737.220	+354.173	-48.424	-1894.190	-.378	-.381
17	+853.360	+406.290	-53.283	-2189.206	-.447	-.478

SPIN SUSCEPTIBILITY AND MOMENT FORMATION IN METALLIC SOLIDS

In this second example of the use of local approximations, linear response theory and the density functional techniques are used to obtain the basic expressions for the static spin susceptibility χ_s of metallic materials as a local property $\chi_s(r)$[6].

The basic idea of several types of condensed matter calculations is that the electronic structure in a region of a material can be computed using a self consistent potential and the appropriate boundary conditions for the wave functions. This is also the idea of cluster methods for a real space approach or of the band structure methods in the reciprocal space treatment. The spin susceptibility as well as other properties are then computed per atom, per formula unit or per cell. The extension discussed here consists in considering small regions of space, actually reduced to a point, as those for which the spin susceptibility is computed. We should then find which ones will present a larger tendency to acquire a magnetic moment or even a magnetic instability leading to a permanent moment behavior. The numerical analysis of the consequences of these ideas, after computing the electonic structure of some materials within the spin restricted method, using cluster multiple scattering techniques, we analyze[6] the results for $\chi_s(r)$ of some examples to show that a paramagnetic material (Nb) will present no regions of instability, a localized moment ferromagnetic material (bcc-Fe) shows where localized moment will develop and an itinerant ferromagnet ($TiBe_2$), where the difuse regions of magnetic instability are to be expected.

All these results are obtained within the linear response theory, the actual magnetization of the material cannot, of course, be accurately computed without a fully selfconsistent treatment.

The basic formulae used for the calculation of the local spin susceptibility are:

$$\chi_s(r) = 2\mu_B^2\, \gamma(r)/(1-\gamma(r)I(r))$$

where the density $\gamma(r)$ of electrons with energy equal to the Fermi level energy E_f is

$$\int \gamma(\underline{r})\,d\underline{r} = N(E_f), \quad \gamma(r) = \sum_{\ell m} \gamma_{\ell m}(r)$$

$N(E_f)$ is the electron density of states at E_f, μ_B the Bohr magneton and the enhancement factor $(1-\gamma(r)I(r))^{-1}$ contains the local, Curie-Weiss, exchange field

$$I(r) = \sum_{11'} \gamma_1(r)\gamma_{1'}(r)\,|K(r)|/\gamma^2(r)$$

computed from $|K(r)|$, the second derivative with respect to the local magnetization $m(r)$ of the electronic exchange-correlation potential $E_{xc}(r)$, as $K(r) = \partial^2 E_{xc}/\partial m(r)^2$

The approach followed here is parallel to that used by Janak[7] (and Gunnarsson) for the study of the uniform spin susceptibility.

The usual magnetization experiments only measure the total susceptibility[8] $\chi = \chi_S + \chi_C + \chi_D + \chi_L$ which is the sum of the average value

$\chi_s = \langle\chi_s(r)\rangle$ of the spin enhanced susceptibility, the core electron diamagnetic susceptibility χ_C , the conduction electron diamagnetic susceptility χ_D and the localized magnetic moments paramagnetic susceptibility χ_L. But, as we will discuss further below, the properties of transition, actinide and rare earth metals can strongly depend on the local spin enhanced susceptibility $\chi_s(r)$.

LOCALIZED ELECTRON ORBIT - ITINERANT ELECTRON SPIN INTERACTIONS IN HEAVY FERMION INTERMETALLICS

In many studies of the properties of valence and conduction electrons in the different materials, an approximate theoretical result to fit experimental data is obtained through the introduction of an effective mass m*, using a nearly free electron picture and formulae with the mass m parametrized accordingly. In most metals, where the ratio m*/m varies in the range of $0.1 < m^*/m < 10$, this value of the parameter m* can be explained considering an actual electron density of states N(E) modified from the free electron value $N_o(E) = aE^{\frac{1}{2}}$ for two main reasons[9] : resonant states corresponding to s,p,d,f bands of atomic origin and, second, splitting of those bands by crystal field, ligand field or chemical bonding interactions in general. Each one of these contributions will tend to create a sub-band, generally narrow, which will change m*(E). A transition metal for example, will have, from resonant d atomic scattering, a large N(E) in the energy range corresponding to the d band, that d band will be split, in cubic crystals, into E_g and T_{2g} sub-bands, corresponding to the crystal field symmetry, and, finally, each sub-band will in turn be split into (a series) of bonding and antibonding sub-bands from the chemical bonding with the rest of the material, mainly the nearest neighbours. Nevertheless the sharpest of those sub-bands will not explain an effective mass larger than m* = 20 m.

Heavy electron systems (see for example Ott, Rudigier, Fisk & Smith)[10] have been known in the last ten years, since the discovery of the unusual properties of $CeAl_3$. They show from measurements of the specific heat C_p , a large effective mass ranging from 150 m for UAl_2 to 1180 m for UBe_{13} ; simultaneously a very large magnetic susceptibility ranging from a spin fluctuation behavior (where the susceptibility is at least two orders of magnitude higher than the transition metal usual spin enhanced susceptibility) to a high temperature local moment behavior, with a tipical Curie-Weiss law and a localized moment corresponding to an orbital and spin parts very similar to the electron atomic values, and finally, a resistivity maximum at low temperatures, suggesting a Kondo effect contribution. Even more surprising was the finding that these heavy electron systems, being spin fluctuators, could become superconducting as in the case of $CeCu_2Si_2$, UPt_3 , UBe_{13} , URe_2 , showing unconventional superconductivity; the experimental data suggesting from the very beginning the possibility of s = 1 and p-pairing. A theoretical discussion can be found, for example, in Ott, Rudigier, Rice, Ueda, Fisk and Smith[11], where the analogy suggested by Pines of a behavior similar to superfluid liquid ^3He is further analyzed.

At this point we want to stress that the heavy fermion systems[12] are related to a very large spin enhanced susceptibility or localized magnetic moments, very narrow bands of elementary excitations at the Fermi level, and a new type of pairing, in the case where they become superconducting, at low temperatures even if they were spin fluctuators above the critical superconducting transition temperature T_s.

In our studies of the f bands, in terms of the actual f-wave functions in the atomic and in the condensed matter case (C. de Teresa,

J. Keller and J. Schoenes[13]) we found that the width of the band is related to a distortion, in real space, of the radial part of the f-wave function to adapt to the condensed matter potential and boundary conditions: if the distortion is large, (αCe for example) the band width is also large, otherwise if the f-wave function remains very close to atomic like the band width is smaller (γCe, for example). Then it is possible that the peculiar behavior of heavy fermion intermetallics is related to a mechanism which will contract the f-wave functions to an atomic, core like form.

The question of these materials being spin fluctuators on the other hand, similar to the problem of the itinerant $TiBe_2$ is related to both a large conduction electron density of state at the Fermi level and to a large enhancement of the spin susceptibility. The approach we have followed for these materials is to analyze the spin susceptibility density in real space to show those regions of the material with a large electron density of electrons at the Fermi level $\gamma(r)$ and the local enhancement those electrons will have in that point. The spin fluctuators are those materials for which β is large either in an atomic cell (with localized moment behavior) or in the interstitial region (with itinerant magnetism behavior like in $TiBe_2$). We have found that the rare earth and actinides compounds of Ce and U present a large spin susceptibility in the boundary regions between atomic cells, similar to early transition metals, but larger.

Suppose now that some mechanism, which will be explained in the self consistent approximation below, contracts the f-wave functions of the rare earth or actinides metals to the point where the orbital angular momentum is not quenched,as in broad d-band transition metals, then the orbital magnetism will act as a strong local magnetic moment which will induce a (local) spin magnetic moment of the enhanced susceptibility conduction band electrons. This will contribute an additional energy $\Delta E_{O-S} = -\frac{1}{2}\chi_S H_O^2$ to the system, with the following consequences: first an additional, magnetic, scattering of the conduction electrons will be found, increasing the resistivity in a Kondo like mechanism, second, the susceptibility to an external field will have two contributions, one from the spin enhanced susceptibility and the second from the giant localized moments consisting of the f-electron moment and the induced local spin moment. Third, at low temperatures, the additional ion conduction electron magnetic scattering, for electrons of the same spin, will add a p-pairing (because of the total spin s of the conduction electrons being s = 1) to the weak, normal, s-pairing of the Cooper mechanism. It is known that when p-pairing is present the s-pairing will tend to be negligible. Then, a mechanism which changes the hibridized f-electron wave functions back into atomic like will, in a high spin susceptibility material, induce the observed behavior now known as heavy electrons in metals, if at the same time the coupling, through magnetic interaction, of the electrons at the Fermi level and the electrons in the f-band, is accounted for.

Because of the coupling between Fermi level conduction electrons and the f-band, allowing the hoping between both types of states, it could be thought that the one electron f band will always be at the Fermi level, but it is known that this is not the actual case because the occupation of the f-band cannot be deduced from the energy eigenvalue (because a removal of an f-electron will drastically change the electronic structure and the relaxation corrections are large and cannot be neglected). Then, the narrowed f-band occupation has to be found from the study of the lowest total free energy configurations.

The mechanism we propose in the present paper is just a self consistent analysis of all the previous considerations and their introduction into the, otherwise tipical, calculational procedure.

The magnetic coupling between conducting electrons and f-electrons will modify the kinetic energy of the f electron. For this the conduction electrons have to be thought as the source of the magnetic field and the field momenta incorporated into the Schroedinger equation

$$+\frac{1}{2m}\left(p_i - \frac{e}{c}A_i\right)^2 \approx +\frac{1}{2m^\bullet}p_i^2 \quad \text{with } m^\bullet - m \approx \left[\frac{\Delta E_{o-s}}{E_k}\right]^{1/2} 2$$

The effective magnetic field H, from the orbital part onto the conductions electrons, can be approximated from the spin orbit splitting because $\Delta E_{SO} \approx gH$. It should also be remembered that the effective field could be larger because the spin orbit interaction, in the lowest state, alignes the spin of the f-electron antiparallel to the orbital magnetic moment, and the Curie field, from the exchange forces between the f-electron spin and the conduction electron spin, will add to the dipole field of the orbital moment. To ilustrate the effect of this effective mass we present the change in the f-band width of Ce in γCe as a function of the effective mass, Table 2 and Fig. 1.

Table 2. The change in the energy of the center of the 4f and 5d bnads, $\overline{\varepsilon}_f$ and $\overline{\varepsilon}_d$ respectively, and the band widths Γ_f and Γ_d as a function of the electron effective mass $m^\bullet$ in γ-cerium metal, from the electron orbit-conductions electron spin magnetic interaction. Energy in Rydberg units. For $m^\bullet/m > 1$ the state becomes almost atomic like, the f bands, for different J = L + S values do not overlap in energy, $\Delta E_{SL} >> \Gamma_f$.

$m^\bullet/m$	$\overline{\varepsilon}_f - V_{MTO}$	Γ_f/Ry	$\overline{\varepsilon}_d - V_{MTO}$	Γ_d/Ry
0.98	1.048	0.099	0.90	0.48
0.99	0.845	0.045	0.84	0.45
1.00	0.619	0.016	0.77	0.42
1.01	0.362	0.003	0.71	0.38
1.02	0.072	0.001	0.64	0.35

We found the contraction of the f-wave functions to an atomic like character. This has an important consequence, because we see the onset of a cooperative effect, the orbital magnetic moment to spin conduction band electrons interaction, will generate an effective mass for the f-electrons which, in turn, will, by contraction of the f-wave function enhance the atomic like character and then enhance the full process, hence the process will not be smooth or present linear behavior; the process will tend to present a sudden change from a band character behavior of the f-electrons to an almost atomic like, then from the coupling to the electrons at the Fermi level, a heavy fermion behavior. This many-body effect is responsible for further scattering between the conduction electrons and the rare earth or actinide ions which of course will show into magnetoresistance, Kondo like behavior and a very special electron pairing possibility, with total spin S = 1, which at low temperatures could be responsible for the very special superconductivity

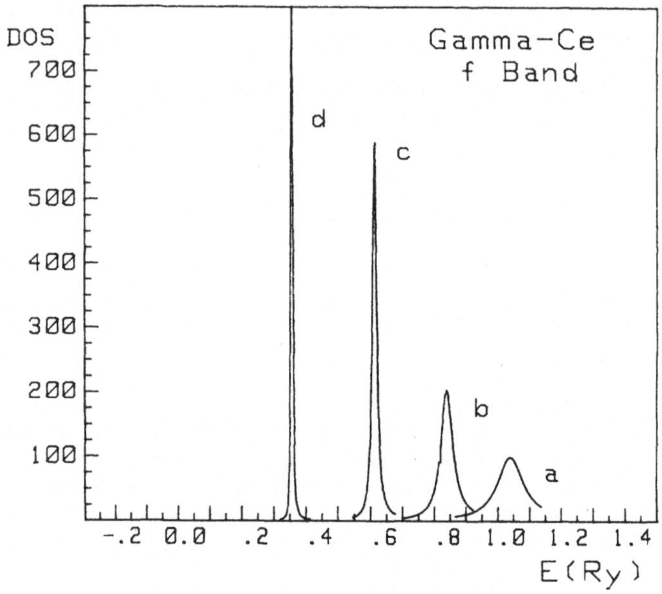

Figure 1. The cerium f-bands as a function of the effective mass m* in the kinetic energy term of the hamiltonian a) m* = 0.98 m, b) m* = 0.99 m, c) m* = 1.0 m and d) m* = 1.01 m. In crystalline γ-Ce, DOS is the electronic density of states.

behavior of some of these materials with the appropriate crystalline symmetry (see for example P.W. Anderson[14] or J.P. Rodriguez[15]).

ACKNOWLEDGEMENT

This work was partially supported by CONACYT, México, proyect PCCBBNA-022702.

REFERENCES

1. P. Hohenberg and W. Kohn, Phys. Rev. B 136, 864 (1964).
2. W. Kohn and L.J. Sham, Phys. Rev. 140, A1133 (1965).
3. M. Levy, J.P. Perdew and V. Sahni, Phys. Rev. A 30 #5, 2745-48 (1984).
4. J.F. Janak, Phys. Rev. B 18, 7165-68 (1978).
5. L. Eyges, Phys. Rev. 111, 683 (1958); K.H. Johnson, J. Chem. Phys. 45, 3085 (1966); K.H. Johnson, in Advances in Quantum Chemistry, edited by P.O. Löwdin, Vol. 7, pp. 143, Academic, New York, 1973; K.H. Johnson and F.C. Smith, Phys. Rev. B 5, 831 (1972); J. Keller, Int. J. Quantum Chem. 9, 583 (1975). Paper presented at the Sanibel Symposia (1973).
J. Keller, Computational Methods for Large Molecules and Localized States in Solids, edited by F. Herman, A.D. McLean and R.K. Nesbet, 341-56, Plenum Press 1973 and J. Physique, 33, C3, 241 (1972); J. Keller, Hyperfine Interact. 6, 15 (1979).
J. Keller, J. Fritz and A. Garritz, J. Physique, 35, C4, 379 (1974); J. Keller and J. Fritz, Proceedings of the V Int. Conf. on Amorphous and Liquid Semiconductors 1973; A. Garritz and J. Keller, in Proceedings of the Int. Conf. on the Electronic and Magnetic

Properties of liquid Metals, University of Mexico publica tions 1978. M. Castro, J. Keller and P. Rius, Hyperfine interactions 9, (1982). J. Keller and C. Amador, Lecture Notes in Physics, edited by J.G. Zabolitzky, M. de LLano, M. Fortes and J.W. Clark, 142, 364 (1981).
J. Keller and J.L. Gazquez, Phys. Rev. A 20, 1289 (1979); J.L. Gazquez and J. Keller, Phys. Rev. A 16, 1385 (1977); J.L. Gazquez, E. Ortiz and J. Keller, Int. J. of Quantum Chemistry, Quantum Chemistry Symposia, edited by P.O. Løwdin and Y. Ohrn 13, 377 (1979). Lecture Notes in Physics 187, Edited by J. Keller and J.L. Gazquez, 1983.

6. J. Keller, C. Amador and C. De Teresa, Rev. Mex. Fis. 30, No. 3,447 (1984).
7. J.F. Janak, Phys. Rev. B16, 255 (1977), and references therein.
8. J. Keller, Hyperfine Interact. 6, 15 (1979).
9. J. Keller, J. of Molecular Structure 93, 93 (1983). C. De Teresa, C. Amador and J. Keller, to be published Physica B (1985). A. Pisanty, E. Orgaz, C. De Teresa and J. Keller, Physica 102B, 78 (1980).
10. H.R. Ott, H. Rudigier, Z. Kisk and J.L. Smith, Physica 127B, 359 (1984). K. Andres, J.E. Graebner and H.R. Ott, Phys. Rev. Lett. 35 1779 (1975). F. Steglich, J. Aarts, C.D. Bredl, W. Lieke, D. Meschede, W. Franz and H. Schafer, Phys. Rev. Lett. 43, 1892 (1979).
11. H.R. Ott, H. Rudigier, T.M. Rice, K. Ueda, Z. Fisk and J.L. Smith, Phys. Rev. Lett. 52, 1915 (1984).
12. H.R. Ott, H. Rudigier, Z. Fisk and J.L. Smith J. Appl. Phys. 57 (1), 3044 (1985). Z. Fisk, J.L. Smith, H.R. Ott and B. Batlogg, to be appear in J. of Magnetism and Magnetic Materials (1985).
13. C. De Teresa, J. Keller and J. Schoenes, to be appear in Solid State Comm. (1985).
14. P.W. Anderson, Phys. Rev. B30, 4000 (1984).
15. J.P. Rodriquez, Phys. Rev. Lett. 55, 250 (1985).

LOCAL-DENSITY-DEPENDENT DIELECTRIC FUNCTION FOR ELECTRONS IN METALS

Ralph J. Harrison

Mechanics and Structural Integrity Laboratory
Army Materials and Mechanics Research Center
Watertown, MA 02172

INTRODUCTION

It has seemed useful to begin this discussion of the dielectric function for electrons in metals by reviewing some aspects of how the well known behavior of the static dielectric constant in ordinary materials provides guidelines which must be met by any valid quantum mechanical description. Thus I shall first introduce the dielectric constant in terms of the behavior of point charges in a dielectric, describing how they are screened and interact with each other, and how such features may be described by the quantum mechanical many body theory. The concept of a dielectric function, both local and non-local, will then be introduced and a very brief review of what has been done in developing the theory of the static dielectric function for application to problems in metals will be given. The dielectric function determines the screening of ions in metals and consequently also their interaction energy. The usual theories have the density of electrons as parameter in the dielectric function, with the density defined in terms of a global average of the number of electrons per unit volume; however for physical reasons one expects that there must be some upper limit to the averaging volume over which the average density of electrons will affect the screening behavior at a point. This physical notion provides support for the idea that a more correct dielectric function must be a local-density-dependent one. A way to modify the usual derivation of the dielectric function to justify this is suggested. The relevance of the use of local-density-dependent dielectric functions to the "compressibility problem", that of preserving the identity of the compressibility when computed by the two methods, that of uniform deformation and the "method of long waves" is emphasized.

CLASSICAL DIELECTRIC CONSTANT FROM A QUANTUM VIEWPOINT

Consider the classical point charge q in a dielectric medium characterized by dielectric constant ε, with the electrostatic potential at a distance r given by $q/(\varepsilon r)$. For the corresponding case of two point charges q_1 and q_2 a distance apart r_{12}, their interaction energy is $q_1 q_2/(\varepsilon r_{12})$.

For the single point charge the many body wave function of the system of n electrons and N nuclei can be written as $\Psi(q,r;\rho_1,\ldots\ldots\rho_n,R_1,\ldots\ldots R_N)$.

The charge q and its position r enter as classical parameters in the Hamililtonian describing the Coulomb interaction with the electrons and nuclei. For the second case there exists a corresponding wave function $\Psi(q_1,r_1,q_2,r_2;\rho_1,\ldots\rho_n,R_1,\ldots R_N)$. Here we can, for example, write the Hamiltonian as

$$H = H_0(\rho_1,\ldots\ldots\rho_n,R_1,\ldots R_N) - eq_1\sum^{n}|\rho_i-r_1|^{-1} - eq_2\sum|\rho_i-r_2|^{-1} + q_1q_2|r_1-r_2|^{-1} + eq_1\sum^{N}Z_j|R_j-r_1|^{-1} + eq_2\sum^{N}Z_j|R_j-r_2|^{-1} \qquad (1)$$

The requirement for reduction to the classical electrostatics description is that the net charge enclosed by spheres drawn about the charge q or the charges q_1 and q_2 must be independent of radius (as long as the radius is greater than atomic dimensions.) The "screening charge" enclosed in the small sphere about q must be $-q(1 - (1/\varepsilon))$, with corresponding values of screening charge about q_1 and q_2. The quantum mechanical charge density operator

$$O_{c.d.} = - e\sum^{n} \delta(\rho_i-r) + e\sum^{N} Z_j\ \delta(R_j-r) \qquad (2)$$

is used to obtain the charge density at a point r.

The change in this charge density associated with the change in the wave function due to the presence of the external charge or charges is the screening or "induced" charge and is then

$$\int (|\Psi|^2 - |\Psi_0|^2)O_{c.d.}\,d\rho_1 \ldots d\rho_n\, dR_1 \ldots dR_N \qquad (3)$$

The wave function contains both the electronic and nuclear coordinates, and the corresponding charge density operator also contains these coordinates. In solids where the polarization occurs almost entirely due to the change in the electron wave function it is simplest to deal only with the one-electron charge density, compensated for by a "smeared-out" nuclear charge, taking into account that the electron coordinates are usually defined with respect to the center of mass of the nuclear coordinates, and that the total nuclear and total electronic charges are almost equal. There are cases when the polarization due to the nuclear motion cannot be neglected and it is necessary to use the full charge density operator. In order to reduce to the classical result the screening charge (3) must describe localized screening charges of magnitude $-q_1(1 -(1/\varepsilon))$ localized at r_1 and $-q_2(1 - (1/\varepsilon))$ localized at r_2. One may have to perform at least two averaging operations to avoid dependence on the exact microscopic positions at which the external charges q_1 and q_2 are placed and at which the induced charge density is sampled. In any case if the stated limiting results are obtained the system can be said to describe an isotropic linear classical dielectric. Since the total charge will be conserved this change in the local regions about the external charges must be compensated for elsewhere. Taking the classical results as a guide, the compensating charge will be at the surface of the solid. This is a cue that in any formal computational theory to obtain the dielectric screening in a solid, one may have to deal with the surface boundary conditions quite carefully in order to get the proper surface charge.

It may be noted parenthetically that in quantum electrodunamics the concept of <u>renormalized</u> charge is used to describe vacuum polarisation effects on the force between "bare" charges in the vacuum. The renormalized charge would correspond to a reduction of the bare charge q by a factor $\varepsilon_v^{1/2}$ where ε_v is the "dielectric constant of the vacuum." This renormalized charge therefore is not equivalent to the screened charge q/ε_v. Kohn (1958), for example, has discussed the ordinary dielectric in terms of this renormalized charge. The renormalization concept appears to be a more symmetric description and in the vacuum one cannot tell whether the interaction energy between two charges q_1 and q_2 is that of a test charge q_2 in a field $q_1/(r^2\varepsilon_v)$ or of a test charge $q_2/\varepsilon_v^{1/2}$ in an electric field $q_1/(r^2\varepsilon_v^{1/2})$. For the ordinary dielectric one can resolve the ambiguity by measuring the value of charge outside the dielectric and trusting it to be the same when placed inside.

In the classical case of the two charges one may wonder for a moment why, if the screening charges exist, they don't contribute to the interaction energy, making this a total of $q_1q_2/(\varepsilon^2 r_{12})$. However a little thought shows that the conventional result measures the work done on the second charge q_2 as it is brought from infinity, subject to the electrostatic field due to q_1 and to the screening charge $-q_1(1 - (1/\varepsilon))$, giving $q_1q_2/(\varepsilon r_{12})$. It will be seen below that the quantum mechanical Hellmann-Feynman theorem yields the same results.

The application of the Hellmann-Feynman theorem (Hellmann,1937; Feynman, 1939) to the example of the charge interaction described above is both instructive and simple. This theorem, for the many body problem involving the n electron coordinates $\rho_1, \ldots, \rho_n$ and the N nuclear coordinates $R_1,..R_N$, states that the partial derivative with respect to a parameter q, of the energy E of an eigenstate Ψ is given by

$$(\partial E/\partial q) = (\Psi,(\partial H/\partial q)\Psi), \tag{4}$$

where H is the Hamiltonian operator of the problem. If we replace q by q_1 and then take a further derivative of the energy with respect to a second parameter q we obtain

$$\begin{aligned}(\partial^2 E/\partial q_1 \partial q_2) &= (\Psi,(\partial^2 H/\partial q_1 \partial q_2)\Psi) + ((\partial\Psi/\partial q_2),(\partial H/\partial q_1)\Psi) + \\ &\quad (\Psi(\partial H/\partial q_1),(\partial\Psi/\partial q_2)) \\ &= (\Psi,(\partial^2 H/\partial q_1 \partial q_2)\Psi) + ((\partial H/\partial q_1),(\partial/\partial q_2)|\Psi|^2), \end{aligned} \tag{5}$$

where the Hamiltonian H is as defined before in (3) and the matrix elements represent integrals over the n electronic and N nuclear coordinates. Since the left hand side of (5) is just the coefficient of the biquadratic terms (in q_1 and q_2) in a Taylor expansion of the energy, the right hand side can be identified with the coefficient of q_1q_2 in the interaction energy of point charges in the linear dielectric. The first term is obviously, from (1), just the coefficient $1/r_{12}$ of the interaction energy of the bare charges. If the localization of screening charge behaves classically the second term on the right of (5) is just the derivative with respect to q_2 of the interaction energy of the screening charge at r_2 with a unit charge at r_1, thereby contributing the factor $q_1q_2(1 - (1/\varepsilon))$ to the bilinear part of the interaction energy. (The energy associated with the surface part of the screening charge may be neglected in the limit of large dimensions.)

The generalization to the interaction between ions in metals is obvious (e.g., Harrison and Paskin,1963) and in the pseudopotential formulation for example one calculates the electrostatic energy of interaction between the bare pseudocharge of one ion and the screened pseudopotential of the other (Harrison,W. A.,1966).

DIELECTRIC FUNCTION OF METALS

It has been seen that the dielectric <u>constant</u> represents a situation where the screening charge follows the external charge point by point, except for surface effects, with factor $-(1 - (1/\varepsilon))$.

A <u>local dielectric function</u> $\varepsilon(q,\omega)$ represents the analogous relationship in wave number and frequency space, namely

$$\rho_{\text{screening}}(q,\omega) = -(1 - (1/\varepsilon(q,\omega))\, \rho_{\text{ext}}(q,\omega) \tag{6}$$

(The left hand side is also commonly referred to as the induced charge.)

It is apparent that for metals the dielectric constant itself is not useful except as a limiting concept. One may regard a metal as a material for which ε is infinite and therefore the screening charge is exactly -q, with the screened charge identically zero. Again this means there is a localized excess of electronic density made up for by a deficiency at the surface of the metal. This is the conventional "Faraday Cage" effect in metal shielding. If a metal shield is thin compared to the dimension over which the details of the screening charge are important, the Faraday cage effect will not be complete (Harrison, R. J., 1966). It is here as well as in many other applications that one finds the dielectric concept to be useful. The lack of strict localization of the screening charge is conveniently represented in terms of the wave number-dependent dielectric function. Although both the frequency and wave number dependencies of the dielectric constant are important in general, in the present context only the dependence upon the wave number in the limit of zero frequency is emphasized. For optical phenomena, where plasmon excitation, for example, is important, both frequency and wave number dependence must be taken into account explicitly.

There have been some recent reviews covering the theory of the dielectric function of metals (Hanke,1978; Gorobchenko and Maksimov, 1980) but we shall give a brief discussion of some of these developments as introduction to the title problem.

A simple example of a local dielectric function is that arising from the linearized Thomas-Fermi treatment as given, e.g., by Mott and Jones (1936)which has proved quite useful in metal theory. An even more realistic dielectric function for metals is the Lindhard function (Lindhard,1954), corresponding to the self-consistent Hartree screening or RPA approximation, and which leads to an oscillatory screening charge density (Friedel oscillations) with the inverse of the Fermi wave vector describing a length scale. This was generalized to include band structure effects by Ehrenreich and Cohen (]959). Improvements in the dielectric functions to include additional many body effects by taking into account corrections for exchange and correlation have been made. (Geldart and Taylor, 1970; Gorobchenko and Maksimov,1981). Some work on higher order corrections to take non-linear effects into account have improved the reliability of the calculated charge density, primarily at distances close to the point charge. Other types of improvement in the theoretical treatment of screening have been the consideration of effects due to finite temperature and to the finite mean free path of conduction electrons. Both of these reduce the sharpness of the Fermi surface and consequently introduce a damping into the long-range oscillatory screening charge distribution. The temperature and the mean free path of the electrons enter as parameters as well as does the original charge density.

An important contribution to the theory has been the incorporation of concepts using the density functional theorem of Hohenberg and Kohn (1964) and related developments(Kohn and Sham, 1965) into the theory of the dielectric function(Yang et al.,1972,1975). The Kubo response theory has also been used to discuss the dielectric function within the framework of many body theory.

Another direction that has been taken in regard to the screening problem has been with regard to the inclusion of effects associated with inhomogeneities in the medium. Work of this nature was carried out long ago in connection with corrections to the Thomas-Fermi theory and also has been recently extended to use in the density functional development. Two types of inhomogeneity have been considered; first the case of a nearly homogeneous electron gas with a fairly small and uniform density gradient for which the general functional relationship may be expected to be expandable in terms of kernels involving as parameters the average density and gradient of density of the electron gas; and second the case of abrupt changes in density such as those due to the presence of surfaces. Some of the latter developments have utilized non-local dielectric functions such that the relation between induced charge and the external charge densities is given by a linear integral relationship, replacing Eq. (6) and utilizing a dielectric function $\varepsilon(q,q')$. For example this kind of relationship would be expected to arise if the dielectric material is inhomogeneous and the screening extends over a region which is comparable to the scale of the inhomogeneity. In a crystalline material, for example, the response might depend upon exactly where with respect to the exact atomic positions one places the test charge. Close enough to a surface the polarizability might be different enough to require a non-local description. Near a surface one also gets image-charge effects which arise even where the conventional dielectric constant description is valid. Treatments of surface effects include those of Yang et al. (1972),(1975) and Rasolt and Perrot (1973).

The local-density-dependent dielectric function can be regarded as a special case of the non-local dielectric function which it is convenient to regard as dependent not upon two wave numbers q and q', but upon one wave number q and one spatial variable r'. That is, the test charge, or at least the screening charge is spread out over some finite region so that it is convenient to describe the response in terms of the wave number q. However the respons, (except for the distance surface charging effect), is localized over a volume over which the material can be regarded as homogeneous. Over larger distances where this assumption of material inhomogeneity cannot be made one uses a dielectric function appropriate to the corresponding local neighborhood. This concept appears to be qualitatively plausible and in fact seems required if the macroscopic description of an inhomogeneous dielectric material holds.

However the formal theory of the dielectric response in the literature presents the dielectric function as one which depends upon the global electron density. I don't know of any treatment which specifically introduces any characteristic dimension marking the transition between local and global averaging. I shall return to this point in the concluding section where it is suggested that perhaps the mean free path may be such a characteristic dimension.

COMPRESSIBILITY PROBLEM

This problem has to do with the fact that although the compressibility of a metal ought to be the same, in a correct calculation, no matter how it is calculated. In terms of the bulk modulus $B = 1/\kappa = (1/3)(c_{11} + 2c_{13})$, the "method of uniform compression" obtains it from

$$B = V\ (\partial^2 B/\partial V^2),$$

whereas the "method of long waves" obtains the elastic constants from calculation of the dispersion relation for phonons, in turn which follows from the solution of the dynamical matrix. However Soma et al. (1984) quote a discrepancy of 10-15% for the compressibility of aluminum while others (Taole and Glyde, 1979; Finnis, 1974) suggest that the discrepancy can be 50% or even greater. Brovman et al. (1969) suggested that the inconsistency arises because the dynamical matrix is not computed to high enough order, and showed that inclusion of fourth order terms in the pseudopotential will reduce the discrepancy. Similarly Wallace (1970) suggests that an infinite order computation must give complete agreement between the two methods. Taole and Glyde have emphasized that the discrepancy arises because the dynamical matrix utilizes ion-ion interactions which assume screening at constant density whereas the uniform compression method explicitly accounts for the density dependence of the screening. They therefore eliminate the discrepancy in compressibility calculation by assuming a completely local-density-dependent screening, and taking local averages of electron density as a function of wavelength for longitudinal waves. They note that this method makes a correction in the dispersion curves not only at the long wavelength limit, but may be significant over approximately the lower tenth of the Brillouin zone for longitudinal waves.

AVERAGING LENGTH FOR LOCAL-DENSITY-DEPENDENT DIELECTRIC FUNCTION

As has been pointed out there apparently exists no theory for a characteristic averaging length for local density. If one uses the approach of Brovman et al. to resolve the compressibility paradox one avoids the problem provided that one can calculate the three and four-ion terms and if these are large enough to make the two computations of compressibility agree. If not, one has to try to compute higher terms. The alternate approach of Taole and Glyde also avoids the necessity for computing an averaging length by defining it in accord with the relevant wave vector.

Our own hypothesis is that the electronic mean free path forms at least an upper bound for the length over which one must average in order to form a definition of local density. The qualitative justification for this is that one can make a parallel derivation of the SCF dielectric function, using the first-order self-consistent perturbation theory technique, which avoids the fact that in the usual derivation where one starts with plane waves the normalization which naturally comes into the problem is the entire space over which the plane waves are defined. If it were reasonable to take localized wave functions as the choice of unperturbed functions, the result would immediately fall out that the proper averaging region is that region over which the wave functions are localized, provided that within this region these functions approximated plane waves. Since it is probably not justifiable to take localized plane waves to describe the wave functions near the Fermi surface in a metal one has to argue that it is sufficient if the wave functions were good approximations to plane waves within a region of extent of a mean free path, and outside this region did not necessarily fall off in amplitude, but they no longer were coherent with the plane wave inside this region. This will effectively cause the wavefunctions to no longer be good momentum eigenfunctions and one can then renormalize within each local region by matching to an appropriate set of momentum states whose spacings reflected the effect of this dephasing.

It would be of interest if there were an experimental way of determining the averaging length. It is possible that an accurate determination of the phonon dispersion curves, coupled with a good theoretical pseudopotential, could be helpful in this regard. Another possibility is an accurate determination of the approach to the compressibility limit in the low angle scattering observations of the density-density correlations in liquid metals. It is hoped that the present analysis, although admittedly only qualitative, will stimulate more detailed quantitative treatments of this problem.

ACKNOWLEDGMENTS

Some of the ideas regarding the classical dielectric constant description trace back to earlier work in collaboration with Dr. Arthur Paskin (Harrison and Paskin, 1963). I also have benefited from discussions with Dr. Paolo Ascarelli regarding the compressibility problem. I should also like to thank Dr. H. R. Glyde for discussions and for calling my attention at this workshop to the work of Taole and Glyde (1979) and of Finnis (1974).

REFERENCES

Brovman, E. G., Kagan, Yu., and Kholas, A., 1970, The compressibility problem and violation of the Cauchy relations in metals, Zh. Eksp. Teor. Fiz. 57:1635.

Brovman, E. G., Kagan, Yu. M., 1974, Phonons in nontransition metals, Usp. Fiz. Nauk 112:369.
Feynman, R. P., 1939, Forces in molecules, Phys. Rev. 56:340.
Usp. Fiz. Nauk 112:369.
Feynman, R. P., 1939, Forces in molecules, Phys. Rev. 56:340.
Finnis, M. W., 1974, Energy and elastic-constants of simple metals in terms of pairwise interactions, J. Phys. F. 4:1645.
Ehrenreich, H. and Cohen,M. H.,1959, Self-consistent field approach to the many-electron problem, Phys. Rev. 115:786.
Gorobchenko, V. D. and Maksimov, E. G., 1980, The dielectric constant of an interacting electron gas, Usp. Fiz. Nauk 130:65.
Gorobchenko, V. D. and Maksimov, E. G., 1981, Dielectric matrix of Bloch electrons with account for exchange-correlation effects, Zh. Eksp. Teor. Fiz. (USSR) 81:1847.
Geldart, D. J. W. and Taylor, R., 1970, Wave-number dependence of the static screening function of an interacting electron gas,I and II, Can. J. Phys. 48:155,167.
Hanke, W., 1978, Adv. in Physics 27:287.
Harrison, R. J. and Paskin, A. 1963, Many-electron description of dielectric polarization, Bull. Amer. Phys. Soc. 8:255.
Harrison, R. J., 1966, Non-classical electrostatic screening in metals, Bull. Amer. Phys. Soc. 11:273.
Harrison, W., 1966, "Pseudopotentials in the Theory of Metals", Benjamin, New York.
Hohenberg, P. and Kohn, W., 1964, Inhomogeneous electron gas, Phys. Rev. 136:B864.
Hellman, H., 1937,"Einfuhrung in die Quantenchemie", Sec. 54, Franz Deuticke, Leipzig.
Kohn, W. and Sham, L. J., 1965, Self-consistent equations including exchange and correlation effects, Phys. Rev. 145:561
Lindhard, J.,1954, On the properties of a gas of charged particles, K. Danske Vidensk. Selsk.,mat.-fys. Medd. 28:No.8
Rasolt, M. and Perrot, F.,1983, Pair potentials on metal surfaces, Phys. Rev. B28:6749.
Soma, T.,Itoh, T., and Kagaya, H.-Matsuo,1984,Lattice dynamics of aluminum, Phys. stat. sol. (b) 123:463.
Taole, S. H. and Glyde, H. R.,1979, Volume forces in simple metals, Can. J. Phys. 57:1870.
Sham, L. J. and Kohn, W. ,1966, One-particle properties of an inhomogeneous electron gas, Phys.Rev. 145:561.
Wallace, D. C. ,1969, Pseudopotential calculation of the elastic constants of simple metals, Phys. Rev. 182:778.
Yang, S. C., Smith, J. R., and Kohn, W., 1972, Self-consistent screening of charges embedded in a metal surface, J. Vac. Sci. Technol. 9:575.
Yang, S. C., Smith, J. R., and Kohn, W., 1975, Density-functional theory of chemisorption on metal surfaces, Phys. Rev. B11:1483.

CONVERGENCE PROPERTIES OF AN EXACT BAND THEORY*

R. G. Brown and M. Ciftan

Physics Department
Duke University
Durham, NC 27706

INTRODUCTION

Green's function band theory (GFBT) has advanced considerably in its attainable precision and range of applicability since its development by Korringa[1] and by Kohn and Rostoker[2]. The Korringa-Kohn-Rostoker (KKR) equations can be applied to virtually any crystalline structure and yield reasonably accurate bands. They have substantial computational advantages, including the separation of the effects of structure and potential in the secular equation and rapid convergence with a relatively small secular determinant.

In order to achieve this separability the KKR equations require the potential to be muffin-tin (MT) restricted, that is, defined to be non-zero only within the largest sphere that could be inscribed completely inside a unit cell of the crystal. To simplify the equations still further, the potential is restricted to be spherically symmetric within that cell. The rapid convergence is a consequence of using an angular momentum expansion of the wavefunctions within the cell; the angular momentum cutoff then limits the contribution from terms with large angular momenta.

While the KKR method is capable of evaluating many crystal bands with great accuracy, particularly when the actual crystal potential is nearly of the MT form, in many cases of interest the potential is not of the MT form. The potential may be anisotropic both inside the MT sphere (breaking the condition of spherical symmetry) and in the interstitial region. In these crystals, MT approximation may be a poor one.

Furthermore, in any calculation of the single-electron energy bands of a crystal, it is desirable to make the calculation self-consistent[3] to correct for many body effects like electron correlation which are known to be important in, for example, the transition metals. When one begins a self-consistent calculation with a potential of the MT form, there is no reason to assume that the potential will remain in that form through the self-consistent cycle. Attempting to enforce the MT condition artificially leads to an unpredictable error in the final self-consistent state.

For these reasons it is desirable to eliminate the MT approximation

*This work was supported by the Army Research Office.

from the Green's function band theories, and yet retain as much as possible its principal attractive features. The restriction to spherically symmetric potentials was eliminated by Evans and Keller[4], who allowed the potential to mix different angular components within the MT sphere. Keller[5] also tried to go beyond the MT restriction by placing additional MT spheres in the interstitial region (reducing the total interstitial volume) but the computational effort this required prevented it from becoming a popular approach.

A number of researchers[6-10] tried to take a non-zero interstitial potential into account, but succeeded only at the expense of the separability. Another approach is to extend one's basis or do a perturbative calculation based on a KKR calculation begun with a MT potential[11]. Giving up separability or doing a multistep calculation both substantially increase the difficulty and expense of a calculation.

Ziesche[12] analyzed the Green's function method from the multiple-scattering point of view. He showed that if one breaks up the single cell scattering into incoming and outgoing scattered waves and then uses the asymptotic form of those scattered waves in the interstitial region to derive a secular determinant, then an error is made if the potential is not of the MT form.

The error arises because the asymptotic form of the radially scattered wave is only an approximation to the true scattered wave in the region between the bounding sphere and the MT sphere of the central WS cell. Ziesche introduced a term to explicitly account for the scattering in this region and called it the "near field" correction because of its volume of origin. Because this term involves a lattice sum over nearest neighbor cells, it effectively destroys the separability of the theory.

At the same time, Williams and van W. Morgan[13] published a paper in which they derived a generalized, non-MT band theory from the multiple scattering point of view. In their derivation they introduced a basis different from that used in previous attempts and used a three-center expansion theorem for the Green's function[14] to obtain a separable theory. Unfortunately, the expansion theorem was in error and their basis functions, based on the phase functional method of Calogero[15], turned out to be incomplete in the outer part of the cell.

As a result, their method also contained a "near field" error term, though their numerical tests of convergence (done on the empty lattice) demonstrated that it was certainly a small one. Faulkner[16] derived a near field correction term for multiple-scattering band theories that differed somewhat from that of Ziesche but still destroyed the separability of the theory.

The purpose of the present paper is to study the convergence properties of a generalized, non-MT Green's function band theory we derived[17], using the theory of Williams and van W. Morgan[13] as a starting point. This theory, like theirs, is based on the use of the phase functional method to obtain suitable basis functions in terms of which to express the solution. However, we prove that the basis we use is complete and rapidly convergent in the cell. In addition, we do not use the three-center expansion for the Green's function to obtain our secular equation.

This theory retains all the formal simplicity of KKR, that is, the secular equation still separates into the (now non-diagonal) scattering matrix of a single cell of the crystal and the usual structure constant matrix of KKR, which can be independently and efficiently evaluated[18,19]. Because the potential is not MT approximated, the method is suitable for use in covalent crystals and in applications where self-consistency is desired.

In this paper we wish to answer two questions: First, is our theory in fact an exact solution to the band structure problem in the sense that it converges to the correct energy bands for non MT potentials? Second, is the theory practically applicable, that is, does it converge fast enough to be of some use and is it relatively inexpensive to apply when compared to other methods of comparable precision?

Both of these questions can be at least partially answered by applying the theory to the empty lattice problem. The empty lattice is a simple non-MT problem in which the exact eigenvalues are known. We thus compare the results produced by the application of our method to the exact results and directly measure the error. We can directly study the convergence properties of the theory by observing the systematic behavior of the error as the number of expansion functions is increased.

The empty lattice problem is a non-trivial problem in GFBT. The unit cellular potential is a step function equal to one only inside the WS cell of the crystal. It is non-spherically symmetric and thus mixes different angular components of the radial wavefunction in the cell. The potential has sharp edges where it abruptly goes to zero. In a typical real metal, the potential outside the MT region is much smaller in magnitude than the depths of the empty lattice potential we study here[20,21], which means that the errors we obtain should be reasonable worst case estimates. In addition we study the empty BCC lattice since it produces a larger error in most applications than the FCC lattice.

The empty lattice test is not the only test a candidate high precision band theory has to pass. Ultimately the theory must be successfully applied to real crystals. The empty lattice test serves as a useful benchmark in the design of these applications and is thus a necessary first step.

As we have noted before in an earlier, less extensive test of the method[22], our theory has the further advantage of handling core states on exactly the same footing as the valence bands; it requires no orthogonalization procedure and does not require a "valence potential". This is important in applications where the valence electron penetrates deep into the cell. It is also important in transition metals where total self-consistency is required to study (for example) their ferromagnetic properties.

To demonstrate that the phase functional basis (which is a scattering basis) can handle the "bound" core states we have applied the phase functional method to several simple potentials where the actual solution is analytically known. The phase functional method successfully reproduces the known eigensolutions to 6 figures. This shows that it can be used to integrate the wavefunctions and find the eigenenergies of bound states as well as scattering states.

This aspect of the phase functional basis is a potentially unifying factor across several disciplines. The phase functional basis can be used in problems in nuclear physics (the context in which the phase functional method was originally applied) and atomic and molecular physics, as well as condensed matter physics. It is, in a sense, an "atomic orbital" basis where the atomic orbitals can respond self consistently to changes in the potential because it actually solves Schrödinger's equation explicitly for the potential of interest. We feel that the PF basis is the "natural" basis in terms of which to expand almost any problem because of these unifying aspects and its extreme accuracy.

In this paper we will review the theory of the phase functional basis. We will then briefly rederive the equations of Green's function band theory. This derivation is simpler than that presented in our previous papers and it does not use a variational step like that used by Kohn and Rostoker[2]. This variational step, while perfectly legitimate, has recently been questioned by Faulkner[23]. The derivation we present here should make it clear that our secular equation in no way depends on this step.

We will then show the results of the application of generalized GFBT to the BCC empty lattice. In particular we will present the lowest bands on the principal symmetry axes to demonstrate that the theory works at a high level of attainable precision. We will study the rms splitting of the degenerate energies produced by truncating the phase functional basis at several symmetry points as the size of the basis is increased. We will also study the rms splitting at the symmetry points as the depth of the empty lattice potential is varied. Finally we will compare the rms splitting produced by our theory at the symmetry points to that produced by Williams and van W. Morgan's[13] theory at $\ell_{max} = 6$ in the scattered wave.

THE PHASE FUNCTIONAL BASIS

The phase functional basis is defined by the integral equation solutions to the time-independent Schrödinger equation,

$$[\nabla^2 + \kappa^2]\ \Psi(\vec{r}) = V(\vec{r})\ \Psi(\vec{r}) \tag{1}$$

that behave like a regular free spherical wave

$$J_L(\vec{r}) = j_\ell(\kappa r) Y_L(\theta,\phi) \tag{2}$$

at the origin. In this expression and throughout, $L \equiv (\ell,m)$ and $\kappa^2 = E_0$ (the kinetic energy of the free wave). $V(\vec{r})$ is the potential formally defined throughout all space. These solutions can be written

$$\Phi_L(\vec{r}) = \int_{\mathbb{R}^3} G_0(\vec{r},\vec{r}_0) V(\vec{r}_0) \Phi_L(\vec{r}_0) d^3 r_0 \tag{3}$$

where

$$G_0(\vec{r},\vec{r}_0) = -\frac{\cos(\kappa|\vec{r}-\vec{r}_0|)}{4\pi|\vec{r}-\vec{r}_0|} \tag{4}$$

is the stationary wave Green's function.

The phase functional solutions (3) are usually desired in some domain Ω with bounded support. Frequently the potential $V(\vec{r})$ also has bounded support, whereupon the solutions (3) take on an asymptotic (scattering) form far from Ω. We find it convenient to break the integral in (3) into two pieces: one over the interior and one over the exterior of the sphere S, which is defined to be the smallest sphere completely containing Ω.

$$\Phi_L(\vec{r}) = \int_{\mathbb{R}^3 - S} G_0(\vec{r},\vec{r}_0) V(\vec{r}_0) \Phi_L(\vec{r}_0) d^3 r_0 + \int_S G_0(\vec{r},\vec{r}_0) V(\vec{r}_0) \Phi_L(\vec{r}_0) d^3 r_0 . \tag{5}$$

The Green's function (4) can be expanded as

$$G_0(\vec{r},\vec{r}_0) = -\kappa \sum_L J_L(\vec{r}_<) N_L(\vec{r}_>)^* \tag{6}$$

where

$$N_L(\vec{r}) = n_\ell(\kappa r) Y_L(\theta,\phi) \tag{7}$$

is an irregular free spherical wave and the argument of $J_L(\vec{r})$ is restricted to be strictly less than the argument of $N_L(\vec{r})$. Equation (5), with $\vec{r}$ restricted to lie in S (which contains Ω) can thus be written in the form

$$\Phi_L(\vec{r}) = \sum_{L'} C^S_{LL'}(\infty) J_{L'}(\vec{r}) + \int_S G(\vec{r},\vec{r}_0) V(\vec{r}_0) \Phi_L(\vec{r}_0) d^3r_0 \ , \tag{8}$$

where

$$C^S_{LL'}(\infty) = -\kappa \int_{\mathbb{R}^3-S} N_L(\vec{r}_0)^* V(\vec{r}_0) \Phi_L(\vec{r}_0) d^3r_0 \, . \tag{9}$$

The solution to (8) for arbitrary $\vec{r} \in S$ is defined to be

$$\Phi_L(\vec{r}) = \sum_L \left\{ C^S_{LL'}(r) J_L(\vec{r}) + S^S_{LL'}(r) N_L(\vec{r}) \right\} \tag{10}$$

with

$$C^S_{LL'}(r) = C^S_{LL'}(\infty) - \kappa \int_r^\infty N_L(\vec{r}_0)^* V_S(\vec{r}_0) \Phi_L(\vec{r}_0) d^3r_0 \tag{11a}$$

and

$$S^S_{LL'}(r) = -\kappa \int_0^r J_L(\vec{r}_0)^* V_S(\vec{r}_0) \Phi_L(\vec{r}_0) d^3r_0 \, . \tag{11b}$$

In equations (11), $V_S(\vec{r})$ is just $V(\vec{r})$ restricted to be zero outside of S. This effectively bounds the region of integration in (11). In the event that the support of $V(\vec{r})$ is Ω the restriction is, or course, not necessary.

The $C^S_{LL'}(r)$ and the $S^S_{LL'}(r)$ are the non-diagonal equivalent of the cosine and the sine of the partial wave phase shifts of scattering theory. The phase shifts are (in Caloger's original view[15]) *functions* of r, so the $C^S_{LL'}(r)$ and $S^S_{LL'}(r)$ are then *functionals* of the phase shifts, hence the terminology phase functional basis. However, as we shall see, it is never necessary to extract the phase and amplitude of the partial waves; in fact, when the partial waves are mixed by a non-spherical potential the concept of partial wave phase shift itself is a fuzzy one.

Any practical application of the phase functional basis requires the evaluation of the phase functionals $C^S_{LL'}(r)$ and $S^S_{LL'}(r)$ (which we obtain as functions of r). This is most easily done by integrating the coupled ordinary differential equations implied by (11) from the origin outward. Since $\Phi_L(\vec{r})$ at $\vec{r} = 0$ is defined to be $J_L(\vec{r})$, equations (11) become

$$C^S_{LL'}(r) = \delta_{LL'} + \kappa \int_0^r N_{L'}(\vec{r}_0)^* V_S(\vec{r}_0) \Phi_L(\vec{r}_0) d^3r_0 \tag{12a}$$

and

$$S^S_{LL'}(r) = -\kappa \int_0^r J_{L'}(\vec{r}_0)^* V_S(\vec{r}_0) \Phi_L(\vec{r}_0) d^3r_0 \, . \tag{12b}$$

These equations can be integrated from the origin out to r for any potential less singular than r^{-2} at the origin. If one integrates them to $r = \infty$ (beyond the range of the potential) then one obtains the asymptotic forms $C^S_{LL'}(\infty)$ and $S^S_{LL'}(\infty)$ used in scattering theory, from which the scattering matrix and cross section can be directly extracted.

The proof that the phase functional basis is complete (subject to certain restrictions) is quite simple. It is clear that any solution to Schrödinger's equation that is regular at the origin can be written for $\vec{r} \in S$ as

$$\Psi(\vec{r}) = \sum_{L'} C^{S}_{\Psi L'}(\infty) J_{L}(\vec{r}) + \int_{S} G_0(\vec{r},\vec{r}_0) V_S(\vec{r}_0) \Psi(\vec{r}_0) d^3 r_0 \; . \tag{13}$$

If the $\Phi_L(\vec{r})$ form a basis, then $\Psi(\vec{r})$ can be exactly expanded in the form

$$\Psi(\vec{r}) \;=\; \sum_{L} a_L \Phi_L(\vec{r}) \; ; \tag{14}$$

that is, there exists a set of a_L's such that this equation is satisfied. Substituting (14) into (13) and using the definition of the $\Phi_L(\vec{r})$ (equation (8)) we obtain

$$\sum_{L'} C^{S}_{\Psi L'}(\infty) J_{L'}(\vec{r}) \;=\; \sum_{L} a_L \sum_{L'} C^{S}_{LL'}(\infty) J_{L'}(\vec{r}) . \tag{15}$$

This simply states that the homogeneous term of the integral equation (13) must equal its expanded form. From this we can project out the L'th term from both sums at any $\vec{r} \in S$ such that $j_{\ell'}(\kappa r) \neq 0$, which leads to the following set of simultaneous equations for the a_L's:

$$\sum_{L} C^{S}_{LL'}(\infty) a_L \;=\; C^{S}_{\Psi L'}(\infty) \; . \tag{16}$$

Clearly a non-trivial solution exists if the det $|C^{S}_{LL'}(\infty)| \neq 0$.

Equation 16 is an important result to remember. The band theory equations follow directly from this equation and the definitions of the $C^{S}_{LL'}(\infty)$ and the $C^{S}_{\Psi L'}(\infty)$ (as we shall see in the next section). The homogeneous (boundary) term of a Fredholm integral equation of the second kind (like equation 13) uniquely determines the solution, except at eigenvalues[24]. In order for the integral equation solution (13) to equal its expansion in integral equation solutions (8), they must have equal homogeneous terms. Since the homogeneous terms in (15) have the same form, (16) just says that we can obtain equality by equating coefficients. The determinantal condition is just the restriction on uniqueness at eigenvalues in Fredholm's theorem.

The angular part of the integration in equations (12) can be done by decomposing $V(\vec{r})$ and $\Phi_L(\vec{r})$ in spherical harmonics and doing the resulting integral of three spherical harmonics

$$I(L',L'',L''') \;=\; \iint Y_{L'}(\theta,\phi) Y_{L''}(\theta,\phi) Y_{L'''}(\theta,\phi) \sin\theta d\theta d\phi . \tag{17}$$

The I(L',L",L"') are coefficients with strong symmetry properties. They can be evaluated from a product of Clebsch-Gordan coefficients and tabulated for use in a given calculation,

The remaining radial integrals,

$$C^{S}_{LL'}(r) = \delta_{LL'} + \kappa \int_0^r n_{\ell'}(\kappa r) \sum_{L''} V_{SL''}(r) \sum_{L'''} \Phi_{LL'''}(r) I(L',L'',L''') r^2 dr \tag{18}$$

and

$$S^S_{LL'}(r) = -\kappa\int_0^r j_{\ell'}(\kappa r) \sum_{L''} V_{SL''}(r) \sum_{L'''} \Phi_{LL'''}(r) I(L',L'',L''') r^2 dr$$

are one dimensional and can easily be evaluated with standard ODE solvers.

To demonstrate the accuracy and potential utility of the phase functional basis in bound state problems, we used equations (18) to compute the phase functional basis for several simple trial potentials for which the analytic results were known. We solved equations (18) for $V(\vec{r}) = -\frac{1}{r}$ (the hydrogen atom, $V(\vec{r}) = r^2$ (the 3 dimensional simple harmonic oscillator) and $V(\vec{r}) = r^2 + 0.1r^4$ (the anharmonic oscillator) and obtained the bound state wave functions $\Phi_L(\vec{r})$ for several values of (n,ℓ). The results for the Hydrogen atom are plotted in Figure 1.

In the first two applications (hydrogen and the 3D SHO) the error in the calculated wavefunctions at the known eigenenergies was never more than 10^{-6}. This gives us a fair degree of confidence that the AHO wavefunctions, which are not analytically known and are very difficult to calculate using the usual approximate techniques of quantum mechanics were also accurately obtained.

The eigenvalues themselves can be found from the phase functional method as well. At an eigenvalue, the solution $\Phi_L(\vec{r})$ as $r \to \infty$ must approach zero asymptotically. At an energy slightly above or below an eigenvalue, the solution diverges to either + or $-\infty$. As the energy sweeps through the eigenvalue, the solution changes the sign of its divergence, that is, if it diverged to $+\infty$ it diverges to $-\infty$ and vice versa. The scattering matrix evaluated from the $C^S_{LL'}(\infty)$ and the $S^S_{LL'}(\infty)$ becomes singular at the eigenvalue. This behavior can be used to obtain eigenvalues to any desired degree of precision for an arbitrary potential.

The eigenenergies are real numbers with an infinite number of significant figures. Any deviation from the exact number results in a solution that ultimately diverges from a bound state solution, thought it may follow the eigensolution for a long ways out from the origin. Since computers necessarily discretize real numbers and the integration procedure, the cumulative numerical error always ultimately pushes the solution off the eigenvalue; the numerical integration of eigensolutions is unstable when the solutions themselves get small.

The consequence of this is that, even at the true eigenvalue, the $n = 1$ wavefunction in figure 1 becomes unstable and ultimately diverges. This does not prevent one from evaluating the wavefunction out to where it is essentially zero and obtaining the bound state energy to a high degree of precision.

From this we may conclude the following things. First, the phase functional method can easily and accurately integrate Schrödinger's equation for reasonably physical potentials including those singular at the origin (atomic potentials). Second, in spite of its being a "scattering" theory it can also be used to find bound state eigenfunctions and eigenvalues. Third, it can be applied in problems where the classical analytic and perturbative techniques (such as the Born approximation) are either ineffective or extremely difficult to use as easily as it can be applied to a trivial problem. These three things, taken together, indicate that the phase functional basis is potentially an extremely powerful tool for doing quantum mechanics.

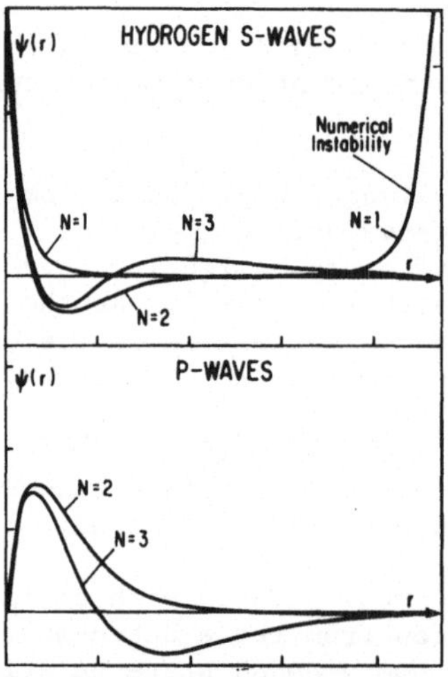

Fig. 1. Bound state s and p wavefunctions of hydrogen obtained using the phase functional method.

Once the properties of the phase functional basis are established, it is fairly straightforward to derive a non-MT GFBT in terms of them. We begin with the Lippmann-Schwinger equation for the crystal

$$\Psi(\vec{k},\vec{r}) = \int_{\mathbb{R}^3} G_0(\vec{r},\vec{r}_0)V(\vec{r}_0)\Psi(\vec{k},\vec{r}_0)d^3r_0 \ . \tag{19}$$

In this equation $V(\vec{r}_0)$ is the periodic crystal potential and $G_0(\vec{r},\vec{r}_0)$ is the usual stationary wave Green's function.

We break the integral into two pieces over the interior and exterior of the smallest sphere S bounding the central WS cell Ω of the crystal,

$$\Psi(\vec{k},\vec{r}) = \int_{\mathbb{R}^3-S} G_0(\vec{r},\vec{r}_0)V(\vec{r}_0)\Psi(\vec{k},\vec{r}_0)d^3r_0 + \int_S G_0(\vec{r},\vec{r}_0)V(\vec{r}_0)\Psi(\vec{k},\vec{r}_0)d^3r_0 \tag{20}$$

and express (20) in the form (for $\vec{r}\in S$)

$$\Psi(\vec{k},\vec{r}) = \sum_{L'} C^S_{\Psi L'}(\infty)\, J_{L'}(\vec{r}) + \int_S G_0(\vec{r},\vec{r}_0)V(\vec{r}_0)\Psi(\vec{k},\vec{r}_0)d^3r_0 . \tag{21}$$

The definition of the $C^S_{\Psi L'}(\infty)$ follows from (20) (and the expansion of the Green's function, equation (6))

$$C^S_{\Psi L'}(\infty) = -\kappa\int_{\mathbb{R}^3-S} N_{L'}(\vec{r}_0)^* V(\vec{r}_0)\Psi(\vec{k},\vec{r}\ )d^3r_0 \ . \tag{22}$$

We wish to expand the desired solution, $\Psi(\vec{k},\vec{r})$, in the phase functional basis $\Phi_L(\vec{r})$ for $r\in\Omega\in S$:

$$\Psi(\vec{k},\vec{r}) = \sum_L a_L(\vec{k})\Phi_L(\vec{r}). \tag{23}$$

We can do this whenever equation (16) is satisfiable, that is, when

$$\sum_L a_L(\vec{k})\ C^S_{LL'}(\infty) = C^S_{\Psi L'}(\infty) = -\kappa\int_{\mathbb{R}^3-S} N_{L'}(\vec{r}_0)^* V(\vec{r}_0)\Psi(\vec{k},\vec{r}_0)d^3r_0 \ . \tag{24}$$

We rewrite the integral in (24) to involve the exterior region of the cell Ω instead of the sphere S and collect terms on one side:

$$\sum_L a_L(\vec{k})C^S_{LL'}(\infty) - \kappa\int_{S-\Omega} N_{L'}(\vec{r}_0)^* V(\vec{r}_0)\Psi(\vec{k},\vec{r}_0)d^3r_0 + \kappa\int_{\mathbb{R}^3-\Omega} N_{L'}(\vec{r}_0)^* V(\vec{r}_0)\Psi(\vec{k},\vec{r}_0)d^3r_0 = 0. \tag{25}$$

We use Bloch's theorem, the symmetry of $V(\vec{r})$ for the crystal and the definition of the structure constants of KKR to write (for r_0 strictly less than the nearest neighbor distance)

$$\kappa\int_{\mathbb{R}^3-\Omega} N_{L'}(\vec{r}_0)^* V(\vec{r}_0)\Psi(\vec{k},\vec{r}_0)d^3r_0 =$$

$$= \frac{1}{\kappa} \sum_{L''} B_{L'L''}(\vec{k})\, \kappa \int_{\Omega} J_{L''}(\vec{r}_0)^* V(\vec{r}_0)\Psi(\vec{k},\vec{r}_0)d^3r_0 \ . \qquad (26)$$

Substituting (26) into (25), substituting (23) for all occurances of $\Psi(\vec{k},\vec{r}_0)$ (which is now allowed because $\vec{r}_0 \in \Omega$) and collecting terms, we get the secular equation,

$$\sum_L a_L(\vec{k}) \left\{ \sum_{L'} \left[C^{\Omega}_{LL'}(\infty) + \frac{1}{\kappa} \sum_{L''} B_{L'L''}(\vec{k}) S^{\Omega}_{LL''}(\infty) \right] \right\} = 0 \qquad (27)$$

where

$$C^{\Omega}_{LL'}(\infty) = C^{S}_{LL'}(\infty) - \kappa \int_{S-\Omega} N_{L'}(\vec{r}_0)^* V(\vec{r}_0)\Phi_L(\vec{r}_0)d^3r_0 \qquad (28a)$$

$$= \delta_{LL'} + \kappa \int_{\Omega} N_{L'}(\vec{r}_C)^* V(\vec{r}_0)\Phi_L(\vec{r}_0)d^3r_0$$

and

$$S^{\Omega}_{LL'}(\infty) = -\kappa \int_{\Omega} J_{L'}(\vec{r}_0)^* V(\vec{r}_0)\Phi_L(\vec{r}_0)d\ r_0 \ . \qquad (28b)$$

$B_{LL'}(\vec{k})$ is the usual structure constant matrix of KKR. Equation (27) is a set of linear equations for the $a_L(\vec{k})$,

$$\sum_L M_{LL'}(E,\vec{k})a_L(\vec{k}) = 0 \qquad (29)$$

with

$$M_{LL'}(E,\vec{k}) = \left[C^{\Omega}_{LL'}(\infty) + \frac{1}{\kappa} \sum_{L''} B_{L'L''}(\vec{k})\ S^{\Omega}_{LL''}(\infty) \right] \qquad (30)$$

The band energies are the $E(\vec{k})$ where the det $|M_{LL'}(E,\vec{k})| = 0$ and equation (29) has a non-trivial solution.

Note that the matrices $C^{\Omega}_{LL'}(\infty)$ and $S^{\Omega}_{LL'}(\infty)$ are evaluated from equations (28). The differentials to be solved are thus the same as those given by equations (18) for $C^{S}_{LL'}(r)$ and $S^{S}_{LL'}(r)$ with two important differences. First, $V(\vec{r})$ in the equations for $C^{\Omega}_{LL'}(\infty)$ and $S^{\Omega}_{LL'}(\infty)$ is restricted to the WS cell Ω. One thus decomposes the potential in the unit cell Ω, not the sphere S, when integrating (28). In cubic lattices the cellular potential can be decomposed in kubic harmonics which greatly simplifies the coupling of the problem via the $I(L,L'',L''')$ and reduces the number of terms to be evaluated. In lattices with inversion symmetry it is possible to use real spherical harmonics and positive kinetic energies (as described below) and thus avoid entirely the burden of complex arithmetic even for negative total energies.

Second, since $\Phi_L(\vec{r})$ is expressed in terms of the $C^{S}_{LL'}(r)$ and $S^{S}_{LL'}(r)$ (which are evaluated separately) the differentials in (28) decouple. $C^{\Omega}_{LL'}(\infty)$ and $S^{\Omega}_{LL'}(\infty)$ are evaluated by solving a matrix of *decoupled* ordinary integrals where $\Phi_L(\vec{r})$ is *already* evaluated using (10) and (12)!

We should note that $M_{LL'}(E,\vec{k})$ is usually written[27]

$$M_{LL'}(E,\vec{k}) = \left\{ \eta^{\Omega}_{LL'}(\infty) + B_{LL'}(\vec{k}) \right\} \qquad (31)$$

where

$$\eta^{\Omega}_{LL'}(\infty) = (-1)^{(\ell+\ell')} \kappa \sum_{L''} C^{\Omega}_{L''L}(\infty) \left[S_{L''L'}(\infty)\right]^{-1} \tag{32}$$

is the scattering matrix evaluated using the complete phase functional basis. The phase in (32) makes $\eta^{\Omega}_{LL'}(\infty)$ compatible with the $B_{LL'}(\vec{k})$ as defined elsewhere.[18,19]

The derivation presented here is done slightly differently from the way it is presented in our previous work[17,22]. We work only with the homogeneous term of the integral equation and we use projection to obtain equation (16) for an arbitrary $\vec{r} \in S$. This equation is general and is derived out of the context of band theory entirely. We do this to emphasize that the derivation of equation 16 is quite independent of the particular application of equation (16) to GFBT.

The advantage of deriving the secular equation directly from (16) (which always holds) is that the problem of a three center expansion for the Green's function (which orginally led to the MT approximation) *never arises in any form!* A completely straightforward algebraic step converts equation (24) to (25); the addition theorem for Neumann functions, Bloch's theorem and the definition of the structure constants then leads to the secular equation (27) with only the rather weak restriction that the sphere S must not contain any nearest neighbor centers (which is not desirable from the point of view of integrating the singularity there as well!)

This method of derivation also allows us to avoid the use of a variational step. As we have noted before[17,25], the variational step in our former derivation is quite harmless, quite valid, and yields useful information about the convergence properties and expected error. Nonetheless, it has been criticized[23] and we wish to make it clear that it is absolutely inessential to our results.

If one uses the basis functions of Williams and van W. Morgan[13] (or any similar set) instead of ours in step (23) of this derivation, one obtains a separated secular determinant with the form of equation (27), showing that the possible divergence predicted by Ziesche does not arise (since the three center Green's function is never used in this derivation). However, since these $\Phi_L(\vec{r})$ do not correctly propagate between S and Ω, they still make the near field error. This error can be viewed as resulting solely from the use of the wrong basis functions in equation (23), which do not correctly include the effects of scattering from the near field region.

RESULTS

In the case of the empty lattice (i.e.--a lattice with a constant, non-zero potential $V(\vec{r}) = \Delta$) the phase functional basis functions are given by

$$\Phi_L(\vec{r}) = \left[\frac{E}{E-\Delta}\right]^{\ell/2} j_\ell\left(\sqrt{E-\Delta}\, r\right) Y_L(\theta,\phi). \tag{33}$$

In the empty lattice, the correct $\Phi_L(r)$ are just the $J_L(\vec{r})$ themselves, scaled to the appropriate zero of total energy. In this case, the $\Phi_L(\vec{r})$ *manifestly* form a basis in terms of which the exact solution can be expanded, and the exact solution is in fact a linear combination of these $\Phi_L(\vec{r})$, with known coefficients.

The energy bands are efficiently obtained by evaluating $C^{\Omega}_{LL'}(\infty)$, $S^{\Omega}_{LL'}(\infty)$ and $\eta^{\Omega}_{LL'}(\infty)$ for a given fixed (positive!) kinetic energy E_0. The structure constants $B_{LL'}(E_0,\vec{k})$ are evaluated for the same E_0. The total energy $E = E_0 - \Delta$ is then varied by changing Δ and re-evaluating $\eta^{\Omega}_{LL'}(\infty)$ and $M_{LL'}(E,\vec{k})$ until the determinant is as close to zero as desired. Restricting E_0 to be positive eliminates (for lattices with inversion symmetry) the need for complex arithmetic even at negative total energies.

Because the determinant is generally a very large, highly non-linear function of Δ for large ℓ_{max} (in the scattered wave), it is best to examine the pivots of the determinant (obtained, say, from Gauss-Jordan elimination) individually to detect a zero. We found it fairly easy (depending of the degeneracy of the eigenvalue sought) to obtain five significant figures in our computation of the eigenenergies.

The BCC lattice has full cubic symmetry, therefore the lattice potential can be expanded in kubic harmonics instead of spherical harmonics. This effectively eliminates one index and simplifies the radial integrals (see equations (28) and (18). The $I(L', \ell'', L''')$ coefficients (where ℓ'' is the index of the kubic harmonic) have the property that the ℓ quantum numbers must triangle and their sum must be even for the coefficient to be non-zero. As a consequence, the largest ℓ'' coupled to a given ℓ_{max} (in the scattered wave, i.e.--in $C^{\Omega}_{LL'}(\infty)$ or $S^{\Omega}_{LL'}(\infty)$ is $2.\ell_{max}$. As a further consequence, even and odd parity terms do not mix in the scattering matrix.

We thus see that in order to estimate the expected degree of convergence we need to look at the relative sizes of the kubic components of the unit cell potential. In figure 2 we present $rV_{\ell''}(r)$ for $\ell'' \leq 12$ (which is the largest term coupled to $\ell_{max} = 6$ in the scattered wave) between r_{MT} (the radius of the MT sphere) and r_{BS} (the radius of the bounding sphere, S). The factor of r is part of the volume element in (28) or (18) and eliminates the singularity in atomic potentials.

The striking feature of these terms is the $\ell'' = 10$ term is surprisingly large; it is bigger than the $\ell'' = 8$ term and comparable in magnitude to $\ell'' = 6$. We might therefore expect that its inclusion will greatly improve the convergence of the odd parity eigenvalues. On the other hand, $\ell'' = 12$ is quite small and we might therefore expect its effect to be more ambiguous.

In the empty lattice the energies at the H-point and the second Γ-point are six and twelve-fold degenerate, respectively. In a numerical calculation of the band energies at these points [$E(H) = 1.0$ and $E(\Gamma) = 2.0$ in dimensionless units (d.u.)] the degeneracy is split by several things including numerical error, the error created by truncating the basis, the error created by truncating the expansion of the unit cell potential (which can be distinct if all the coupled terms in the potential are not kept) and any method error. The dimensionless units used throughout this study are related to Rydbergs by a factor of $(2\pi/a)$. For many metals this factor is nearly one.

One measure of the precision of candidate high precision band theory is, therefore, the rms splitting of the degenerate energies at the symmetry points. We can get some idea of the convergence properties of a given theory by studying the systematic behavior of the rms splitting as convergence-related properties are varied. Muffin tin theories are known to

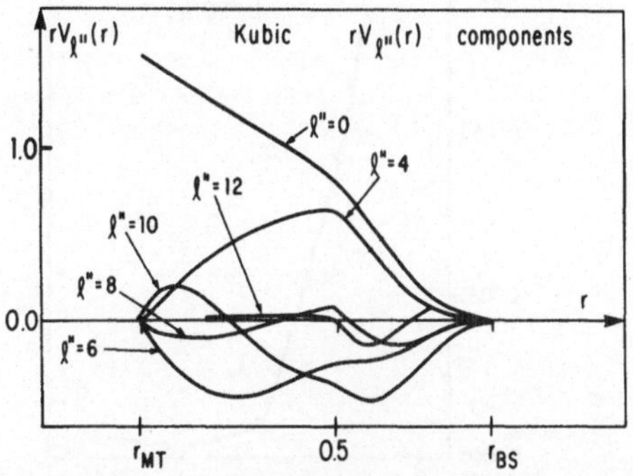

Fig. 2. Spherically decomposed components of the unit cellular potential (in units of a=1, for the BCC lattice) times the radius, between the muffin-tin (MT) sphere radius and the bounding sphere radius of the unit cell.

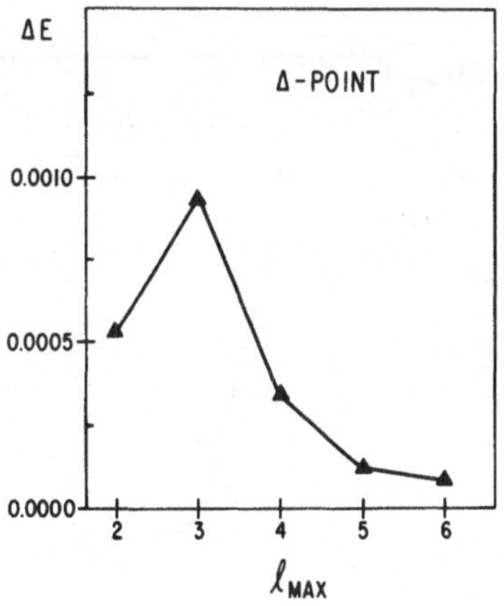

Fig. 3. The dependence of the rms splitting on ℓ_{max} in the scattered wave of the Δ point of the Brillouin zone, produced by the theory of present authors; Δ, the depth of the potential, is -0.25.

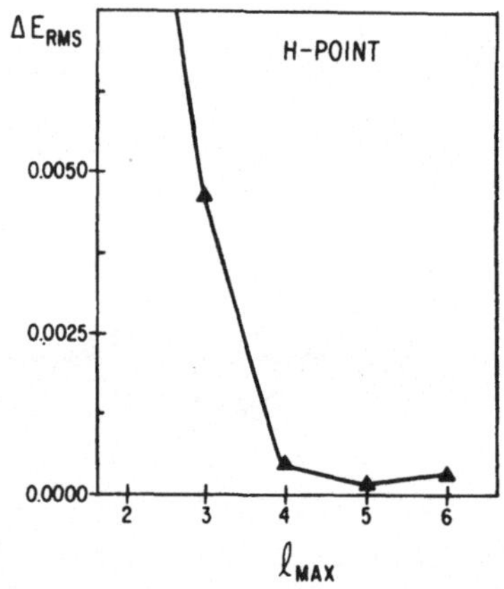

Fig. 4. The dependence of the rms splitting on ℓ_{max} in the scattered wave of the H-point of the Brillouin zone, produced by the theory of present authors; Δ, the depth of the potential, is - 0.25.

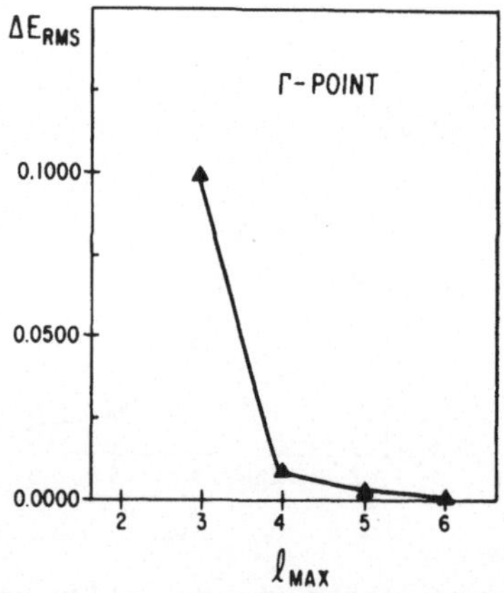

Fig. 5. The dependence of the rms splitting on ℓ_{max} in the scattered wave of the Γ-point of the Brillouin zone, produced by the theory of present authors; Δ, the depth of the potential, is - 0.25.

produce rather laıge rms erros (~0.01 - 0.10 d.u.) in the empty lattice case[23] and we can therefore appreciate the improvement in the calculated bands resulting from inclusion of non-MT effects.

In figures 3, 4 and 5 we present the change in the rms splitting produced by our theory when ℓ_{max} in the scattered wave is varied from 2 (3 in the case of the Γ-point) to 6 at a Δ-point ($\vec{k}$ = 0.6, 0, 0)), the H-point and the second Γ-point. Δ (the depth of the empty lattice potential) is fixed at -0.25. The error generally decreases as ℓ_{max} is increased, though there are exceptions. We can understand the exceptions by examining the behavior of the individual energies at these points as Δ is varied for fixed ℓ_{max} (see figure 6). The degree of convergence ultimately attained is remarkably high; the errors at ℓ_{max} = 6 on the order of 0.0001 - 0.001 d.u.

In figure 6, we show the behavior of the individual eigenvalues at the H-point as Δ is varied for ℓ_{max} = 4 and ℓ_{max} = 6 in the scattered wave. The eigenvalues (or the error) vary approximately quadratically in Δ, with the error always approaching zero as Δ approaches zero. From this figure we see why the convergence of the H-point fails to improve in figure 4 above; the Δ = -0.25 is located at a point where the error in H_{12} is coincidentally approaching zero because of the quadratic behavior. The error in H_{15} dramatically decreases as the ℓ'' = 10 term in the potential is coupled in and the rms error decreases. But when the ℓ'' = 12 term is coupled (as ℓ_{max} goes up to 6) the *form* of the H_{12} eigenvalue improves (it flattens out) but its *value* does not, and so the rms splitting actually grows a little worse.

This illustrates one danger in using rms splitting as a measure of a theory. One must separate out coincidentally good performance (illustrated in figure 7 at Δ ≈ -0.35) from systematically good performance. Figure 6 shows that the behavior of the eigenvalues *systematically* improves as ℓ_{max} is increased through the range of Δ.

It also illustrates another feature of the secular determinant; the importance of coupling. H_{12} is an ℓ = 2 state that is not directly coupled to the ℓ = 0 state. Including additional even-parity coupling terms in the potential like ℓ'' = 12 radically improves $E(H_1)$ (which is virtually exact across at ℓ_{max} = 6) but does not affect the more weakly coupled H_{12} eigenvalue as much. This effect is even more pronounced in the case of the Γ-point. The twelve eigenvalues at the Γ-point cannot be adequately represented at all unless ℓ_{max} is at least 3, since three of them couple to ℓ = 3, and higher odd terms.

Figure 7 displays the rms splitting at the H-point and the Γ-point as functions of for ℓ_{max} = 4 and ℓ_{max} = 6. In addition it displays the rms splitting produced by the theory of Williams and van W. Morgan when identically implemented in code.

At the H-point we observe the rms effect of the coincidental behavior of the individual eigenvalues in figure 6. The second zero in the quadratics produces a dip in the rms error around Δ = -0.35 in the ℓ_{max} = 4 results. This produces a rms splitting that is locally better than the ℓ_{max} = 6 results, in spite of their systematically superior performance. Because of this we feel that the best test of a theory's Δ convergence behavior is to look at the second derivative of the (approximately quadratic)

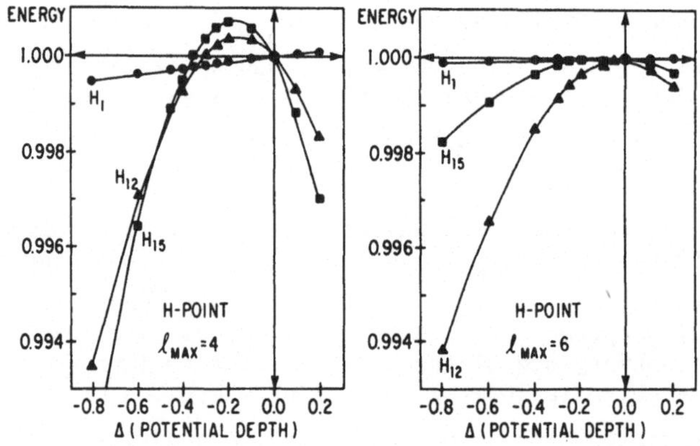

Fig. 6. The behavior of individual eigenvalues at the H point as the depth of potential Δ is varied, for ℓ_{max} = 6 in scattered wave, in dimensionless units.

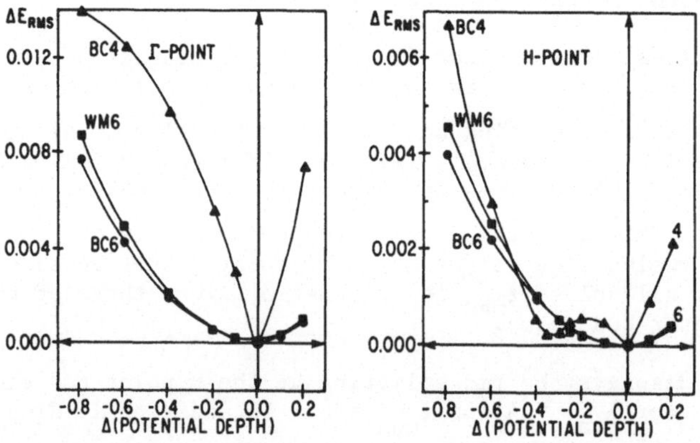

Fig. 7. Comparison of rms splittings obtained from the theory of the present authors (BC4 for ℓ_{max} = 6 in dimensionless units.

mean error. The flattest curve (that goes through zero) has the smallest second derivative and will produce the smallest rms splitting overall, although it may not be the best at any particular point.

The results at the Γ-point illustrate the poor convergence at $\ell_{max} = 4$ resulting from inadequate representation of the irreducible representations corresponding to individual eigenvalues. Since there are five distinct eigenvalues (with additional internal degeneracy) and only five ℓ-values represented, each eigenvalue is rather minimally represented. Increasing ℓ_{max} dramatically improves the behavior of the eigenvalues. As ℓ_{max} increases (and an additional term in the potential is coupled in) one can see the odd and even parity eigenvalues improve in alternate steps. At $\ell_{max} = 6$ they are sufficiently well represented that the error curves in figure 7 at the Γ and H-points are virtually scaled versions of each other.

Note finally that our theory produces systematically better results than that of Willians and van W. Morgan's at $\ell_{max} = 6$. Our curve is flatter and is lower at nearly all values of Δ. On the other hand, the difference is not very great, indicating that the near field error may be very small indeed.

Finally, in figure 8 below we present the results of applying our theory at many points on the principal symmetry axes to extract the bands. The values of Δ used varied between -.1 and -.25, which is a little larger than would be expected in a typical application. We used $\ell_{max} = 6$ to obtain the highest possible degree of precision. The largest error observed was less than 3×10^{-3} d.u., the smallest was around 1×10^{-4} d.u. On the scale of the actual bands plotted, the average error is (much) less than the width of the drawn line. This serves to emphasize that this really is a high precision band theory.

To summarize the numerical results, the theory exhibits a strong convergence to the correct energies as the size of the basis (scattering matrix) is increased. The actual size of the error depends on Δ (which is directly proportional to the neglected potential) approximately quadratically. The rms errors are much better behaved at $\ell_{max} = 6$ than at $\ell_{max} = 4$, though the degree of improvement depends on the details of the coupling of the different representations. Overall the study predicts errors on the order of mRydbergs or less in the application of the theory to real metals at these levels of convergence.

CONCLUSION

It is impossible to prove a theory correct or incorrect on the basis of numerical evidence alone. The most one can do is apply an exact theory to a problem where the exact answer is known and see how close the theory comes. Since even an exact theory becomes approximate on application due to the truncation of its (infinite) expansion set and numerical error, it is not correct to assume that the error produced by an exact theory should be zero.

What one is really testing is whether or not the results thus obtained are consistent with the theory being exact. If the results are in error by an inexplicably great amount, if they are diverging from the correct result when they should be converging, if the equations turn out to be unsolvable, then there may be a problem with the theory. On the other hand, if the theory converges to the correct answer as the number of basis functions is increased and if the absolute error produced is very small at a resonable

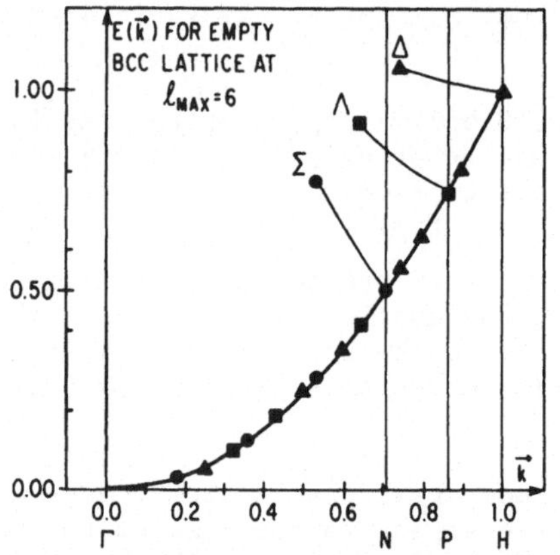

Fig. 8. Parts of the energy bands of the empty lattice obtained from the theory of the present authors demonstrating the high precision obtained, in dimensionless units. The error in energy is less than the width of the penline.

level of representation, then the theory is behaving correctly. If it produces converged answers that are better than those produced by a theory known to contain an additional approximation, then the theory is also behaving correctly.

We believe that the results that we present here are consistent with our theory being, in fact, an exact solution to the Schrödinger equation for the infinite lattice. Furthermore, we believe that the theory has certain features inherent to GFBT that will ultimately make it the most attractive band structure calculation scheme available to those interested in high resolution band structure and self-consistent applications. It is rapidly convergent, inexpensive to apply, and produces extremely high precision results. Future "fast band"[26] versions of the theory based on fitting curves through the matrix elements of $C^{\Omega}_{LL'}(\infty)$ and $S^{\Omega}_{LL'}(\infty)$ as functions of the depth of the potential will make it much less expensive to run without greatly compromising precision.

We are currently working on the application of this method to real crystals. We hope to present the results at some future date.

ACKNOWLEDGEMENTS

The authors wish to express their gratitude to H. L. Davis who provided them with listings of his programs that greatly facilitated this work.

REFERENCES

1) J. Korringa, Physica 13, 392 (1947).

2) W. Kohn and N. Rostoker, Phys. Rev. 94, 1111 (1967).

3) D. D. Koelling, Rep. Prog. Phys. 44 139, 187 (1981).

4) R. Evans and J. Keller, J. Phys. C 4, 3155 (1971).

5) J. Keller, J. Phys. C 4, L85 (1971).

6) H. Bross and K. H. Anthony, Phys. Status Solidi 22, 667 (1967).

7) F. Beleznay and M. J. Lawrence, J. Phys. C 1, 1288 (1968).

8) A. R. Williams, Phys. Rev. B 1, 3417 (1970).

9) M. A. Ball, J. Phys. C 5, L23 (1972).

10) A. R. Williams and J. van W. Morgan, J. Phys. C 5, 1293 (1972).

11) A. R. Williams, J. F. Jana-, and V. L. Moruzzi, Phys. Rev. B 6, 4509 (1972).

12) P. Ziesche, J. Phys. C 7, 1085 (1974).

13) A. R. Williams, and J. van W. Morgan, J. Phys. C 7, 37 (1974).

14) A. R. Williams, S. M. Hu and D. W. Jepsen, Recent Developments in KKR Theory, in *Computational Methods in Band Theory*, P. M. Marcus, J. F. Janak, and A. R. Williams, eds., Plenum Press, New York, 1971.

15) F. Calogero, *Variable Phase Approach to Potential Scattering*, Academic Press, New York, 1967.

16) J. S. Faulkner, Phys. Rev. B 19, 6186 (1979).

17) R. G. Brown and M. Ciftan, Phys. Rev. B 27, 4564 (1983).

18) B. Segall and F. S. Ham, *Methods in Computational Physics*, Academic Press, New York, 1968, Vol. 8, p. 251.

19) H. L. Davis, Efficient Numerical Techniques for the Calculation of KKR Structure Constants, *Computational Methods in Band Theory*, P. M. Marcus, J. F. Jank and A. R. Williams, eds., Plenum Press, New York, 1971.

20) G. S. Painter, J. S. Faulkner and G. M. Stocks, Phys. Rev. B 9, 2448 (1974).

21) N. Elyoshar and D. D. Koelling, Phys. Rev. B 13 5362 (1976).

22) R. G. Brown and M. Ciftan, A Generalized Non-Muffin-Tin Theory of Band Structure, *International Journal of Quantum Chemistry, Proceedings of Quantum Chemistry Symposium* No. 18, 1984, Per-Olov Löwdin, John R. Sabin, Michael C. Zerner, eds., John Wiley, p. 87.

23) J. S. Faulkner, Phys. Rev. B 32 (comments) 1339 (1985).

24) This is Fredholm's theorem; See for example, R. Courant and D. Hilbert, *Methods of Mathematical Physics*, Interscience, New York, 1953, Vol. I, pp. 112-122.

25) R. G. Brown and M. Ciftan, Phys. Rev. B 32 (response to ref. 23) 1343 (1985).

26) J. S. Faulkner and T. T. Beaulac, Phys. Rev. B 26, 1597 (1982).

27) This necessitates a redefinition of $a_L(\vec{k})$ in equation (29); see ref. 17.

CORRELATIONS IN FRACTIONAL HALL EFFECT

A. Kallio, P. Pollari, J. Kinaret and M. Puoskari

University of Oulu
Department of Theoretical Physics
SF-90570 Oulu, Finland

INTRODUCTION

Although most of the experimental facts [1-4] about the fractional quantum Hall effect can be understood on the basis of Laughlin wave function [5] or its generalizations [6-10], there still remains questions connected with the correlations and the thermodynamic limit. Many calculations have been done for few electron systems [11-13]. They are all restricted to the lowest Landau level, the argument also used by Laughlin [5] to set up his function. Can this restriction be relaxed? With a notable exception [15] most of the calculations in the thermodynamic limit use the analogy with the classical one-component plasma [5,16-18] to show e.g. that the Laughlin state lies lower in energy than the CDW (Charge-Density Wave)-states. The connection between a classical system and a physically completely different system is always interesting, but the restriction to only two particle correlations usually sets the limit to the practical usefulness of the analogy.

The purpose of the present paper is to apply the formalism of hypernetted chain for inhomogeneous quantum system. We derive Euler-Lagrange equation for two dimensional electron gas in magnetic field. This takes into account the effect of higher Landau levels in addition of the Laughlin-Jastrow correlations in the fractional 1/m-effect. We present an explicite proof for quantization of conductivity.

FULL AND FRACTIONALLY FILLED LANDAU LEVEL

We use the symmetric gauge $\vec{A}=\frac{1}{2}B(y\hat{x}-x\hat{y})$ and the lengths and energies are measured in the units $a=\sqrt{\hbar c/eB}$ and $\hbar\omega_c = \hbar(eB/m_e c)$. The Hamiltonian is

$$H = \frac{1}{2m_e} \sum_{i=1}^{N} [\vec{p}_i - \frac{e}{c}\vec{A}_i]^2 + \sum_{i<j} \frac{e^2}{r_{ij}} \tag{1}$$

The single particle states and energy of the kinetic energy part of the Hamiltonian are given for the lowest Landau level as

$$\varphi_m(z) = (2^{m+1}m!\pi a)^{-1/2} z^m e^{-|z|^2/4}, \quad m = 0,1,\ldots,N-1 = M$$

$$E_o = \frac{1}{2}\hbar\omega_c$$

with $z = (x + iy)a$ in the plane perpendicular to the magnetic field. When the lowest Landau state is filled one obtains by summing the squares of the single particle wave functions the density

$$\rho_1(r) = (2\pi a^2)^{-1} \sum_{m=0}^{N-1} \frac{(r^2/2)^m}{m!} e^{-r^2/2} \tag{2}$$

$$= (2\pi a^2)^{-1}[1-\xi_N(r)\ e^{-r^2/2}]$$

which is uniform up to R, the radius of the circular sample. This comes out of the study of the remainder $\xi_N(r)$. Within this radius we have N electrons with uniform density $(2\pi a^2)^{-1}$, hence

$$N = \pi R^2(2\pi a^2)^{-1} = \frac{1}{2}\frac{R^2}{a^2}. \tag{3}$$

In the thermodynamic limit the system is therefore clearly uniform.

The corresponding Slater determinant of aligned electrons made of the single particle states above is of the Vandermonde type and hence can be expanded in the product form

$$\Phi_1 = \prod_{i<j} (z_i - z_j)\ e^{-\sum_i |z_i|^2/4} . \tag{4}$$

It is manifestly antisymmetric and can easily be shown also explicitely to be eigenfunction of the kinetic energy part of the Hamiltonian

$$H_o = \sum_{k=1}^{N} \frac{1}{2m_e} (-i\hbar \vec{\nabla}_k - \vec{A}_k)^2 \tag{5}$$

Clearly then by using the analogy to the classical plasma at $\Gamma = 2$ the wave function (4) also minimizes the Coulomb and the total energy

$$\frac{E_o}{N} = \varepsilon_o + \frac{1}{2}\rho_1 \int_0^\infty d^2r(g_1(r)-1)\frac{e^2}{r} \tag{6}$$

The radial distribution function $g_1(r) = 1-e^{-r^2/2}$ can in principle be calculated from the HNC (Hypernetted Chain) theory

$$g_1(r) = f^2(r)\ e^{N(r)+E(r)} = e^{\ln r^2+N(r)+E(r)} \tag{7}$$

In the Fourier space the nodal function N(k) and the liquid structure factor $S_1(k) = 1-e^{-k^2/2}$ are related by the Örnstein-Zernicke relation

$$N(k) = \frac{(S-1)^2}{S} . \tag{8}$$

Since $S(k) \sim k^2$ for small k the logarithm in eq. (7) gets cancelled by a corresponding term in N(r). One notices that although the wave function (4) is of Jastrow form the correlation factor does not possess the usual property

$$\lim_{r\to\infty} |f^2| = 1.$$

Nevertheless the HNC-equations do possess a well defined solutions provided that we use in the Fourier transforms the small k - large r connections $k^{-2} \sim - \ln r$, $k^{-1} \sim r^{-1}$ etc. Throughout this paper we neglect the elementary diagrams E(r).

Suppose now that we diminish the number of electrons N_e<M but keep the ion disc radius R fixed. Amongst the many degenerate Slater determinants only one is again of the Vandermonde type except now the last occupied single particle state has maximum $m = N_e - 1$. Nevetheless it can be expanded in the product form (4) and by equation (2) the density becomes constant $(2\pi a^2)^{-1}$ in the thermodynamic limit. If we call this determinant Φ one may multiply it with a symmetric correlation factor $F(z_1, z_2, \ldots, z_{Ne})$ such that it gives an exact eigenstate of the full Hamiltonian. This way we can avoid taking complicated mixtures of Slater determinants as is necessary in the few electron calculations. Simultaneously with the particle numbers N_e< M we can look for an improved eigenfunction also for the full lowest Landau level, because after all the wavefunction (4) is uncorrelated. Hence our trial wave function would be for both cases written in the form

$$\Psi = F \prod_{i<j}^{N} (z_i - z_j)\, e^{-\sum |z_i|^2/4} \tag{9}$$

Once the wavefunction is fixed, one may compute the density distributions $\rho_1(r)$, $\rho(r)$ and radial distribution functions $g_1(1,2)$ and g(1,2) corresponding to the two cases from the usual definitions

$$\rho(r_1) = N \frac{\int d\tau_1 |\Psi|^2}{\int d\tau |\Psi|^2} \tag{10}$$

$$g_{12}\rho(r_1)\, \rho(r_2) = N(N-1) \frac{\int d\tau_2 |\Psi|^2}{\int d\tau |\Psi|^2} \tag{11}$$

In eq.(10) $\vec{r}_1$ is left unintegrated and in (11) both $\vec{r}_1$ and $\vec{r}_2$.

If in the thermodynamic limit we manage to get constant density and translationally invariant g_{12} the total potential energy will be given by the same expression as in eq. (6). The kinetic energy due to the correlations can be easily calculated and we obtain for the total energy the expression

$$E = N\varepsilon_o + \frac{N}{2}\rho\int(g(r)-1)\,\frac{e^2}{r}\,d^2r \tag{12}$$

$$- \frac{\hbar^2}{4m_e}\frac{N}{Q}\int d\tau\,[F^*\vec{\nabla}_1^2 F + F\vec{\nabla}_1^2 F^*]|\phi|^2$$

$$- \frac{\hbar^2}{2m_e}\frac{N}{Q}\int d\tau\,[(F^*\vec{\nabla}_1 F)(\phi^*\vec{\nabla}_1\phi) + (F\vec{\nabla}_1 F^*)(\phi\vec{\nabla}_1\phi^*)]$$

$$+ \frac{\hbar^2}{2m_e}\frac{N}{Q}\int d\tau\,\frac{eA_1}{\hbar c}\,[F^*\vec{\nabla}_1 F - F\vec{\nabla}_1 F^*]|\phi|^2,$$

where $Q = \langle\psi|\psi\rangle$.

Suppose now that $F(z_1,z_2,\ldots,z_N)$ is analytic in each variable z_i in the complex plane so that the Cauchy-Riemann equations are satisfied. It is then easily seen that the extra kinetic energy given by eq. (12) vanishes. In what follows we will restrict ourselves to the case

$$F = \prod_{i<j}^{N} f(z_i - z_j)$$

where the correlation factor $f(z)$ is analytic. This way we need to consider functions of only one complex variable. Further restrictions come from the antisymmetry of function (9) which requires $f(z)$ to be symmetric. Clearly $f(z)$ has to be free of branch points and poles, i.e. an entire function. We can therefore write

$$f(z) = e^{q(z)} = e^{u(x,y)+i\,v(x,y)}\ . \tag{13}$$

In the thermodynamic limit $|f(z)|^2$ and hence $u(x,y)$ must be a function of the distance r alone in order to obtain a uniform density. Requiring also the Cauchy-Riemann conditions one learns that $f(z) = z^p$ is the only possibility. All the other entire functions lead to CDW-states. For the latter ones we have also shown that only potential energy needs to be calculated if $f(z)$ is an entire function. For the uniform case we have shown that the Laughlin wave function is the only possibility in this class. We may now summarize both cases in the form

$$\psi_m = \prod_{i<j}^{N}(z_i - z_j)^m\, e^{-\frac{1}{4}\sum|z_i|^2}\ , \quad m = 1,3,5\ldots \tag{14}$$

where m=1 for the full Landau level, and odd for any fractionally filled state. Physically it's clear that the density for state ψ_m is smaller than the one for m=1 since the particles are pushed away from each other also by correlation function f.

SCALING OF THE DENSITY, ANGULAR MOMENTUM AND CONDUCTIVITY

Although the scaling of densities $\rho_m = \rho_1/m$ for the states of eq. (14) can be proven by the one-component plasma argument we want to do it also explicitely by resorting to the BBGKY-integral equation for the densities. Both wave functions ψ_m are of the form

$$|\psi_m|^2 = \prod_{i<j} f^2(r_{ij}) \prod_k R^2(r_k). \tag{15}$$

Using the definitions in eqs. (10)-(11) one may derive the following integral equation [19] for the densities (BBGKY-eq.)

$$\vec{\nabla}_1 \ln \frac{\rho(\vec{r}_1)}{R^2(\vec{r}_1)} = \int d\vec{r}_2 \rho(\vec{r}_2) g_{12} \vec{\nabla}_1 \ln f_{12}^2 \tag{16}$$

which is exact for any wave function of type (15). This equation has been found useful in the treatment of impurities and quantum surfaces [20-22].

Next we write for the two cases in eq. (13)

$$g_m(1,2)=1+h_m(1,2),$$

take the difference of the two respective eqs. (16) to eliminate the common factor $R^2(r_1)$, and finally get

$$\vec{\nabla}_1 \ln \frac{\rho_1(\vec{r}_1)}{\rho_m(\vec{r}_1)} = \int_A d^2r_2[\rho_1(\vec{r}_2)-m\rho_m(\vec{r}_2)]\vec{\nabla}_1 \ln r_{12}^2 \tag{17}$$

$$+ \int_A d^2r_2[\rho_1(\vec{r}_2)h_1(1,2)-m\rho_m(\vec{r}_2)h_m(1,2)]\vec{\nabla}_1 \ln r_{12}^2$$

Here the integration is performed over a finite domain A before going to the thermodynamic limit. If within this domain ρ_1 and ρ_m are constant then the first term on the right hand side of eq. (17) gives

$$2\pi(\rho_1-m\rho_m)\vec{r}_1 = 0, \tag{18}$$

since the angular integration on second line of eq. (16) vanishes when also the radial distribution functions become translationally invariant. Hence we obtain the correct scaling $\rho_m = m^{-1}\rho_1$.

For purposes of applying eq. (16) to a calculation of CDW states one may write it with $\rho_m(r) = \rho_o^m + \delta_m(r)$ thereby eliminating the R^2-factor against the diverging first term in eq. (16):

$$\vec{\nabla}_1 \ln \rho_m(\vec{r}_1) = \int d^2r_2\, \delta_m(r_2)\, \vec{\nabla}_1 \ln f_{12}^2 \tag{19}$$

$$+ \int d^2r_2\, \rho_m(r_2)\, h_m(1,2)\, \vec{\nabla}_1 \ln f_{12}^2 ,$$

where all terms are finite in the thermodynamic limit. The scaling equation (16) shows the intimate interplay between the correlation factor $|f|^2 = r_{12}^{2m}$ and the Gaussian factor $R^2 = e^{-r_1^2/2}$, which in fact makes it permissible to use such improper correlation factors from the point of view of hypernetted chain theory.

In passing one notices that the Slater determinant with m=1 in eq. (14) carries the minimum angular momentum $M_1 = N(N-1)/2$, whereas the Laughlin state ψ_m carries the angular momentum mM_1, so that $L_+ = \pi(z_i - z_j)$ is an angular momentum uppering operator for the states with constant density [5,15].

Based on the previous discussion one can also understand the quantization of the conductivity [24] σ_{xy} in the case of Laughlin wave function. Two essential ingredients are needed: The first is the connection of σ_{xy} with the linear density response function [25,26] $\chi(k)$ and the second is the small k expression for $\chi(k)$ in terms of the structure function [27] $S(k)$. It has been shown by McMullen[25] that for the 2D Hall system one obtains the following connection

$$\sigma_{xy} = - \frac{e^2 \omega_c}{2} \lim_{q \to 0} \frac{\partial^2 \chi(q)/\partial q^2}{1 - v_0(q)\chi(q)} \qquad (19a)$$

where the $v_0(q)$ is the coulomb interaction. On the other hand applying the HNC theory for mixtures, in zero concentration limit, one obtains the expression

$$\chi(q) = - \frac{4m_e \rho}{\hbar^2} \frac{S^2(q)}{q^2 a^2} \approx \frac{4m_e \rho}{\hbar^2} b^2 q^2 a^2 \qquad (19b)$$

where for the Laughlin state ψ_m one obtains by eqs. (7) and (8) $b = (4\pi m\rho)^{-1}$. Hence by eq. (19) one obtains, since the denominator becomes 1 for 2D Coulomb force, the expression

$$\sigma_{xy} = \frac{4m_e \rho}{\hbar^2} \frac{e^2 \omega_c}{2} \frac{2a^2}{16\pi^2 m^2 \rho^2} = \frac{1}{m} \frac{e^2}{h} .$$

This is the correct quantization rule. In what follows we show that the real correlations will not change the coefficient b. Furthermore adding impurities would simply be contained in δv_{ext} of ref. 25 and hence change nothing except when their concentration becomes so large that their effect can no longer be treated by linear response theory.

Finally since both formulas (19a) and (19b) are valid also for more general fractional states $\nu = P/m$ we can make prediction for the small q-behaviour of their liquid structure factor

$$S(q) \sim \frac{(qa)^2}{(4\pi\nu^{-1}\rho_\nu)a^2}$$

This should also be valid for normal QHE when ν is integer. Clearly this requires detailed wave function to be known. Approximate radial distribution function for $\nu = 2/5$ given by MacDonald et al. [28] satisfies this condition.

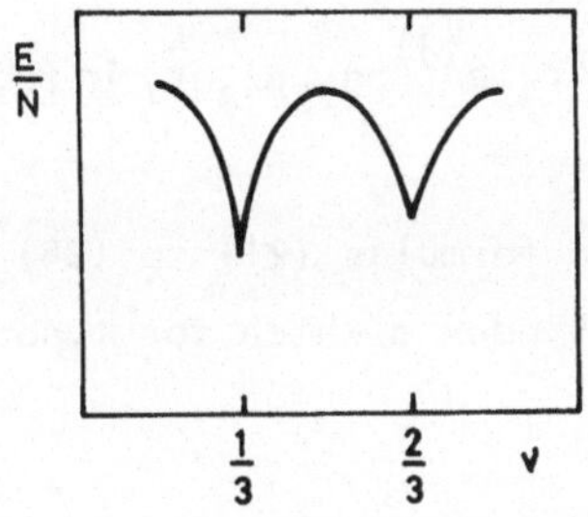

Fig. 1. Fractional state has no neighbouring states, hence compressibility is infinite. The state with $\nu = 2/3$ is the particle-hole conjugate [28] of the Laughlin state with $\nu = 1/3$.

OPTIMIZED CORRELATIONS WITH REAL FUNCTIONS

It was shown before that the Laughlin state exists as an infinitely sharp state below the sea of CDW states. Since there are no nearby states at all, the fluid is incompressible as is illustrated qualitatively in fig. 1. Can one obtain new nearby constant density states by adding on the top of previous complex correlations also real ones? To this end we call the previous Laughlin wave functions Φ_m and multiply them by a two-body correlation factor

$$|\Psi_m|^2 = \exp\left(\sum_{i<j} u_{ij}\right) |\Phi_m|^2 = |F|^2 |\Phi_m|^2, \quad m=1,3,... \tag{20}$$

with $u_{ij} = u(r_{ij})$ a real function. For normal electron systems such a wave function is known to produce a satisfactory accuracy. On the other hand if F is mearly an analytic function of its arguments z_i then again the kinetic energy vanishes and Ψ remains an eigenfunction of H_o so that the effect of higher Landau levels is not taken into account.

We may again use the formula (12) which now gives us the following kinetic energy expression:

$$\langle T\rangle = -\frac{\hbar^2 \rho N}{8 m_e}\left\{\int d^2r\, [g(r)\nabla^2 u + g(\vec{\nabla}u)\cdot\vec{\nabla}\ln(r^{2m})]\right. \tag{21}$$

$$\left. +\rho\int d^2r_{12} d^2r_{13} [g^{(3)}(1,2,3) - g_{12}g_{13}]\, (\vec{\nabla}_1 u_{12})\cdot\vec{\nabla}_1 \ln(r_{13}^{2m})\right\}$$

Here we have used the two lowest BBGKY-equations to reduce the 3-body term into Jackson-Feenberg form in eq. (21). We write it in the form which holds also for the CDW states:

$$g_{12}\, \vec{\nabla}_1 \log \frac{g_{12}}{f_{12}^2} = \int \rho_3 d^2 r_3 [g^{(3)} - g_{12} g_{13}] \vec{\nabla}_1 \ln f_{13}^2 \tag{22}$$

Here $f_{12}^2 = r_{12}^{2m} e^{u(r_{12})}$. In formulas (21) to (24) all functions g_{12}, $g^{(3)}$ and $u(r)$ should carry a subindex m which for typographical reasons has been dropped.

The best way to optimize is to derive Euler-Lagrange equations [20-22] for g(r). For normal quantum fluids this can be done by eliminating u with HNC-connection

$$g = e^{u + \ln r^{2m} + N(r)} \tag{23}$$

$$N(r) = \frac{1}{(2\pi)^2 \rho} \int d^2k\, e^{i \vec{k}\cdot\vec{r}} \frac{(S-1)^2}{S} \tag{24}$$

from the energy expression. Equation (24) is solution of the Örnstein-Zernicke equation. Here this approach is hampered by the existence of the logarithmic term as was discussed before. It is, however, evident from eq. (21) that one actually needs derivatives and for them the Fourier transforms do exist. Take, for example, the full Landau level with $g_1(r) = 1 - \exp(-r^2/2)$. The derivative of N is

$$\frac{dN}{dr} = \frac{r \exp(-r^2/2)}{1 - \exp(-r^2/2)} - \frac{2}{r}$$

and is F-transformable with only 1/r-singularity.

Clearly the 3-body term is intractable in general terms. Let us see if the Euler-Lagrange equation is consistent with the small-k behaviour of S(k) which has to be [14] quadratic by eqs. (23) and (24).

The E-L equation is similar to that of the Coulomb gas

$$-\frac{\hbar^2}{m_e} \nabla^2 \sqrt{g} + \left(\frac{e^2}{r} + W(r) + W_m(r) + W_3(r)\right) \sqrt{g} = 0 \tag{25}$$

$$W(k) = -\frac{\hbar^2 k^2}{4m_e} \frac{(S-1)^2}{S^2} (2S + 1) \tag{26}$$

where W(k) comes from the variation of the first two-body term in the standard way. The induced potential $W_m(r)$ comes from the variation of the second term in eq. (21) A straightforward calculation gives

$$W_m(r) = \frac{\hbar^2}{m_e} \left[\frac{m^2}{r^2} + \frac{m}{2r} \frac{dN}{dr} \right]$$

$$- \frac{\hbar^2 m}{2m_e} \frac{1}{(2\pi)^2 \rho} \int e^{i \vec{k}\cdot\vec{r}}\, \xi_k \frac{S_k^2 - 1}{S_k^2}\, d\, k \tag{27}$$

Here ξ_k is the Fourier-transform

$$\xi_k = \rho \int e^{i\vec{k}\cdot\vec{r}} \left(\frac{1}{r}\frac{dg}{dr}\right) d^2r. \tag{28}$$

Simple inspection shows that the small r behaviour of radial distribution function $g(r) \sim r^{2m}$ remains the same as in the unperturbed state. At large distances the first two terms in W_m cancel the r^{-2} singularity.

For normal 2-dimensional Coulomb gas with $\vec{A} = 0$ one would obtain eq. (25) with $W_m = 0$. At large distances the Coulomb interaction e^2/r has to get cancelled by a counterterm in $W(r)$. That happens if the structure factor behaves like $S \sim k^{3/2}$ for small k. Here the behaviour of S has to be quadratic which means that $W(r)$ has a logarithmic term. That in turn has to be cancelled by a corresponding term in $W_m(r)$. This would require that $\xi_k = \alpha k^2$ for small k. From the definition we get $\xi(k=0) = 2\pi\rho$. Hence it does not even vanish at k=0. We therefore conclude that there is no consistent solution for the Jackson-Feenberg form of kinetic energy without the 3-body term.

In order to get a consistent solution for g we have to take into account also the three-body kinetic energy term. We use the superposition approximation $g^{(3)}_{123} = g_{12}g_{13}g_{23}$ (the other simple possibility, the convolution approximation would lead to a singularity near the origin due to the r^{2m}-behaviour of g). Straightforward variation with respect to g gives us the following additional terms in the induced potential

$$W_3(r) = W_g + W_{ug} + W^{(1)}_{uN} + W^{(2)}_{uN}$$

$$W_g = -\frac{\hbar^2}{2m_e}\rho m \int d^2r_{13}g_{13}g_{23}\left[\frac{\hat{r}_{12}\cdot\hat{r}_{13}}{r_{12}}\frac{du_{13}}{dr_{13}} + \frac{\hat{r}_{12}\cdot\hat{r}_{13}}{r_{13}}\frac{du_{12}}{dr_{12}} + \frac{\hat{r}_{13}\cdot\hat{r}_{23}}{r_{23}}\frac{du_{13}}{dr_{13}}\right]$$

$$W_{ug} = \frac{\hbar^2}{2m_e}\rho m \int d^2r_{13}\frac{\hat{r}_{12}\cdot\hat{r}_{13}}{r_{13}}\frac{d\ln g_{12}}{dr_{12}} g_{13}g_{23}$$

$$W^{(1)}_{uN} = \frac{\hbar^2}{2m_e}\frac{m}{(2\pi)^2}\left[\iint e^{i\vec{k}\cdot\vec{r}}\frac{S_k^2-1}{S_k^2}\xi(k)\,d^2k - \int e^{i\vec{k}\cdot\vec{r}}\frac{S_k^2-1}{S_k}\xi(k)\,d^2k\right]$$

$$-\frac{\hbar^2}{2m_e}\frac{m}{(2\pi)^4\rho}\iint d^2k\,d^2q\; e^{i\vec{k}\cdot\vec{r}}\frac{(S_k^2-1)}{S_k^2}(S_k-1)(S_{k-q}-1)\xi(q)$$

$$W^{(2)}_{uN} = \frac{\hbar^2}{2m_e}\frac{m}{(2\pi)^4\rho}\iint d^2k\,d^2q\; e^{i\vec{k}\cdot\vec{r}}\,\vec{q}\cdot(\vec{k}-\vec{q})\,\frac{(S_k^2-1)}{S_k^2}(S_q-1)(S_{k-q}-1)\,\varphi(q)$$

$$\xi(r) = \frac{d\varphi}{dr}$$

Now we see immediately that the first term of $W^{(1)}_{uN}$ cancels the singular term $(S_k^2-1)\xi_k/S_k^2$ in W_m while the second term cancels the k^2/S^2-singularity in W. The first term of W_g and the term W_{ug} behave as r^{-1} and thus can take care of the Coulomb term in the EL-equations. So we conclude that including the three-body kinetic energy terms in the EL-equation it seems possible to find a consistent solution which has physically meaningful small and large r-behaviours. Without explicitely constructing numerical solution one cannot make any further comments. It is clear that the inclusion of the higher Landau levels should depress the energy and therefore one should have a consistent solution. As mentioned before the liquid structure factor has to behave like $S_k \sim bk^2$ hence the Feynman spectrum would have a gap $E(k) \sim h^2/2mb$ but b is not changed by correlations.

Finally we may ask what interaction v does one obtain if we adopt the present theory with u=0, $W_m=0$. Eq. (25) would now give with the known radial distribution function g(r) of the fractional Laughlin state

$$v(r) = \frac{\hbar^2}{m_e} \frac{\nabla^2 \sqrt{g}}{\sqrt{g}} - W(r)$$

This would give the following behaviours

$$\lim_{r\to\infty} v(r) = A \ln \frac{r}{a}$$

$$\lim_{r\to 0} v(r) = \frac{\hbar^2 m^2}{m_e\, a^2} \left(\frac{a}{r}\right)^2 \quad , \; m=3,5,\ldots$$

For the full lowest Landau state there will be no r^{-2}-term.

REFERENCES

1 D.C. Tsui, H.L. Störmer and A.C. Gossard, Phys. Rev. Lett. 48, 1559 (1982).
2 M.A. Paalanen, D.C. Tsui and A.C. Gossard, Phys. Rev. B25, 5566 (1982).
3 H.L. Störmer, A. Chang, D.C. Tsui, J.C.M. Hwang, A.C. Gossard and W. Wiegman, Phys. Rev. Lett. 50, 1953 (1983).
4 E.E. Mendez, L.L. Chang, M. Heiblum, L. Esaki, M. Naughton, K. Martin and J. Brooks, Phys. Rev. B30, 7310 (1984).
5 R.B. Laughlin, Phys. Rev. Lett. 50, 1395 (1983).
6 B.I. Halperin, Helv. Phys. Acta 56, 75 (1983).
7 F.D.M. Haldane, Phys. Rev. Lett. 51, 605 (1983).
8 A.H. MacDonald, Phys. Rev. B30, 3550 (1984).
9 C. Kallin and B. Halperin, Phys. Rev. B30, 5655 (1984).
10 T. Chakraborty, Phys. Rev. B31, 4026 (1985).
11 R.B. Laughlin, Phys. Rev. B27, 3383 (1983).
12 S.M. Girvin and T. Jach, Phys. Rev. B28, 4506 (1983).
13 F.C. Zhang and T. Chakraborty, Phys. Rev. B30, 7320 (1984).
14 S.M. Girvin, Phys. Rev. B30, 558 (1984).
15 F.D.M. Haldane and E.H. Rezayi, Phys. Rev. Lett. 54, 237 (1985).
16 B. Jancovici, Phys. Rev. Lett. 46, 386 (1981).
17 D. Levesque, J.J. Weis and A.H. MacDonald, Phys. Rev. B30, 1056 (1984).
18 J.P. Hansen and D. Levesque, J. Phys. C14, 1603 (1981).
19 E. Feenberg, Theory of Quantum Fluids, Academic 1969, New York.
20 M. Saarela, P. Pietiläinen and A. Kallio, Phys. Rev. B27, 231 (1983).
21 P. Pietiläinen and A. Kallio, Phys. Rev. B27, 224 (1983).
22 J.M. Caillol, D. Levesque, J.J. Weis and J.P. Hansen, J. Stat. Phys. 28, 325 (1981).
23 S.M. Girvin and T. Jach, Phys. Rev. B29, 5617 (1984).
24 R.B. Laughlin, Phys. Rev. B23, 5632 (1981).
25 T. McMullen, Phys. Rev. B32, 1415 (1985).
26 Qian Niu, D.J. Thouless and Yong-Shi Wu, Phys Rev. B31, 3372 (1985).

[27] A. Kallio, M. Puoskari, L. Lantto, P. Pietiläinen and V. Halonen, Proceedings of "Recent Progress in Many-Body Theories", Altenberg, Germany 1983, Lecture Notes in Physics 198, p. 210.
[28] A.H. MacDonald, G.C. Aers and M.W.C. Dharma-wardana, Phys. Rev. B31, 5529 (1985).

HEAVY FERMION SYSTEMS: FERMI LIQUID ASPECTS AND MODEL CALCULATIONS

Khandker F. Quader

Physics Department
University of Illinois at Urbana-Champaign
Urbana, Illinois 61801

INTRODUCTION

The exotic "heavy fermion" systems (HFS) with enormous effective masses, $m^*/m \sim O(10^2)$ have recently generated a great deal of interest in the condensed matter and many-body community. These f-electron systems are[1] superconducting (s.c.), magnetic, or non-superconducting and non-magnetic (Table 1). Below $T \sim O(10K)$, they exhibit large linear coefficient, γ ($c_v = \gamma T$), and enhancement in magnetic susceptibility, χ, suggestive of strongly interacting Fermi liquids with large many-body renormalizations. Though the nature of superconductivity (for the superconductors) is still an outstanding question, there have been suggestions[1-4] that these may be the first triplet metallic superconductors in nature.

The accumulation of a phenomenal number of experiments in this rapidly developing field have posed a challenge for theorists[1]. Among the intriguing questions are the physical origin of the heavy electrons, the nature of the Fermi liquid and that of the superconducting and the magnetic states, the role of spin fluctuations, spin-orbit interactions, etc. In a fast developing field as this, any attempt at a review is outside the scope of this article; (see Ref. 1 for a current review). Here, we shall focus on the Fermi liquid aspects of these systems, UPt_3 in particular. Since we assume isotropy (though we do not assume Galilean invariance), this could be viewed as a "spherical cow" description. However, given some of the experimental features in UPt_3, especially the existence of a $T^3 \ell n T$ term in C_v (a general property of normal Fermi liquids[5]), it is tempting to apply Fermi liquid theory (FLT) to analyze the experiments.

Table 1. Typical Heavy Fermion Systems and some Properties (sc = superconducting, am = antiferromagnetic)

Systems	γ mJ/mol-K^2	m^*/m	$\chi(T\to 0)$ 10^{-3}mu/mol	T_c (K)
$CeCu_2Si_2$ (sc)	1100	460	7	0.6
UBe_{13} (sc)	1100	300	15	0.9
UPt_3 (sc)	429	178	7	0.5
U_2PtC_2 (sc)	75	47	4.4	1.4
U_6Fe (sc)	24	20	3.1	3.8
U_2Zn_{17} (am)	535	>100	12.5	9.7
UCd_{11} (am)	840	>100	38	5.0
$NpBe_{13}$ (am)	900	230	56	3.4
$CeAl_3$	1600	600	36	---
$CeCu_6$	1600	740	27	---
UAl_2	140	55	4.3	---

The discussion will essentially be in two parts. In the first part, we discuss how we deduce[6] the properties of UPt_3 by analyzing various experiments. The $T^3 \ell nT$ term in specific heat under pressure[7], $C_v(P)$, enables us to obtain the $\ell = 0$ antisymmetric Landau parameter, $F_o^a(P)$. This, together with sum-rule arguments leads us to conclude that UPt_3 is a triplet superconductor of purely electronic origin; our calculated transition temperature as a function of pressure, $T_c(P)$ agrees well with experiments[8]. For consistency, we consider as well the relationship between thermal expansion, α and $C_v(P)$. We also show that FLT extended to finite wavevectors and energies leads to strong spin fluctuation peaks in neutron scattering experiments, and provide an explanation of the results of Aeppli et al.[9]. In passing, we note the shortcomings of "paramagnon" theory and attempts to determine F_o^a from susceptibility. Other heavy fermion systems, UAl_2 and $TiBe_2$ are also briefly considered.

In the second part, we discuss how we apply[10] the "induced" interaction model[11] to calculate the Landau parameters in typical heavy mass systems. From this, we conclude that in systems with large m^*/m, p-wave pairing is favored. Additionally, in the extreme limit of our model, m^*/m diverges while F_o^a saturates to a finite value. This is suggestive of a Mott-like localization (instead of a ferromagnetic divergence) found in Gutzwiller-Hubbard approaches[12].

FERMI LIQUID ASPECTS

The recent experimental[1,7,13] data on the specific heat of UPt_3, UAl_2 and $TiBe_2$ can be fit by an expression containing a $T^3 \ln T$ term

$$C_v = \gamma T + \delta T^3 \ln T + \beta T^3 \tag{1}$$

The work of Pethick and Carneiro[5] in terms of "statistical" quasiparticles showed that long-wavelength fluctuations lead to non-analytic behavior of the quasi-particle interaction, $f_{p,p+q}$ (small q), which then produces a $T^3 \ln T$ term in C_v. Fermi liquid theory enables one to obtain a relationship for the coefficient of this term:

$$\delta = -\frac{3\pi^2}{10}\,\gamma\,\frac{B^s}{T_F^2} \tag{2}$$

where $\gamma = (\pi^2/3)k_B^2 N(0)$ is the linear coefficient in C_v, $N(0) = m^* k_f/\pi^2 = \frac{3}{2}\frac{n}{\varepsilon_F}$ is the density of states, and $\varepsilon_F = k_F^2/2m^*$ is the Fermi energy for the equivalent spherical Fermi surface of radius k_F. B^s is given by the scattering amplitude (hence by $f_{p,p+q}$). For $\ell \leqslant 1$

$$B^s = -\frac{1}{2}\sum_{\lambda=s,a} \omega_\lambda[(A_o^\lambda)^2\{1+A_1^\lambda - \frac{\pi^2}{12}A_o^\lambda\}+(A_1^\lambda)^2\{1 - \frac{\pi^2}{48}A_1^\lambda\} - 2A_o^\lambda A_1^\lambda] \tag{3}$$

where s and a refer to symmetric and antisymmetric terms, $\omega_s = 1$, $\omega_a = 3$, and the Landau scattering amplitudes are

$$A_\ell^\lambda = \frac{F_\ell^\lambda}{1 + F_\ell^\lambda/(2\ell + 1)} . \tag{4}$$

For electron systems, $A_0^s = 1$; hence an appreciable $T^3 \ln T$ term in C_v requires a large negative value of A_0^a. From the experimental results of Brodale et al. for γ and δ as functions of pressure, one obtains the corresponding values of B^s/T_F^2. On adopting the value of $k_F = 1.08\ A^{-1}$, we obtain the results given in Table 2 for and F_0^a. We note that F_0^a is large, negative and decreases with increasing density.

It has been noted by Brown et al. that F_0^a cannot be obtained from the experimental values of the spin susceptibility and the specific heat (as done in liquid ^3He) because in the present case the effective moment of the heavy fermion quasiparticle is not known. Given, however, our deduced values of F_0^a, one can invert this procedure, and determine μ_{eff}; for UPt_3, $\mu_{eff} \approx 0.5\ \mu_B$.

To examine the possible transition to the superconducting state, we use the approximate expression of Patton and Zaringhalam[15], and the s-p approximation expression of Dy and Pethick[16]. For electron systems, since $A_0^s = 1$, the forward scattering sum rule in the s-p aproximation reads,

$$A_1^s + A_1^a = -(1 + A_o^a) \tag{5}$$

on neglecting possible phonon contributions to the scattering amplitudes. For the large negative values of A_0^a obtained here, it is readily seen from Eq. (5) that p-state pairing is preferred.

The transition temperature for our model calculation of triplet pairing is given by

$$T_c = T^* \exp(1/g_1) = \alpha T_F \exp(6/(1 + A_0^a)) \tag{6}$$

where $T^* = \alpha T_F$ is a cut off (analogous to the Debye cutoff introduced in BCS theory) which reflects the fact that the interactions we consider are highly frequency dependent, and the expression used for the effective interaction is valid only for a restricted range of energies, T^*, in the immediate vicinity of the Fermi surface. On substituting the values we have obtained for A_0^a into Eq. (6), on choosing the cutoff α (=0.0103) so as to get the correct value of T_c at zero pressure, and on assuming that α does not vary with pressure, we obtain the results shown in Table 2. Given the approximate nature of our calculations, the agreement with experiment is highly gratifying.

Thermodynamics provides a connection between measurements of thermal expansion and the pressure dependence of the specific heat. The volume coefficient of thermal expansion is

$$\alpha_v = -\left(\frac{\partial s}{\partial P}\right)_T \tag{7}$$

where s is the entropy per unit volume; on making use of Eq. (7), we find for the average linear coefficient of thermal expansion,

$$\alpha \equiv \alpha_v /3 = \gamma_\alpha T + \delta_\alpha T^3 \ln T + O(T^3), \tag{8a}$$

where

$$\gamma_\alpha = -\frac{1}{3}\ (\partial\gamma/\partial P) \tag{8b}$$

and

$$\delta_\alpha = \frac{1}{9}\,\delta\kappa\ [2 - \frac{3}{\kappa\gamma}\frac{\partial\gamma}{\partial P} - \frac{1}{\kappa B^s}\frac{\partial B^s}{\partial P}\ ], \tag{8c}$$

with κ as the compression modulus. Our fit to the thermal expansion[17] yields the values $\gamma_\alpha = 1.12 \times 10^{-6} K^{-2}$ and $\delta_\alpha = 0.765 \times 10^{-8} K^{-4}$; these coefficients are in excellent agreement with the values we calculate from a quadratic fit to the specific heat experimental results of Brodale et al.,[7] $\gamma_\alpha = 1.05 \times 10^{-6} K^{-2}$ and $\delta_\alpha = 0.8 \times 10^{-8} K^{-4}$. The internal consistency of the two experiments is thereby demonstrated.

The recent neutron scattering experiments of Aeppli et al.[9] provide a valuable check on the Fermi liquid picture we have developed for UPt_3. The spin-spin correlation function at finite wavevectors and energies, takes the form

$$\chi^I(q,\omega) = \frac{\chi_o(q,\omega)}{1 - f^a_q\ \chi_o(q,\omega)}\ , \tag{9}$$

where f^a_q is a non-local potential which reduces in the long wavelength limit to the Landau result: $\lim_{q\to 0} f^a_q = F^a_o/N(0)$ and $\chi_o(q,\omega)$ is the Lindhard function for quasiparticles of mass m^*. On taking this

Table 2. Fermi liquid parameters relevant to the $T^3 \ell nT$ terms in the specific heat of UPt_3. The corresponding parameters for UAl_2 and $TiBe_2$ are given for the purpose of comparison.

System	Pressure (kbar)	γ (mJ/mole-K^2)	δ (mJ/mole-K^4)	k_F (Å^{-1})	m^*	E_F (K)	F^a_0	A^a_0	T_c^{Theo} (K)	T_c^{Exp} (K)
UPt_3	0	429	1.90	1.08	178	289	-0.811	-4.29	0.48	0.48
	3.8	378	0.97	1.08	157	328	-0.794	-3.85	0.41	0.43
	8.9	332	0.46	1.08	138	375	-0.773	-3.41	0.32	0.36
UAl_2	0	142	1.4	1.15	55	1054	-0.93	-13.8		
$TiBe_2$	0	56	0.066	1.38	39	1500	-0.91	-11.1		

expression, and neglecting any q dependence in f^a_q, one finds for $S^I(q,\omega) \equiv -(1/\pi)\mathrm{Im}\,\chi^I(q,\omega)$, the dynamic spin-spin structure function, the result shown by the dashed line in Fig. 1 for $q = 1Å^{-1}$. The spin fluctuation excitation peak at 8 meV, which has been calculated with no free parameters from our fit to the specific heat results, is in quantitative agreement with the neutron scattering experiments which show a broad peak at around 10 meV. To take into account experimental resolution, we next introduce Lorentz broadening

$$S^I_L(q,\omega) = \int d\omega'\, S^I(q,\omega')\, \frac{1}{\pi} \frac{\Gamma}{(\omega' - \omega)^2 + \Gamma^2} \tag{10}$$

with the half-width $\Gamma = 4$ meV given by the instrumental resolution of Aeppli, et al.[9]. As evident from the calculated $S^I_L(q,\omega)$, shown by the full line in Fig. 1, the agreement with the experiments is even better, the peak now being at ~ 10 meV. It should be noted that if one neglects the spin-spin restoring force, f^a_q, one obtains only the quite broad

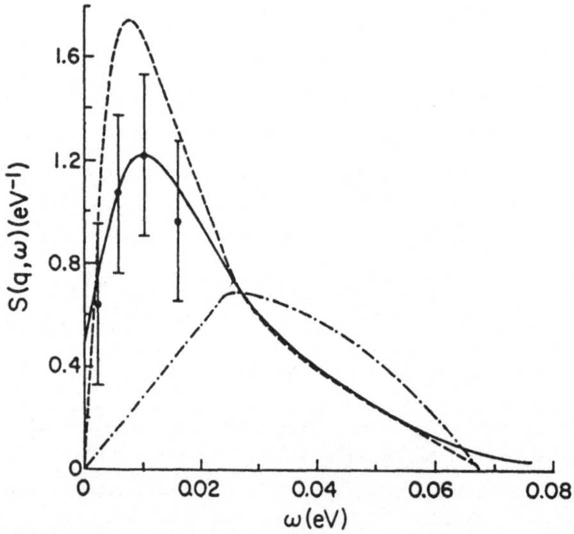

Fig. 1 The calculated dynamic spin-spin structure function for $q = 1Å^{-1}$. The full line is the result including both the effects of the spin-spin restoring force, f^a_q, and Lorentz broadening (discussed in text); the dashed line is that without the broadening, while the dot-dashed line is the corresponding result neglecting f^a_q. The experimental points with error bars are those of Ref. 9.

peak shown by the dot-dashed line in Fig. 1; thus any attempt to explain experiment without invoking f_q^a, requires an effective quasiparticle mass at least twice that derived from specific heat experiments.

Regarding the often used "paramagnon" model, we note that paramagnon calculations are model-dependent, since one must make an assumption about the physical origin of the effective mass. Additionally, as known[5] they can easily lead to an overestimate of the spin fluctuation contribution to the $T^3 \ln T$ term by a factor as large as three owing to the use of "dynamical" quasiparticles rather than the appropriate "statistical" ones.

It is instructive to consider UAl_2 and $TiBe_2$ which also exhibits a pronounced $T^3 \ln T$ term in the specific heat. We find rather negative values for F_0^a (Table 2) so that in these metals one is very close to a ferromagnetic instability - so close indeed, that any application of Fermi liquid theory to calculations of transport properties or the superconducting transition temperature must necessarily involve quasiparticle interactions in many different l channels.

There are a number of features of the experimental results on normal UPt_3, (e.g. the temperature dependence of the resistivity, thermoelectric power, etc.) which deserve examination using Fermi liquid theory, as well as others (e.g. the temperature dependence of the quasiparticle mass, the anisotropic spin susceptibility and the magnetoresistance) which clearly find a physical origin elsewhere. We have demonstrated that Fermi liquid theory offers a useful starting point for understanding the low temperature properties of UPt_3; it will be interesting to see whether it is equally useful for the other heavy fermion superconductors.

THE INDUCED INTERACTION MODEL FOR HEAVY FERMIONS

Our objective is to construct a simple soluble Fermi liquid theory (FLT) for heavy fermions, starting from a microscopic description and using some empirical imput. It is known[11] that a consistent FLT cannot be formulated in terms of short-range effective interaction, alone; collective excitations generated by these must be exchanged between the quasiparticles.

Here the "induced" interaction model[11] of interacting quasiparticles and collective excitations is adapted[10] to the heavy fermion case. The basic idea is that it is possible to separate the quasiparticle interaction $f_{pp'}^{\sigma\sigma'}$ into two parts (see Fig. 2):

$$f_{pp'}^{\sigma\sigma'} = d_{pp'}^{\sigma\sigma'} + I_{pp'}^{\sigma\sigma'}\left(f_{pp'}^{\sigma\sigma'}\right) \tag{11}$$

where the induced part, $I_{pp'}^{\sigma\sigma'}$ is particle-hole reducible in the crossed channel whereas the "direct" part, $d_{pp'}^{\sigma\sigma'}$ is not particle-hole reducible. Physically, $I_{pp'}^{\sigma\sigma'}$ contains the exchange of virtual collective modes, e.g. density, spin-density, current, and spin-current modes, between the quasiparticles. These arise on demanding that the scattering amplitudes, $A_{pp'}^{\sigma\sigma'}$ are properly antisymmetrized (Fig. 2). The driving term, $d_{pp'}^{\sigma\sigma'}$ contains the scattering of two quasiparticles without the intervention of another. Thus, it has information about the underlying potential, and can be short-ranged.

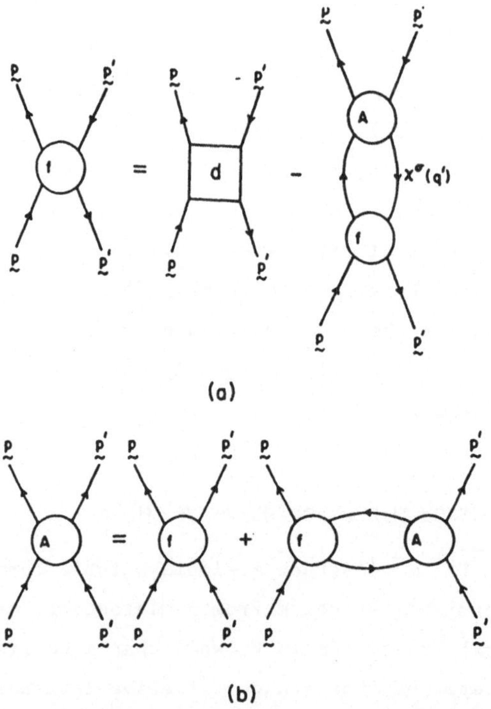

Fig. 2 (a) The particle-hole interactions $f_{pp'}^{\sigma\sigma'}$ with $\chi(q') = N(0)\ U(q')$ [Eqs. (11)-(13)]. (b) The scattering amplitudes $A_{pp'}^{\sigma\sigma'}$; the induced term contained in the first term is the exchange of the second term here.

In accordance with the form of Eq. (11), and with $F^a_{pp'}\sigma.\sigma' = F_{pp'} + F^a_{pp'}\sigma.\sigma'$, $F^{\sigma\sigma'}_{pp'} = N(0)\, f^{\sigma\sigma'}_{pp'}$, the induced interaction model equations are:

$$F^s_{pp'} = D^s_{pp'} + \frac{1}{2}(\gamma^s(q') + 3\,\gamma^a(q')) + \frac{1}{4}(\beta^s(q') + 3\,\beta^a(q')) \tag{12}$$

$$F^a_{pp'} = D^a_{pp'} + \frac{1}{2}(\gamma^s(q') - \gamma^a(q')) + \frac{1}{4}(\beta^s(q') - \beta^a(q')) \tag{13}$$

where

$$\gamma^{s,a}(q') = \frac{(F_o^{s,a})^2 U_o(q')}{1+F_o^{s,a}\, U_o(q')}$$

and

$$\beta^{s,a}(q') = (1+x)\,\frac{(F_1^{s,a})^2 U_1(q')}{1+F_1^{s,a}\, U_1(q')}$$

with $q'^2 = 2k_F^2(1-x)$, $x = \cos\theta_L$. $U_o(q')$ and $U_1(q')$ are the Lindhard functions[11]. We have truncated $I^{\sigma\sigma'}_{pp'}$ at $\ell = 1$. Since $F^{s,a}_{pp'} = \sum_\ell F^{s,a}_\ell\, P_\ell(\hat{p}\cdot\hat{p}')$, the Landau paramters can be obtained by projection.

The effective mass merits some discussion. For the heavy fermion systems (HFS) which are not translationally invariant, $m^*/m \neq 1 + F^s_1/3$. However, for an isotropic system with the "dynamic" mass[18], $m_\lambda(\neq m)$ containing static lattice, phonons, etc renormalizations, the ratio of quasiparticle current to velocity gives $m^*/m_\lambda = 1 + F^s_1/3$, i.e.

$$\frac{m^*}{m} = \frac{m_\lambda/m}{1 - \frac{1}{3}\overline{F}^s_1} \tag{14}$$

where $\overline{F}^{s,a}_\ell = \overline{N}(o)\, f^{s,a}_\ell$ $\overline{N}(o) = k_F m_\lambda/\pi^2$. m_λ, could in principle be determined from the penetration depth for Landau superconductors[18]. Since m_λ is not yet known for these systems, our results will be given in terms of m^*/m_λ (Table 3). We do expect m_λ/m to be large compared with unity for narrow f-band metals as the HFS. Thus, we are starting with large masses even before additional Fermi liquid effects. The general behavior in Eq. (14) should be noted: $m^*/m \to \infty$ when $\overline{F}^s_1 \to 3$.

In our model, hybridization of the f-ortibals with the non-f ligands (conduction electrons) delocalizes the f-electrons sufficiently to form coherent heavy electron states, "h", characterized by a Fermi surface and mass, m_λ. At low temperatures the ground state of the formed Fermi liquid is taken to have the minimal Kramer's doublet; the "h" quasiparticles are characterized by pseudospin $j = 5/2$, $m_j = \pm j$ and effective magnetic moment, μ_{eff}. We calculate the Landau interaction between these "h" quasiparticles, the fermion propagators having m_λ before they are dressed. Though there is no translational invariance, we have assumed isotropy and a positive ionic background.

Our characterization of the physical systems is based on our belief that there are roughly four energy scales in the problem:

(0) Same site electron-electron interaction, $u \sim O(10^4\text{-}10^5K)$

(1) Nearest neighbor interaction, $\alpha u \sim O(10^3\text{-}10^4K)$

(2) Fermi energy, $\varepsilon_F \sim O(10^2K)$

(3) Spin Fluctuation Energy $\sim O(10K)$

Based on this, we model the "direct" part of our interaction (Eqs. 11-13) by a finite ranged renormalized potential. Thus with $U = N(0)u$,

$$D_o^{\uparrow\downarrow} = U \quad ; \quad D_o^{\uparrow\uparrow} = -\alpha U \tag{15}$$

With $D_\ell^{\sigma\sigma'}(\ell \geq 2) = 0$ this gives $D_o^s = (1-\alpha)\, U/2$, $D_o^a = -U/2$, $D_1^s = D_1^a = \alpha U/4$ when $D_{pp'}^{\sigma\sigma'}$ is antisymmetrized. $D_o^{\uparrow\downarrow}$ represents the strong (not infinite) intra-atomic repulsion between electrons of opposite spin, as in the Hubbard model. $D_o^{\uparrow\uparrow}$ is the near-neighbor correction to the Hubbard term. We have taken $\alpha = 0.1$; different choices of $\alpha(\alpha \neq 0)$ do not qualitatively change our results.

We now solve our model Eqs.(12-13) to obtain the various Landau parameters, $F_\ell^{s,a}$ and $\overline{F}_\ell^{s,a}$; (see Table 3). Since α and m_λ is essentially fixed by empirical imput, the adjustable parameter is $\overline{U} = \overline{N}(0)u$. The results are given for various $\overline{U}$'s and in terms of m^*/m_λ, since m_λ is unknown at present. However, we believe that the set with $m^*/m_\lambda \approx 5$ is appropriate to UPt_3. The resulting $F_o^a = -.88$, given the simplicity of our model, is in fair agreement with that obtained above from the $T^3 \ell n T$ analysis.

The general feature of our model is evident: F_o^s, $F_1^{s,a}$ diverge as $m^*/m_\lambda \to \infty$, while $F_o^a \to -0.9$ independent of α or m_λ. The "barred" quantities ($\overline{N}(0)$ as the density of states) have interesting limiting behavior: $\overline{U} \to \overline{U}_{crit} = 9/2\alpha$, $\overline{F}_o^a \to 0$, $\overline{F}_o^s \to (1-\alpha/6)\, \overline{U}_{crit}$, $\overline{F}_1^s \to \alpha\overline{U}_{crit}/3$, $\overline{F}_1^s \to 3$. Thus the susceptibility, $\chi/\chi_{free} \to \infty$ and the compressibility, $\kappa/\kappa_{free} \to$ constant. If $\alpha \to 0$, $\overline{U}_{crit} \to \infty$, $\overline{F}^s_o \to \infty$, but m^* remains finite.

With the saturation of F^a_o the contribution to the effective mass, coming from the $\gamma_1{}^a$ term in $F_1{}^s$, Eq. (12), is modest for the heavy fermion systems. However, they make a significant contribution to the scattering amplitude. This in turn can be used to estimate the pairing interaction, g_1. We find that in the large effective mass limit triplet pairing is favored while singlet pairing is suppressed. The pairing interaction is close to its limiting value for $m^*/m_\lambda \gtrsim 5$.

In many ways our results are similar to the Gutzwiller solutions however, there are important differences. F^s_o diverges as $(m^*/m)^2$ as $U \to U_{crit}$ in the Gutzwiller model. Then the compression modulus κ, which goes as F^s_o/m^* for large m^*, goes to ∞ as $U \to U_{crit}$. In our model κ remains finite as $m^* \to \infty$. Physically, this would imply that electrons in the Gutzwiller case localize on a static incompressible lattice. In our model, the electrons localize on a dynamic compressible lattice.

Our calculations suggest that though the heavy fermion systems bear some semblance to liquid ^{3}He there are subtle and important differences. These are connected with the lack of translational invariance, band structure, spin-orbit effects and possible anisotropy of the Fermi surface. However for some bulk properties, the predictions of our model are consistent with the properties of the heavy fermion systems.

It should be emphasized that the Fermi liquid approach is one of several approaches to this rather complex issue. There are numerous

Table 3 Calculated values of the Fermi liquid paramters, $F^{s,a}_\ell(\ell=0,1)$, and m^*/m_λ for increasing U's. $\overline{F}^{s,a}_\ell$ and $\overline{U}$'s are obtained by dividing out by m^*/m_λ.

U	$\overline{U}$	F^a_o	$\overline{F}^a_o$	F^s_o	$\overline{F}^s_o$	F^a_1	$\overline{F}^a_1$	F^s_1	$\overline{F}^s_1$	m^*/m_λ
150	30.11	-.879	-.176	149	29.95	4.3	0.87	12	2.4	5
1000	41.61	-.900	-.037	985	40.97	32.7	1.36	69	2.87	24
4500	44.14	-.904	-.009	4426	43.42	149.3	1.46	302	2.97	102
9000	44.60	-.905	-.004	8851	43.86	299.4	1.48	602	2.98	202
25000	44.85	-.905	-.002	24585	44.10	832.7	1.49	1669	2.99	557
10^8	44.99	-.906	$-4x10^{-7}$	$9.8x1^7$	44.25	$3.3x10^7$	1.5[a]	$6.7x10^6$	3.00[a]	$2.2x10^6$

questions in this new field. We have tried to address a few, but many remain unanswered; we have barely touched on the superconducting aspect. Our hope is that this will provide a suitable starting point for further work. However, it probably will be sometime before we have all the answers. In the meantime, this field should continue to be a fascinating one for theorists and experimentalists alike.

Much of the work on the first part has been done in collaboration with C. J. Pethick, D. Pines, K. S. Bedell and G. E. Brown, and in the second with K. S. Bedell; I wish to thank them. I would also like to thank D. Hess for various discussions. This work has been supported in part by NSF DMR 82-15128.

References

1. For a review of experimental and theoretical work, see G. R. Stewart, Rev. Mod. Phys. 56, 755 (1984).
2. G. R. Stewart, Z. Fisk, J. O. Willis and J. L. Smith, Phys. Rev. Lett. 52, 679 (1984).
3. H. R. Ott, H. Rudigier, Z. Fisk and J. L. Smith, Phys. Rev. Lett. 50, 1595 (1983); J. J. M. Franse, P. H. Frings, A. de Visser, A. Menovsky, J. J. M. Palstra, P. H. Kes and J. A. Mydosh, Physica B26, 116 (1984).
4. P. W. Anderson, Phys. Rev. B30, 1549 (1984).
5. C. J. Pethick and G. M. Carneiro, Phys. Rev. A7, 304 (1973).
6. C. J. Pethick, D. Pines, K. F. Quader, K. S. Bedell, and G. E. Brown, submitted to Phys. Rev. Lett.
7. G. E. Brodale, R. Fisher, N. E. Phillips, G. R. Stewart, and A. L. Giorgi, preprint (1985).
8. J. O. Willis, J. D. Thompson, Z. Fisk, A. de Visser, J. J. M. Franse, and A. Menovsky, Phys. Rev. B31, 1654 (1985).
9. G. Aeppli, E. Bucher and G. Shirane, BNL preprint 36067.
10. K. S. Bedell and K. F. Quader, Phys. Rev. B32, 3296 (1985).
11. S. Babu and G. E. Brown, Ann. Phys. 78, 1 (1973); K. F. Quader and K. S. Bedell, J. Low Temp. Phys. 52, 89 (1985).
12. J. M. Rice, K. Udea, H. R. Ott, and H. Rudigier, Phys. Rev. B31, 594 (1985); W. F. Brinkman and J. M. Rice, ibid. 4302 (1970); D. Vollhardt, Rev. Mod. Phys. 56, 99 (1984); J. Hubbard, Proc. R. Soc. London, Ser. A276, 238 (1963); 281, 401 (1964).
13. G. R. Stewart, A. L. Giorgi, B. L. Brandt, S. Foner and A. J. Arko, Phys. Rev. B28, 1524 (1983); G. R. Stewart, J. L. Smith, and B. L. Brandt, ibid B26, 3783 (1982).
14. G. E. Brown, K. S. Bedell and K. F. Quader, preprint (1985).
15. B. Patton and Z. Zaringhalam, Phys. Rev. Lett. A55, 329 (1975).
16. K. S. Dy and C. J. Pethick, Phys. Rev. 185, 373, (1969).
17. A. de Visser, J. J. M. Franse, and A. Menovsky, J. Phys. F15, L53 (1985).
18. A. J. Leggett, Ann. Phys. 46 76 (1968).

STABILITY OF RAPIDLY SOLIDIFIED QUASICRYSTALS

Marko V. Jarić*

Department of Physics
Harvard University
Cambridge, Massachusetts 02138

INTRODUCTION

One of the firmest prejudices of solid state physics was shattered in recent experiments by Shechtman and his coworkers[1] when they discovered an aluminum-manganese alloy whose electron diffraction pattern showed an icosahedral arrangement of sharp Bragg peaks. The old prejudice would assert that a Bragg diffraction pattern is produced by a *periodic* lattice of atoms which, by necessity, excludes a possibility of the lattice having a five-fold symmetry (one cannot tile a plane by pentagons!). It is, therefore, not surprising that this discovery generated great excitement and activity among condensed matter theorists and experimentalists creating an instanteneous new field. There are already reported so many experimental and theoretical investigations that a need was felt for an early review.[2] In this paper I shall review current state of the art with an emphasis on theoretical investigations of thermodynamic stability of quasicrystals.

EXPERIMENTAL EVIDENCE

Last November Shechtman et al.[1] investigated rapidly solidified alloys of Al with 10-14 at.% Mn, Fe, or Cr, which they obtained by melt-spinning. The electron diffraction from grains of up to 2μm in size revealed sharp Bragg peaks, like in the case of ordinary crystals, which showed the "forbidden" icosahedral symmetry. The sharpness of the peaks implied a long-range positional ordering but icosahedral symmetry ruled out simple periodicity. Although icosahedral diffraction patterns[3] as well as sharp peaks arising from nonperiodic structures[4] were both observed in the past, they were of substantially different nature. Namely, the diffraction peaks from incommensurate structures showed no icosahedral symmetry while the icosahedral patterns originated from ordinary, multiply twinned crystals. Shechtman et al. ruled out a possibility of twinning by obtaining the same diffraction pattern from various regions of a grain and by showing that the x-ray diffraction pattern cannot be indexed to any Bravais lattice. Similarly, previously known aperiodic (incommensurate) crystals produce Bragg peaks which can be indexed by two or more incommensurate

*On leave from Department of Physics, North Dakota State University, Fargo, ND 58105.

reciprocal lattices. In such aperiodic crystals the ratio of (incommensurate) periodicities is a "continuous" function of temperature. In contrast, in the new icosahedral quasicrystals the incommensurability is purely geometric, that is, the periodicities are locked to certain numbers determined by the geometric, symmetry requirements.

Since the original discovery, several groups confirmed the interpretation that the icosahedral symmetry was inherent to structure of these and few other alloys. The electron diffraction experiments by Field and Fraser[5] confirmed the finds of Shechtman et al. Similar confirmation came from experiments by Kelton and Wu.[6] A high-resolution x-ray powder diffraction measurements were recently reported by Bancel et al.[7] who found icosahedral order not only in Al-Mn alloys but also in Al-Pd, Al-Pt, and Al-Ru alloys. They were able to rule out the multiple twinning hypothesis. Moreover, they succeeded in resolving the width of the peaks and they concluded that the range of the positional ordering is of the order of 100-300 Å while the bond orientational ordering is long range.

Icosahedral quasicrystalline structure was observed by Chen and Chen[8] in Al_6Mn with few percent of Si. The best composition was $Al_{74}Mn_{20}Si_6$.

The most convincing evidence that quasicrystals do not arise from multiple twinning comes from direct lattice-imaging experiments performed by Knowles et al.,[9] by Shechtman et al.,[10] and by Brusill and Lin.[11]

In order to elucidate the structure of quasicrystals several experiments which probe local sites were conducted. Swartzendruber et al.[12] investigated Mössbauer spectra of icosahedral $Al_6(Mn_{0.85}Fe_{0.15})_{1.03}$ and $Al_6(Mn_{0.62}Fe_{0.38})_{1.03}$. Assuming that Fe atoms randomly substitute Mn atoms they concluded that, to lowest order, there are two distinct Mn sites of lower symmetry than icosahedral. Similarly, the extended x-ray absorption fine structure measurements by Stern, Ma, and Bouldin[13] show that there are two kinds of Mn sites occurring in the golden mean ratio. The more populated sites are similar to Mn sites in crystalline Al_6Mn. At the noncrystalline sites, which have smaller coordination the bonding length to the nearest Al was found to be about the same but with a larger uncertainty.

That quasicrystals are here to stay, was dramatically confirmed in a recent experiment by Bendersky[14] who discovered another quasicrystalline phase of Al_6Mn, the so-called T phase. This phase is characterized by lattice periodicity in one direction and by ten-fold symmetry around this direction. Therefore, this decagonal phase, which is a crystal with one-dimensional periodicity and planar quasiperiodicity, seems to interpolate between the crystalline and icosahedral Al_6Mn phases.

These discoveries and results pose many interesting questions to the theorists. The questions of the most immediate importance seem to be the following: Can the appearance of icosahedral symmetry and sharp Bragg peaks be reconciled? What is the structure of icosahedral and decagonal Al_6Mn, and is this a good question? What are the conditions for stability of the quasicrystalline phases? In the rest of this paper we shall review what is known about the answers to these questions focusing on the question of stability.

QUASIPERIODICITY

The answer to the first question posed in the previous section was essentially known before the question arose. Namely, it was known that certain one-dimensional quasiperiodic sequences of lattice points produce sharp diffraction peaks (or, have discrete Fourier transform). In particular, a lattice formed by the Fibonacci sequence of short and long intervals was studied extensively.[15] The N-th lattice point of a Fibonacci sequence is explicitly given by

$$x_N = N\left(1 + \frac{1}{\tau^2}\right) - \frac{1}{\tau}\left\{\frac{N}{\tau} + \zeta\right\} \tag{1}$$

where $\tau = (1+\sqrt{5})/2$ is the golden mean, ζ is the parameter specifying a particular sequence, and $\{x\}$ denotes the fractional part of x. The short intervals in this sequence have length 1 while the long intervals have length τ. (This choice of lengths is related to the pentagonal symmetry in two- and three-dimensional generalizations.) One interesting property of the Fibonacci sequences is their selfsimilarity: if one views every long interval followed by a short interval as a new long interval, and every leftover long interval as a new short interval, one again obtains a Fibonacci sequence. The Fourier transform of the density

$$\rho(x) = \sum_N \delta(x - x_N) \tag{2}$$

is discrete, and the Bragg peaks can be indexed with two integers,

$$\tilde{\rho}(q) = \sum_{n,m} A_{n,m}\, \delta(q - q_{n,m}) \tag{3}$$

where $q_{n,m} = 2\pi(n+m\tau)/(3-\tau)$.

While an incommensurate structure is usually described in reciprocal space, as a superposition of two or more incommensurate density waves, the Fibonacci sequence offers a real-space realization for the "atomic" positions. This realization has the important property that no two lattice points will be found at a distance smaller than a given number.

A straightforward generalization to 3-dimensions of this one-dimensional lattice was used by Levine and Steinhardt[16] who considered a set of lattice points given by intersections of fifteen quasiperiodic families of planes perpendicular to the fifteen directions of edges of an icosahedron. They stated that, generically, there will be a smallest distance between two lattice points and that each lattice point is at an intersection of precisely three planes. Therefore, they easily found the Fourier transform which is discrete, since it is a superposition of products of three one-dimensional transforms, and which has icosahedral symmetry.

A two-dimensional analog of a quasi-lattice was invented in 1974 by R. Penrose[17] who discovered aperiodic tilings of a plane by a pair of tiles. The two-dimensional Penrose tiling has macroscopic ten-fold symmetry and a perfect bond-orientational order. Fourier transform of the tiling was first obtained optically by MacKay[18] who found a diffraction pattern characterized by ten-fold symmetry and sharp diffraction spots. MacKay also discovered a generalization of the Penrose tiling to three dimensions, a nonperiodic tiling by a pair of rhombohedra which has macroscopic icosahedral symmetry and a perfect bond-orientational order.

In order to analyze the Fourier transform of the Penrose lattices several authors[19-22] followed the original work of Janner and Jensen[23] on incommensurate structures and of De Bruijn[24] on two-dimensional Penrose tiles and considered a quasilattice as a projection of a slab from a higher dimensional periodic lattice. In this way, they were able to prove that the Fourier transform will be a discrete set of sharp peaks with an overall icosahedral symmetry. The result of Levine and Steinhardt[16] can also be related to the Penrose tiling since their lattice points corresponds to a particular decoration of the tiling .

The techniques for constructing Penrose lattices can be substantially generalized[25,26] to lead to a wide variety of nonperiodic lattices with arbitrary point group symmetry.

Therefore, it is possible to construct a lattice of points with a nonzero minimal spacing which will have a discrete diffraction pattern with icosahedral symmetry. Whole families of such lattices are currently known and investigated. To which extent these lattices are related to the actual structure of Al_6Mn is not clear at present. The present experimental evidence can be at best judged as ambivalent.

STRUCTURE

In the original paper, Shechtman et al.[1] mentioned that the diffraction data can be qualitatively fitted by a model consisting of a random packing of nonoverlapping parallel icosahedron attached by edges. Presumably, each icosahedron would have Mn in its center and Al at its vertices. Although the details of this model are not yet published, it seems unlikely that a random packing model can be consistent with a high degree of coherence observed in quasicrystals.

Another early proposal was advanced by Field and Fraser.[5] Their model is essentially a multiple twinning model of twenty distorted tetrahedra packed into an icosahedron, each tetrahedron being a fragment of a fcc crystal. As we described earlier, this model appears to be ruled out by the experiments.

The first structural model based on three-dimensional Penrose lattice was suggested by Kalugin et al.[19] They decorate the Penrose tiling by putting an Mn atom at each vertex and an Al atom at the center of each bond. However, this model gives the ratio Al:Mn much smaller than 6:1 observed experimentally.

Another recent study of a possible quasicrystalline structure by Nelson and Sachdev[27] applies the hierarchical inflation of tetrahedra proposed by Mosseri and Sadoc.[28] Although this model is not meant to describe the icosahedral Al_6Mn, it is an interesting alternative to the Penrose tiling since it also exhibits long range bond-orientational order and it seems that it leads also to a long range positional ordering (sharp Bragg peaks).

The two most recent studies of the structure of icosahedral aluminum-manganese attempt to elucidate the problem by analyzing the structures of related crystalline aluminum-manganese alloys. Brusill and Lin[11] observed that high resolution electron microscope image of icosahedral Al-Mn viewed along a 5-fold symmetry axis can be matched to a two-dimensional Penrose tiling. They were unable to find an equally good matching pattern built out of the two tiles of the three-dimensional Penrose tiling. Therefore, they looked for a different structural motif which they discovered in truncated icosahedral clusters which they identified as building blocks of the crystalline Al_6Mn. They were then able to construct a two-dimensional aggregate which matches a selected pattern from the electron microscope image. However, it is not clear whether their construction can be extended to a three-dimensional quasicrystal structure.

The approach of Elser and Henley[29] is based on a systematic analysis of $Al_{72.5}Mn_{17.4}Si_{10.1}$ crystal structure, denoted α(AlMnSi), which they believe relates to the icosahedral $Al_{74}Mn_{20}Si_6$ quasicrystal, denoted i(AlMnSi). They observed that α(AlMnSi) can be described as a packing of 54-atom icosahedral clusters which can be decomposed into the two types of rhombohedral Penrose tiles. They further suggest that the α(AlMnSi) is a finite unit cell approximation of the i(AlMnSi). Moreover, although he did not discuss the T phase[14] of AlMn, Henley[30] suggested that the crystalline $Al_{13}Fe_4$ structure may similarly be a finite cell approximation of a two-dimensional Penrose tiling.

Clearly, in order to make definitive, first principle investigations of stability of the icosahedral and the decagonal quasicrystals, it would first be necessary to determine the precise structure of these phases. However, as we are very much aware in the case of crystals, even when a structure is known it is at the present state of the art practically impossible to make a first principle assessment of its stability. Therefore, one needs to rely on more qualitative and phenomenological approaches which will be discussed in the next section.

THERMODYNAMIC STABILITY

Several different approaches to the question of thermodynamic stability of quasicrystals were taken in recent literature. Steinhardt and Levine found in direct simulations that a two-dimensional state with long range decagonal bond-orientational order is at least locally (mechanically) stable for particular binary and ternary mixtures of Lennard-Jones atoms.

A more phenomenological approach was taken by Bak,[31] Kalugin et al.,[19] and by Mermin and Troian,[32] who based their investigations on the Landau theory of solidification as formulated by Alexander and McTague.[33] In order to bypass the original conclusion[33] that a body centered cubic crystalline structure should generally be favored, they had to either include higher order terms in the Landau expansion,[31] or they had to introduce an additional component to the density.[19,32] With some additional implicit assumptions they then concluded that the icosaderal quasicrystalline structure can be stable.[19,31,32]

The central point of the Alexander-McTague theory of solidification[33] is the expansion of the (positional) Landau free energy near the uniform phase into the Fourier components of the density $\rho(\vec{q})$:

$$F_t = F_{t2} + F_{t3} + F_{t4} + \ldots \qquad (4)$$

The quadratic term has the form

$$F_{t2} = \int d^3\vec{q}\, A(q)\, \rho(\vec{q})\, \rho(-\vec{q}) \qquad (5)$$

which is dictated by translational and rotational invariance. Near the transition into a translationally ordered state, the minimum of $A(q)$ selects the magnitude of the fundamental wave vectors. Which particular set of directions $\{\pm \hat{q}_i\}$ will be chosen depends on the third and fourth order terms (we assume here that the fifth and higher order terms can be neglected near the transition). The discrete version of these terms, for a fixed q, is

$$F_{t3} = B \sum \rho(\hat{q}_i)\rho(\hat{q}_j)\rho(\hat{q}_k),\ \hat{q}_i + \hat{q}_j + \hat{q}_k = 0 \qquad (6)$$

and

$$F_{t4} = \Sigma\ C(\hat{q}_i \cdot \hat{q}_j, \hat{q}_j \cdot \hat{q}_k)\rho(\hat{q}_i)\rho(\hat{q}_j)\rho(\hat{q}_k)\rho(\hat{q}_l), \hat{q}_i + \hat{q}_j + \hat{q}_k + \hat{q}_l = 0 \ . \qquad (7)$$

The rotational and translational symmetries are *fully* implemented in these expressions.

In particular, while the rotational invariance leaves only one independent cubic coupling constant B, the number of quartic coupling constants C is restricted to a two-dimensional continuum.[34] As pointed out by Alexander and McTague, because of these degrees of freedom F_{t4} can depend on specific features, such as bond angles, packing considerations and band structure, so that very little can be said regarding the universal trends in solidification. Namely, two (inequivalent) sets of fundamental wave vectors will generally give rise to nonzero terms with *different* couplings and, consequently, either set could be made to have lower energy (if these different couplings are suitably chosen). A simple illustration of this fact is offered by a comparison between the cubic face-, edge-, and vertex-models, identified here according to the star $\{\pm\, \hat{q}_i\}$ and more commonly denoted sc, bcc, and fcc, respectively. There are two relevant coupling constants for the face model, C(1,-1) and C(0,0) four coupling constants for the edge model, C(1,-1), C(0,0), C(1/2,-1/2), and C(0,-1/2) and three coupling constants for the vertex model C(1,-1), C(1/3,-1/3), and C(-1/3,-1/3).

This difficulty is amplified as one includes into the expansion (4) higher and higher terms. In fact, since $\rho(\hat{q})$ spans the regular representation of SO(3) one can show rigorously that if the degree of the expansion is not limited, then for *every* symmetry in SO(3) there will exist a set $\{\pm\, \hat{q}_i\}$ of that particular symmetry and a set of coupling constants such that $\{\rho(\pm\, \hat{q}_i)\}$ minimizes the free energy.

Therefore, different specific assumptions will lead to different, nonuniversal conclusions. The assumption made by Alexander and McTague is that the isotropic (in ρ-space) component of C typically dominates the quartic term. That is, the dominant term is $\bar{C}|\rho|^4$, where $\bar{C}$ is the average $C(x,-x)$ coupling and $|\rho|^2 = \Sigma\ \rho(\hat{q}_i)\rho(-\hat{q}_i)$. In this case, the free energy F_t is minimized by a set $\{\rho(\pm\,\hat{q}_i)\}$ which *maximizes* $|F_{t3}|/|\rho|^3$. Alexander and McTague considered $\{\pm\, \hat{q}_i\}$ parallel to the edges of a triangle, an octahedron (tetrahedron), and an icosahedron. We can add to this list a tetrahedral bipyramid and an idealized pentagonal bipyramid (assumed to be formed of five perfect tetrahedra). These choices correspond to 2D hexagonal lattice, bcc, icosahedral edge model, 3D hexagonal and idealized closed packing of tetrahedra, respectively. The result is $F_t^{bcc} < F_t^{3Dhex} < F_t^{ideal} < F_t^{2Dhex} < F_t^{ie}$. Therefore, the bcc crystalline order is the most favored while the icosahedral edge-model quasicrystalline ordering is the least likely.

In order to stabilize the icosahedral structure Bak[31] extended previous assumptions by adding a fifth degree term to the expansion. He then considered only a contribution arising from those $\hat{q}_i$ which form a regular pentagon. There are no such $\hat{q}_i$ for the bcc set but they exist in the icosahedral set. Therefore this fifth order coupling can be chosen to make $F_t^{ie} < F_t^{bcc}$ (providing the free energy is stabilized with an isotropic, positive sixth degree term). In fact, already at the fourth degree F_t^{ie} can be made smaller than F_t^{bcc} since it contains terms with $C(\tau/2-1,-1/2)$ which are not present in F_t^{bcc} (τ is the golden mean). However, in the light of what was emphasized above this does not seem to be a sufficiently convincing argument for stability of icosahedral quasicrystals.

Mermin and Troian[32] tried to stabilize the icosahedral translational ordering in another way. First, they assumed that C is constant over its entire domain. Then, they introduced a second component (order parameter) $\rho(\vec{k})$ which selects another wave vector magnitude such that $0 < k < 2q$. Next they assumed that the ordering of $\rho(\hat{k})$ is induced by the ordering of $\rho(\hat{q})$ and they effectively integrate out the $\rho(k)$ component. In this way they generate an additional, effective fourth order term of precisely the same form as the one associated with the coupling $C(k^2/2q^2-1,-1)$. Since this coupling is unique to the icosahedral vertex model when $k^2/q^2 = (2 \pm 2/\sqrt{5})$, it is not surprising that the icosahedral ordering is stabilized for sufficiently large $u/A(k)$, where u is the coupling constant for the interaction term linear in $\rho(\hat{k})$ and quadratic in $\rho(\hat{q})$. Note, however, that this is completely equivalent to a theory with a single component q such that $C(x,y)$ is sharply peaked around, e.g., $C(\pm 1/\sqrt{5},-1)$. (Incidentally, this shows that the initial assumption $C=$ constant is not essential as long as $u/A(k)$ is sufficiently large).

In a related approach, Kalugin et al.[19] also considered a two component density. They argued that since unlike ordinary crystals for quasicrystals higher harmonics of $\hat{q}_i$ can be arbitrarily close to the original basic star $\{\pm\hat{q}_i\}$, the quadratic term in the free energy will be unable to distinguish between the basic and higher harmonics. Concretely, it is unjustified to consider the icosahedral vertex star $\{\pm\hat{q}_i\}$ as a primary order parameter since within 5% the second harmonics, the icosahedral edge star, have the wave vectors of equal magnitude. Consequently, they kept both components and showed, to third degree in the free energy, that for a range of parameters the icosahedral structure is more stable than the body centered cubic structure.

Since in all of these approaches one needs to assume a careful balance between infinitely many coupling constants it seems that the quartic, multicomponent cubic, and higher order terms in the free energy cannot be used in a convincing manner to select the preferred translational ordering; at least not until the nonlocal coupling constants can be calculated independently.

A different mean-field appraoch was employed by Sachdev and Nelson.[35] They adapted a molecular-field (self-consistent field) theory of solidification of Ramakrishnan and Yussouff.[36] This approach is equivalent to a Landau expansion being carried to infinite degree with specific expansion coefficients. In fact, the equilibrium densities obtained by Sachdev and Nelson would require inclusion of several hundred harmonics in the Landau theory. Sachdev and Nelson concluded that undercooled liquids are metastable with respect to an icosahedral crystal. They also found that fcc and bcc crystals have a lower free energy. This stability of the icosahedral phase is caused by the short range icosahedral ordering which enters the theory through the assumed liquid structure factor of metallic glasses.

All the above-mentioned approaches are based on an analysis of the translational ordering. In the following section we shall describe a different approach. The main idea of this approach will be to include in the theory an orientational order parameter which stabilizes the quasicrystalline phase.[37] This does not seem unnatural when one recalls the well-known fact that many supercooled liquids and metallic glasses show a short range icosahedral bond-orientational order[38] and that crystal structures of many alloys contain characteristic (nearly) icosahedral clusters. It is perhaps not an accident that the crystalline Al-Mn alloys contain such clusters.[11,29] We shall, therefore, adopt a view of solidification as an interplay between orientational and translational order parameters.[39]

ORIENTATIONAL ORDERING

The main result of our approach[37] is that the icosahedral quasi-crystalline order is stabilized and, in fact, triggered by the long range icosahedral bond-orientational order. Moreover, it is a direct result of our theory that the fundamental icosahedral wave vectors point in the directions of the twelve vertices of an icosahedron or in the directions of its twenty faces. This is in agreement with the experimental observations which definitely rule out the edge model and seem to be best fitted by the vertex model.[27] More generally, while the translational ordering always triggers an orientational ordering, the reverse does not hold: a transition into an intermediate phase with long-range orientational order but with no translational order may precede the complete ordering.

The results were obtained within the context of the Landau theory, which will be described below.[37] We shall first construct the full quartic Landau free energy for the orientational order parameter and we shall determine all stable phases at the transition from the disordered isotropic phase.[40] We shall find that the transition is first order and that the the icosahedral liquid-crystal phase occupies a large portion of the phase diagram. Next, we shall introduce the lowest order, interaction free energy which has to be linear in the orientational order parameter and quadratic in the translational order parameter. Thus, we shall show that the onset of the translational ordering will always trigger simultaneous onset of the orientational ordering, whereas the onset of the orientational ordering may (because the transition is first order!) but need not trigger simultaneous onset of the translational ordering. In this way we shall show that icosahedral positional ordering can be stabilized by the orientational ordering.

Several years ago Nelson and Toner[39] studied cubic bond-orientational order while more recently Steinhardt et al. investigated the short range icosahedral bond-orientational order.[38] Penrose lattices and the experimentally observed quasicrystals both exhibit long range icosahedral translational order. The pure bond-orientational order (at $q=0$) can be characterized by an order parameter $\mathbb{Q}(\hat{n})$ which gives the density of bonds in the direction n. It is convenient to expand $\mathbb{Q}(\hat{n})$ into spherical harmonics $Y_{Lm}(\hat{n})$ and to consider the expansion coefficients Q_{Lm} as the order parameter. The quadratic term of the corresponding orientational free energy has the form $\Sigma_L a_L \Sigma_m |Q_{Lm}|^2$ dictated by the rotational invariance. Generally, if the transition is continuous or nearly continuous Q_{Lm} associated with a single L can be taken as a primary order parameter. Since we want to model the icosahedral ordering and since six is the lowest degree of a nontrivial icosahedral invariant polynomial in $\hat{n}$ we shall take Q_{6m} as the primary order parameter.[41]

The Landau free energy expansion for Q_{6m} has been constructed up to third order terms.[38] However, as pointed out by Mermin and Stare[42] in the context of BCS pairing with $L\neq 0$, it is generally necessary to include the fourth order terms in the expansion. From the Molien generating function calculated[43] for $L=6$, $M(t) = 1 + t^2 + t^3 + 3t^4 + \ldots$, we conclude that there are two nontrivial linearly independent quartic invariants in Q_{6m}. Therefore, the orientational free energy has the form

$$F_0 = a|\mathbb{Q}|^2 + bI_3(\mathbb{Q}) + c_0|\mathbb{Q}|^4 + c_1 I_{41}(\mathbb{Q}) + c_2 I_{42}(\mathbb{Q}) \tag{8}$$

The cubic invariant and the two quartic invariants can be expressed in terms of (the contractions of) the Wigner's 3-j symbols $\begin{pmatrix} 6 & 6 & 6 \\ m_1 & m_2 & m_3 \end{pmatrix}$.[37,40]

One can now use group theoretical methods to show that the only possible low symmetries are the isotropy groups for the $L=6$ representation of SO(3) spanned by the order parameter. Thus, the possible low symmetry phases for $L=6$ are[44,45] D_∞, Y, O, D_6, D_5, D_4, T, D_3, D_2, C_3, C_2 and C_1. If the degree of F_o would be sufficiently large one could always choose the couplings in such a way to stabilize any of these phases.

In order to decide which of the above listed phases can be stabilized with the quartic free energy we must minimize Eq. (8). This is a very complex problem which requires the solutions of thirteen cubic equations in thirteen unknowns and with three free parameters. Fortunately, over the last ten years group theoretical techniques have been developed for minimizing Landau-Higgs potentials of such complexity.[46] By using these exact and numerical techniques we were able to determine all the stable phases which are separated by a first order transition surface from the isotropic SO(3) phase.[40] These phases are D_∞, Y, O, and D_6. However, the icosahedral phase occupies the largest portion of the $(ac_1/b^2, ac_2/b^2, c_0/\sqrt{c_1^2+c_2^2})$ phase diagram. The icosahedral phase even persists in a large portion of the phase diagram when the transition is continuous $(b=0)$. The order parameter is given in the icosahedral phase by

$$\mathbb{Q}^{icos} = -bI_3(\hat{\mathbb{Q}}^{icos})\hat{\mathbb{Q}}^{icos}/2[c_0 + c_1 I_{41}(\hat{\mathbb{Q}}^{icos}) + c_2 I_{42}(\hat{\mathbb{Q}}^{icos})] \tag{9}$$

where, in a given orientation of Y,

$$\hat{\mathbb{Q}}^{icos} = \frac{1}{5}\,(0\ i\sqrt{7}\ 0000\ \sqrt{11}\ 0000\ i\sqrt{7}\ 0) \quad . \tag{10}$$

rotationally and translationally invariant energy[38,39]

$$F_{int} = \int dq \sum_{L,m} \alpha_L(q) \int d^2\hat{q}\ Q_{Lm} Y^*_{Lm}(\hat{q})\,\rho(\hat{q})\,\rho(-\hat{q}) \quad . \tag{11}$$

We note that, to lowest order, ρ does not couple linearly to $\mathbb{Q}$. Consequently the equilibrium ρ need not have the symmetry of the equilibrium $\mathbb{Q}$ even though the structure factor must have this symmetry. This is precisely the case for the Penrose tiles.

If the translational ordering temperature T_t is greater than the orientational ordering temperature T_o then, because the interaction (11) is linear in $\mathbb{Q}$, the ordering of ρ at T_t will necessarily induce an ordering in $\mathbb{Q}$. On the other hand, if $T_o > T_t$ then, since (11) is quadratic in ρ, the ordering of $\mathbb{Q}$ at T_o will have the effect of renormalizing the quadratic coupling $A(q)$ without necessarily inducing an ordering of ρ. However, if the transition at T_o is discontinuous, like in the case of the icosahedral orientational ordering, $A(q)$ might be sufficiently renormalized for ρ to order. Indeed, the renormalized coupling is

$$A'(\vec{q}) = A(q) + \alpha_6(q) \sum_m Q^{icos}_{6m}\, Y^*_{6m}(\hat{q}) \tag{12}$$

and $\min A'(\vec{q}) < \min A(q)$, as can be seen from Eq. (10) which gives $\max \Sigma\, Q^{icos}_{6m} Y^*_{6m} > 0$ and $\min \Sigma\, Q^{icos}_{6m} Y^*_{6m} < 0$. Depending on the sign and magnitude of b the minimum of $A'(\vec{q})$, will correspond to q's pointing at the vertices or faces of the icosahedron. Typical $A'(\vec{q})$ is shown in Fig. 1. Thus, if the coupling α_6 is sufficiently strong ρ might order at T or at some intermediate temperature. Moreover, the icosahedral vertex or face model will be preferred. The ordering at an intermediate temperature would, in principle, be continuous since there are no icosahedral vertex or face invariants of third degree in ρ. Note that in the case of the icosahedral-face orientational ordering, the corresponding translational ordering would compete with a cubic vertex model (fcc lattice) since the face vectors of an icosahedron can be identified with the vertex vectors of five cubes inscribed inside the icosahedron.

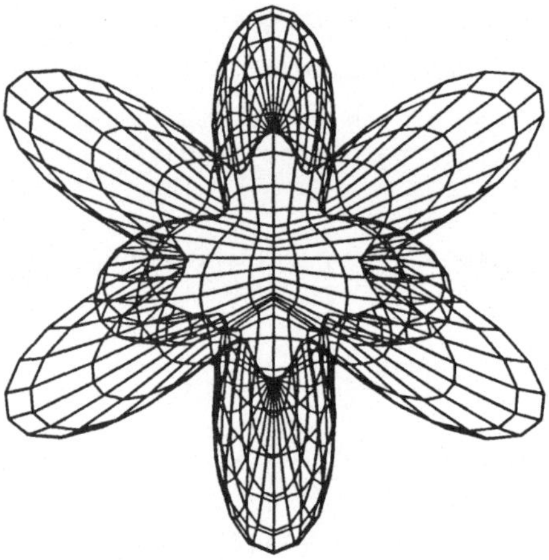

Fig. 1. Polar graph $\sim$ const. + $A'(\vec{q})$ at constant q showing a hemisphere longitude-latitude net for icosahedral symmetry.

Although the original Alexander-McTague theory predicts that the preferred structure of monoatomic crystals should be bcc, the fcc crystal structures are commonly found in nature. Our model offers an avenue for explaining this. Namely, the fcc structure corresponds to the cubic vertex model and, as can be seen from Fig. 2, the cubic bond-orientational ordering leads for $L = 4$ or 6 to the effective quadratic coupling which, depending on the sign of b can have minimum in the directions of the vertices of a cube. Therefore, fcc structure can be stabilized.

As we have seen, an analysis similar to the analysis of cubic-liquid-crystals[39] demonstrates that the icosahedral orientational ordering can induce and stabilize the icosahedral quasicrystalline phase. Although we propose a mechanism which is different from some other recent proposals, some of the general conclusions remain the same. For example, the analysis of elasticity and dislocations in icosahedral quasicrystals by Bak,[21] Levine et al.,[47] and by Lubensky et al.[48] is independent of this mechanism and remains valid. Similarly, the conclusion that a transition from isotropic to icosahedral quasicrystalline phase remains first order even with the inclusion of fluctuations is also still valid.[49] However, the effect

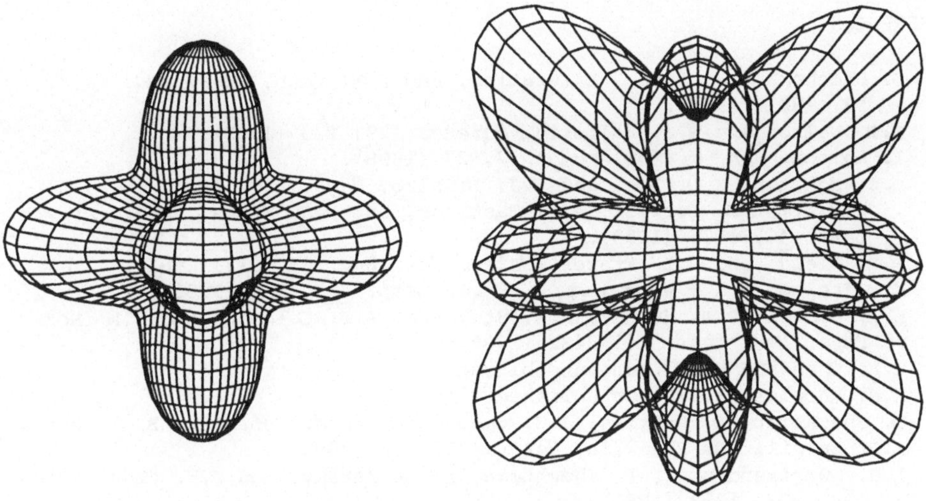

Fig. 2. Polar graph for cubic symmetry at $L = 4$ and 6.

of fluctuations on possible transition between an icosahedral-liquid-crystal phase ($\mathbf{Q} \neq 0$, $\rho = 0$), and between this phase and a (quasi) crystalline phase, remains to be investigated. Hydrodynamics of icosahedral liquid crystals will also have to be investigated.

Our conclusion that the icosahedral edge model is not favored compared to the face or vertex models is in good agreement with the experimental evidence. However, a direct experimental evidence in support of a specific mechanism for quasicrystalline ordering is not available at present. It seems plausible to look in the future experiments for the signs of the icosahedral-liquid-crystalline ordering by analyzing the structure function in the melt just before the formation of the quasicrystals or in the rapidly quenched glass state.

Finally, we shall remark on the stability of the T phase which exhibits decagonal symmetry.[14] Since the lowest nontrivial invariant under D_{10} symmetry is of degree ten,[41] it follows that the lowest appropriate L is $L = 10$. However, for $L = 10$, D_{10} is not a maximal isotropy group so that it is uncertain whether a fourth degree orientational free energy can have an absolute minimum associated with this symmetry. Investigation of the stability of this phase, and of the transition to the icosahedral phase remains to be conducted.

ACKNOWLEDGMENTS

It is a pleasure to thank the condensed matter theory group at Harvard for their hospitality and financial support. This work was supported in part by the NSF through grant DMR 82-07431 and the Harvard Materials Research Laboratory.

REFERENCES

1. D. Shechtman, I. Blech, D. Gratias, and J.W. Cahn, *Phys. Rev. Lett.* 53, 1951 (1984).
2. D.R. Nelson and B.I. Halperin, *Science* 229, 233 (1985).
3. T. Komoda, *Jpn. J. Appl. Phys.* 7, 27 (1968).
4. D.E. Moncton, J.D. Axe, and F.J. DiSalvo, *Phys. Rev. Lett.* 34, 734 (1975); L.M. Corliss, J.M. Hastings, and R.J. Weiss, *Phys. Rev. Lett.* 46, 1135 (1959).
5. R.D. Field and H.L. Fraser, *Mater. Sci. Eng.* 68, L17 (1985).
6. K.F. Kelton and T.W. Wu, *Appl. Phys. Lett.* 46, 1059 (1985).
7. P.A. Bancel, P.A. Heiney, P.W. Stephens, A.I. Goldman, and P.M. Horn, *Phys. Rev. Lett.* 54, 2422 (1985).
8. C.H. Chen and H.S. Chen, unpublished.
9. K.N. Knowles et al., unpublished.
10. D. Shechtman, D. Gratias, J.W. Cahn, *C.R. Acad. Sci.* Ser. B, in press.
11. L.A. Bursill and P.J. Lin, *Nature* 316, 50 (1985).
12. L.J. Swartzendruber, D. Shechtman, L. Bendersky, and J.W. Cahn, *Phys. Rev.* B32, 1383 (1985).
13. E.A. Stern, Y. Ma, and C.E. Bouldin, unpublished.
14. L. Bendersky, unpublished.
15. S. Ostlund and R. Pandit, *Phys. Rev.* B29, 1394 (1984).
16. D. Levine and P.J. Steinhardt, *Phys. Rev. Lett.* 53, 2477 (1985).
17. R. Penrose, *Bull. Inst. Math. Appl.* 10, 266 (1974); M. Gardner, *Sci. Am.* 236, 110 (1977).
18. A.L. MacKay, *Physica* 114A, 609 (1982).
19. P.A. Kalugin, A. Yu. Kitaev and L.C. Levitov, *Sov. Phys. JETP Lett.* 41, 144 (1985).
20. V. Elser, unpublished.
21. P. Bak, *Phys. Rev.* B (to be published).
22. R.K.P. Zia and W.J. Dallas, *J. Phys.* A18, L341 (1985); See Ref. 50.
23. A. Janner and T. Janssen, *Phys. Rev.* B15, 643 (1977).
24. N.G. de Bruijn, Ned. Akad. Weten. Proc. Ser. A43, 39, 53 (1981).
25. M. Duneau and A. Katz, *Phys. Rev. Lett.* 54, 2688 (1985).
26. J.E.S. Socolar, P.J. Steinhardt and D. Levine, unpublished.
27. D.R. Nelson and S. Sachdev, *Phys. Rev.* B32, 689 (1985).
28. R. Mosseri and J.F. Sadoc, *J. de Phys. Lett.* 45, L827 (1984).
29. V. Elser and C.L. Henley, unpublished.
30. C.L. Henley, *J. Non-Cryst. Solids* (in press).
31. P. Bak, *Phys. Rev. Lett.* 54, 1517 (1985).
32. N.D. Mermin and S. Troian, *Phys. Rev. Lett.* 41, 1524 (1985).
33. S. Alexander and J. McTague, *Phys. Rev. Lett.* 41, 702 (1978).
34. The coupling C has an obvious discrete symmetry arising from permutations of $\hat{q}_i$, $\hat{q}_j$, $\hat{q}_k$, and $\hat{q}_l$.
35. S. Sachdev and D.R. Nelson, unpublished.
36. T.V. Ramakrishnan and M. Yussouff, *Phys. Rev.* B19, 2775 (1979).
37. M.V. Jarić, *Phys. Rev. Lett.* (in press).
38. P.J. Steinhardt, D.R. Nelson and M. Ronchetti, *Phys. Rev.* B28, 784 (1983).
39. D.R. Nelson and J. Toner, *Phys. Rev.* B24, 363 (1981).
40. M.V. Jaric and D.R. Nelson, unpublished.
41. M.V. Jaric, L. Michel and R.T. Sharp, *J. Physique* 45, 1 (1984) Appendix D, and references therein.
42. N.D. Mermin and C. Stare, *Phys. Rev. Lett.* 30, 1135 (1973).
43. J. Bystricky et al., *J. Math. Phys.* 23, 1560 (1982).
44. L. Michel, *Rev. Mod. Phys.* 52, 617 (1980).
45. E. Ihrig and M. Golubitsky, *Physica* 13D, 1 (1984).
46. M.V. Jarić, *Phys. Rev. Lett.* 48, 1641 (1982); Lec. Notes. *Phys.* 201, 397 (1984), and references therein.
47. D. Levin et al., *Phys. Rev. Lett.* 54, 1520 (1985).

48. T.C. Lubensky, S. Ramaswamy, and J. Toner, unpublished.
49. S.A. Brazovskii, Zh. Eksp. Teor. Fiz. 68, 42 (1975) [*Sov. Phys. JETP* 41, 85 (1975)]; S.A. Brazovskii and S.G. Dimitriev, Zh. Eksp. Teor. Fiz. 69, 979 (1975) [*Sov. Phys. JETP* 42, 497 (1976)].
50. P. Kramer and R. Neri, Acta Cryst. A40, 580 (1984); P. Kramer, unpublished.

INSTABILITIES AND MODE SELECTION IN EXPLOSIVE CRYSTALLIZATION

Douglas A. Kurtze

Department of Physics
Clarkson University
Potsdam, NY 13676

INTRODUCTION

A great deal of theoretical attention has recently been paid to problems of pattern formation in nonlinear systems.[1] Among the many models in which these phenomena have been studied are several which have arisen from problems in materials processing, especially crystal growth.[2] In these problems, which include directional solidification both of dilute and near-eutectic alloys, and the growth of dendrites, the models investigated admit simple steady states, which describe a solid region advancing at a constant velocity into a melt. Under certain growth conditions, typically amounting to the existence of supercooling, these steady states become linearly unstable and so will not be observed. The problem of theoretical (and, in some cases, practical) interest is to follow the nonlinear developoment of the unstable perturbations of these steady states past the initial stage in which their amplitudes are infinitesimal, and so to predict the form of the solid-liquid interface which is actually observed.

A related problem which has proven particularly susceptible to theoretical analysis is that of explosive crystallization. In this process, a polycrystalline material is grown from the <u>amorphous</u> phase, rather than the liquid. As in crystallization from the melt, the transition taking place is from a high-entropy phase to one of low entropy, and so a latent heat is released. However, the nonequilibrium interface kinetics of explosive crystallization are quite unlike those of growth from a slightly undercooled melt: the former proceeds more rapidly at lower temperatures (higher undercoolings), while crystallization from an amorphous phase takes place more rapidly at <u>higher</u> temperatures, since the metastable amorphous material must surmount an energy barrier in order to crystallize. Thus the role of heat diffusion in explosive crystallization is not to <u>remove</u> latent heat from the advancing crystallization front, but rather to transport it <u>to</u> the front in order to maintain growth. Thus, once the process has been started, the latent heat released when some amorphous material crystallizes then diffuses to neighboring, still-amorphous material, raising its temperature and allowing it to crystallize in its turn and release more latent heat. This can result in a wave of crystallization propagating through the material.

It is well known experimentally that under favorable conditions (usually that the initial temperature be above some threshold value), the release of latent heat may be all that is needed to sustain the advance of the crystallization front.[3] Under such conditions, it is only necessary to start the process by injecting energy locally into the initially amorphous film, say with a laser pulse or by impact with a stylus, and then the progressive release of latent heat will carry the front outward until the entire film has crystallized. Under less favorable conditions, latent heat release alone will not suffice to sustain the reaction, and so the front will eventually stop moving; under such circumstances the crystallization can be maintained by continuous injection of energy from outside, say by scanning a continuous-wave laser across the sample.[4]

It is quite commonly observed that an amorphous film of uniform thickness acquires ridges upon undergoing explosive crystallization,[5] with periodic variations in its thickness, grain size, and degree of crystallinity.[6] Self-sustained crystallization most often produces these ridges in the form of straight, parallel rolls, although wavy structures, in which the rolls are not straight, have also been observed.[7] These ridges are most pronounced when the substrate temperature at which the film was crystallized was near the minimum temperature for self-sustained growth to occur. Similar features are also observed in laser-driven growth, when the substrate temperature is not high enough to allow self-sustained growth.[8-10]

The most successful theoretical model of explosive crystallization was proposed independently by Gilmer and Leamy[11] and Shklovskii.[12] Their model is one-dimensional, and so would be relevant to crystallization of a thin 'fuse', or of a film in which the crystallization front can be expected to be essentially a straight line. The model is based on heat diffusion observed in a moving frame of reference, with linear heat loss to the environment. The structure of the thin zone in which the crystallization is taking place is not resolved; rather the crystallization front is treated as a sharp line of demarcation between the amorphous and polycrystalline regions, at which latent heat is being released. A crucial feature of the model is the inclusion of interface kinetics, in the form of a phenomenological dependence of the local rate of advance of a point on the front on its temperature. The model has steady states[11-12] which describe a crystallization front advancing at a fixed velocity into the amorphous region. The velocity is determined by a balance between the front temperature, as determined from the heat diffusion equation, and the interface kinetics. Shklovskii[12] and van Saarloos and Weeks[13-14] performed linear stability analyses of these steady states, and found that under certain conditions they have oscillatory instabilities, which point to the existence of nonlinear limit cycles in the front motion. It has often been proposed[6,9,13] that the resulting oscillations in front temperature give rise to the observed undulations on the surface of the finally crystallized film: the size of grains produced and the extent of completion of the reaction should depend on the temperature at which the transition took place; the density difference between the amorphous and crystalline phases would then produce a variation of the thickness of the film which mirrors the variation in the temperature at which different points on the film crystallized.

Kurtze, van Saarloos, and Weeks[15] generalized the stability analysis to two dimensions. Following Zeiger et al.,[9] they included a laser slit which can be used to drive the front in cases in which the crystallization is not self-sustained, so that the steady states of the model then represent a straight-line front moving with the laser scan speed. They found that the linear stability of laser-driven growth is very similar to that of self-sustained growth, and that the instabilities are always oscillatory. In addition, they located morphological instabilities, where the straight front is stable against all one-dimensional perturbations but unstable against some having finite wavenumbers transverse to the growth direction.

Zeiger et al.[9] studied one-dimensional laser-driven explosive crystallization numerically, but instead of using nonequilibrium interface kinetics in the general form of the Gilmer-Leamy model, they simplified it by assuming that the interface is always at a fixed temperature (except that it cannot move backwards, i.e., crystalline material cannot become amorphous). They worked with parameter values for which steady-state growth is very unstable, and observed oscillations in which the interface would run ahead of the laser, then stop and wait for the laser to catch up to it. As they increased the substrate temperature in their simulations, they also saw the onset of self-sustained growth, as the front quickly began propagating at a much higher speed than that of the laser. The physical mechanism for the oscillations[9,11] depends crucially on the fact that the front temperature must exceed the substrate temperature for crystallization to proceed. If the front should move ahead of its steady-state position, then it gets ahead of its heat sources, which are the laser (if present) and the previous positions of the front (where latent heat was released in the past). Thus the temperature drops and so the front slows down. If this deceleration is too drastic, then the heat sources catch up and raise the temperature above its steady-state value, forcing the front to move ahead of its steady-state position and beginning the cycle again.

Van Saarloos and Weeks[14] integrated the Gilmer-Leamy model equations for one-dimensional self-sustained growth numerically and found the expected oscillations for substrate temperatures not too far above the threshold for self-sustained growth. In addition, they found a sequence of period-doubling bifurcations.[16] They also performed an analysis (reproduced below in a slightly different form) of a codimension-2 bifurcation, at which the onset of oscillations coincides with the onset of self-sustained growth itself. Their calculations allowed them to find the period of oscillation of the front when the steady state is unstable, and also revealed cases in which the front velocity would perform apparently regular oscillations, but then suddenly decay to zero, so that the transition, although energetically possible, abruptly stops for dynamical reasons.

LINEAR STABILITY ANALYSIS

We will consider a thin film of amorphous material on an inert substrate being crystallized explosively by a laser slit scanned across its surface, as sketched in Fig. 1. The laser defines the y axis, and moves at a speed V_s in the x direction. The crystallization front, at the position $x_b(y,t)$, is driven ahead of the laser at a local normal velocity of $V_b(y,t)$. The temperature far from the laser and the front is T^0. We assume that the temperature field T(x,y,t) determines the behavior of the transition front, and hence the features seen on the finally crystallized film.

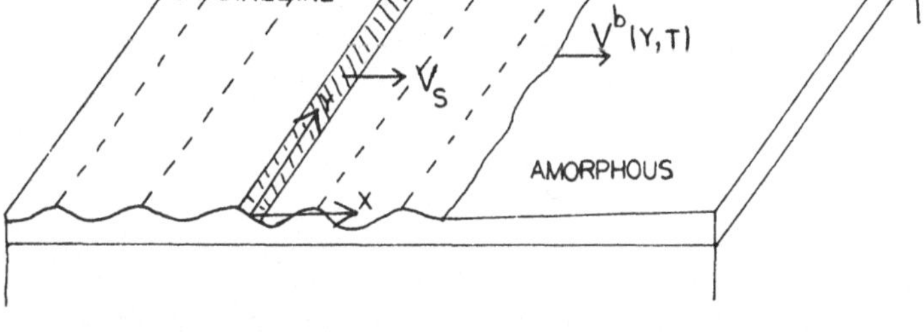

Fig. 1. Schematic view of laser-driven explosive crystallization.

Thus we write the diffusion equation

$$\frac{\partial T}{\partial t} = D\nabla^2 T + V_s\frac{\partial T}{\partial x} - \Gamma(T-T^o) + P(x) + \frac{L}{c}(V_s + \frac{\partial x^b}{\partial t})\delta(x-x^b), \quad (1)$$

where P(x) is proportional to the power density provided by the laser, L is the latent heat of the transition, c is the specific heat of the film, D is the thermal diffusion constant, and Γ is a phenomenological constant which measures the rate of heat loss to the environment. Heat conduction perpendicular to the plane of the film is neglected (actually it is taken to be accounted for by the heat-loss term), and the parameters D and Γ are taken to be the same in both the amorphous and crystalline regions for simplicity. Note that the δ-function term in equation (1) embodies the Stefan boundary condition at the front, which specifies the discontinuity in temperature gradient needed to account for the latent heat release. This model can also be applied to self-sustained growth; in this case P(x) is absent and V_s is the _steady-state_ speed of the front, which must be found as part of the solution. For self-sustained growth, there is an extra translational symmetry, since there is nothing in the problem to define an origin.

We include interface kinetics in the model by giving a phenomenological function, the _growth-rate curve_, relating the temperature T^b at a point on the front to the normal velocity of advance there, namely

$$(V_s + \frac{\partial x^b}{\partial t})[1 + (\frac{\partial x^b}{\partial y})^2]^{-1/2} = V^b(T^b) \quad (2)$$

Such a relation is a continuum approximation to the effect of the microscopic nucleation-and-growth kinetics of the crystallization transition, and so is an intrinsic property of the material. The product consists of grains which are oriented randomly, so the growth-rate curve should be regarded as an average over crystalline orientations; thus there is no dependence of front velocity on orientation. The growth rate will be an increasing function of front temperature such as that plotted as the solid line in Fig. 2, since crystallization of an amorphous material is a thermally activated process; in fact it is typically an Arrhenius law. We will assume further that the relation (2) continues to hold even out of steady state conditions. The model given by (1) and (2) is generally applicable to transitions which are thermally activated and exothermic, and which take place in a zone of negligible width.

The analysis of the model is somewhat simpler if we work in dimensionless variables, so we scale all lengths by the diffusion length $2D/V_s$ and times by the diffusion time $2D/V_s^2$. A simple steady-state solution of the

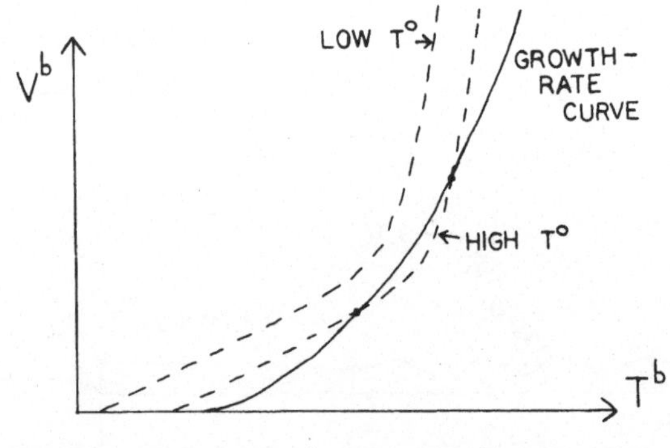

Fig. 2. Growth-rate curve (solid line) and self-sustained steady-state curves (dotted lines).

model, which represents a straight front moving a fixed distance ahead of the laser at the laser scan speed, is given (in dimensionless variables) by

$$x^b(y,t) = x_s^b = \text{constant}, \tag{3a}$$

$$T(x,y,t) = T_s(x) = T^o + T_L(x) + (L\sqrt{\beta}/c)\exp[-(x-x_s^b)-|x-x_s^b|/\sqrt{\beta}], \tag{3b}$$

where we have defined the parameter

$$\beta = \beta(V_s) = V_s^2/(V_s^2 + 4D\Gamma). \tag{4}$$

The last term in (3b), which is the contribution of latent-heat release to the temperature field, is L/c multiplied by the one-dimensional steady-state Green's function for diffusion. This can be used together with the laser power profile P(x) to calculate $T_L(x)$, which is the temperature field produced by the laser alone, and which is absent for self-sustained growth. The parameter β can be thought of as measuring either the front velocity or the unimportance of heat loss (since β increases as Γ decreases). Thus one way of controlling β experimentally would be to vary the thickness of the film; thicker films would not lose heat as quickly as thinner ones, and so would have higher values of β for the same front velocity.

To complete the specification of the steady state, we must determine x_s^b (or V_s for self-sustained growth). For laser-driven growth, the steady-state front temperature T_s^b can be found from the growth-rate curve and the fact that the front velocity is equal to the laser scan speed; evaluating the temperature field (3b) at the front then gives

$$T_L(x_s^b) = T_s^b - T^o - L\sqrt{\beta}/c, \tag{5}$$

which determines x_s^b implicitly. For self-sustained growth, there is no laser to fix the origin and the steady-state front velocity, so x_s^b is arbitrary, but V_s must be found. This is done by evaluating (3b) at the front, which gives a relation between T_s^b and V_s [namely (5) with T_L=0], and to demand that this relation and the intrinsic growth-rate relation be satisfied simultaneously. As shown in Fig. 2, this can occur only for sufficiently high values of the substrate temperature T^o. Thus the model predicts that self-sustained growth is energetically possible only if the substrate temperature is high enough.

The linear stability analysis[12-15] of the steady state proceeds in the standard way: one begins with an infinitesimal perturbation of (3),

$$x^b = x_s^b + \varepsilon\exp(iky + \omega t), \tag{6a}$$

$$T = T_s(x) + \varepsilon\tilde{T}(x)\exp(iky + \omega t), \tag{6b}$$

inserts these into the dimensionless form of equations (1) and (2), linearizes in the small parameter ε, and solves. The result is that the dimensionless wave number k of the perturbation and its dimensionless growth rate ω must satisfy the <u>stability equation</u>

$$[\omega + \alpha(1 + R)](1 + \beta k^2 + 2\beta\omega)^{1/2} = \alpha(1 + \omega). \tag{7}$$

The new parameters α and R appearing in this equation measure the sensitivity of the growth velocity to front temperature and the importance of the laser, respectively; they are given explicitly by

$$\alpha = (L\sqrt{\beta}/cV_s)(\partial V^b/\partial T^b)|_s, \tag{8}$$

$$R = -(c/L\sqrt{\beta})(dT_L/dx)|_s. \tag{9}$$

(The stability equation derived by Kurtze, van Saarloos, and Weeks[15] actually contains another parameter which accounts for the dependence of the growth velocity on interfacial curvature. However, unlike in the case of crystal growth from the melt, in which such a term is necessary to stabilize the system against short-wavelength perturbations, this term has no crucial effect on the stability of explosive crystallization, and so is omitted here.) Once a steady state has been selected, the values of α, β, and R are set; then if there is any value of k for which the stability equation (7) has a solution with the real part of ω positive, then that steady state is unstable.

The results of the analysis of the stability equation are illustrated in Fig. 3, which represents the possible steady states as points in the α–β plane for fixed values of R. The case of self-sustained growth, which is the special case in which R is zero, is shown in Fig. 3a. Points to the left of the solid and dotted lines represent stable steady states. The solid line, which is given explicitly by $\alpha=1/(1-\beta)$, separates steady states which are always unstable, on its right, from ones which may be stable, on its left. Note that in the graphical calculation of the steady-state front velocity for self-sustained growth, there are <u>two</u> intersections of the steady-state and growth-rate curves when the substrate temperature T^0 is sufficiently high, and so there are two possible steady states. At either intersection, the ratio of the slope of the growth-rate curve to that of the steady-state curve is $\alpha(1-\beta)$. Thus it is clear that the <u>slower</u> steady state, at which the growth-rate curve has the greater slope, is represented by a point to the right of the solid line, and so is always unstable, while the <u>faster</u> steady state, at which the steady-state curve is steeper, corresponds to a point to the left of the line, and so may be stable. Thus if self-sustained growth is possible, it will occur at or near the higher steady-state velocity. Along the dashed line in Fig. 3a, an <u>oscillatory</u> instability of the one-dimensional steady state sets in: on this line, which is given explicitly by $\beta=(\alpha^2-1)/4\alpha$, there are solutions of the stability equation for k=0 having purely imaginary ω. This signals the onset of oscillations, so that we should expect steady states to the right of this curve to be unstable, but to have stable limit cycles associated with them. Finally, the dotted line marks the onset of oscillatory <u>morphological</u> instabilities, those having imaginary ω for some <u>finite</u> wavenumber k. This line is given by $\alpha=4\beta$, and the wavenumber of the perturbation which is marginally unstable on it obeys $k^2=(2\beta-1)/\beta$. Steady states to the right of this line will also be unstable, and should give way to a <u>patterned</u> front whose velocity oscillates as the front advances. Fig. 3b shows the situation for laser-driven growth with R nonzero but rather small. Again, points

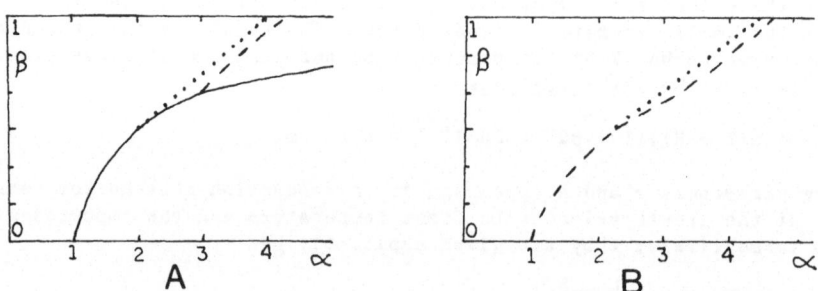

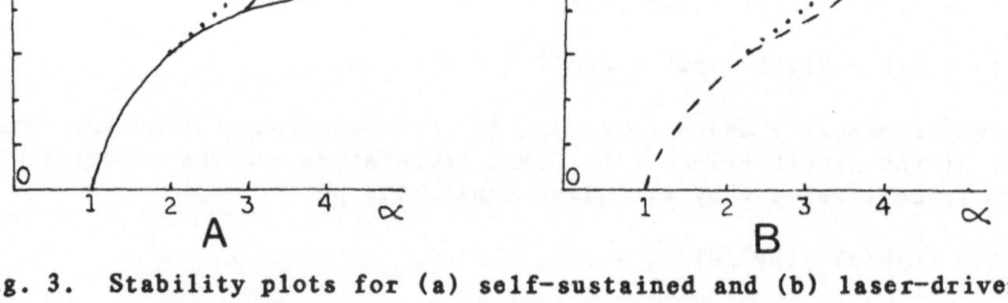

Fig. 3. Stability plots for (a) self-sustained and (b) laser-driven steady states.

to the left of the dashed and dotted curves represent stable steady states. The dotted curve, given by $\alpha=4(1+R)\beta$, again marks the onset of morphological instabilities. The dashed curve shows the onset of oscillatory instabilities for the one-dimensional system; as R goes to zero, it converges to the solid and dashed lines of Fig. 3a. On the section of this curve having $\alpha \lesssim 3$, the growth rate of the unstable perturbation is small for small R, and is in fact given explicitly by

$$\omega \sim i[2R/\beta(2-3\beta)]^{1/2}. \tag{10}$$

This fact will be quite important for the nonlinear analysis below.

One should note that once a sample has been prepared, so that $D\Gamma$, L/c, and the growth-rate curve are fixed, then both α and β are fixed functions of the front velocity, which can itself be adjusted either directly (for laser-driven growth) or by setting the substrate temperature T^o (for self-sustained growth). Thus all the possible steady states for a given sample are represented by points which lie along a single curve in the α-β plane. For example, if the growth-rate curve is a simple Arrhenius law,

$$V^b(T^b) = V_\infty \exp(-E/T^b), \tag{11}$$

then this 'sample curve' is given explicitly by

$$\alpha = (L\sqrt{\beta}/4cE)\{\ln[V_\infty^2(1-\beta)/4D\Gamma\beta]\}^2. \tag{12}$$

For laser-driven growth, we may travel upward along this curve by increasing the laser scan speed (adjusting T^o or the laser power then changes R), while for self-sustained growth we may move outward from the intersection of this curve with the curve $\alpha=1/(1-\beta)$ by increasing the substrate temperature.

NONLINEAR ANALYSIS

It is possible to perform a nonlinear analysis of the one-dimensional instabilities of the steady state and calculate the resulting oscillations of the crystallization front, provided we look at a region of parameter space in which the time evolution of the front (viewed in the frame of reference moving with the steady-state front velocity V_s) is slow. This is the case near the curve $\alpha=1/(1-\beta)$ for R small or zero. There is a natural physical interpretation of this parameter range: recall that if the substrate temperature T^o is barely high enough for self-sustained growth to occur, then the growth-rate and steady-state curves are tangent at their intersection, and so we have $\alpha=1/(1-\beta)$. Thus points in the α-β plane near this curve represent steady-state growth at a front temperature and velocity at which self-sustained growth is either barely possible or almost possible. In the former case we are considering self-sustained growth near its onset, and so set R=0; in the latter the laser is necessary to maintain growth, but only a small contribution is needed from it, so that R will be small.

We now outline the analysis of the slow nonlinear oscillations. Since the time evolution of the front position is slow, we write

$$x^b(t) = x_s^b + \xi(\Omega t) = x_s^b + \xi(\tau), \tag{13}$$

where Ω is a small parameter measuring the slowness of the time evolution of the front, chosen so that the correction ξ to the front position varies on the slow time scale τ, i.e., derivatives of ξ with respect to τ are of the same order as ξ itself. We next use the one-dimensional time-dependent Green's function to write the latent-heat contribution to the front temperature, $T_{lh}(x^b,t)$, using this form for the front position, and expand systematically in powers of Ω. A tedious calculation finally yields

$$T_{1h}(x^b,t) = (L\sqrt{\beta}/c)\{1 + \Omega(1-\beta)\dot{\xi} - (\beta\Omega^2/2)[(2-3\beta)\ddot{\xi} + 3(1-\beta)\dot{\xi}^2]$$

$$+ (\beta\Omega^3/2)[\beta(3-5\beta)\dddot{\xi} + 3\beta(4-5\beta)\dot{\xi}\ddot{\xi} - (1-\beta)(1-5\beta)\dot{\xi}^3]$$

$$- (\beta^2\Omega^4/8)[5\beta(4-7\beta)\ddddot{\xi} + 5\beta(5-7\beta)(3\ddot{\xi}^2+4\dot{\xi}\dddot{\xi})$$

$$- 6(4-35\beta+35\beta^2)\dot{\xi}^2\ddot{\xi} - 5(1-\beta)(3-7\beta)\dot{\xi}^4]$$

$$+ \ . \ . \ .\}, \tag{14}$$

where dots denote differentiation with respect to the slow time τ. We also expand the growth-rate curve about the steady state: for the present analysis, it is convenient to invert the growth-rate condition, and so think of the front temperature T^b as a function of velocity V^b. If the front position is given by (13), then the velocity V^b is $V_s(1+\Omega\dot{\xi})$, and so we may expand T^b about its steady-state value to obtain

$$T^b = T^b_s + (L\sqrt{\beta}/c)[\alpha_{-1}\Omega\dot{\xi} + \alpha_{-2}(\Omega\dot{\xi})^2 + \ . \ . \ .], \tag{15}$$

where we have defined the dimensionless derivatives

$$\alpha_{-n} = (cV^n_s/L\sqrt{\beta}n!)[d^nT^b/d(V^b)^n]|_s. \tag{16}$$

Note that α_{-1} is equal to $1/\alpha$. We derive an evolution equation for ξ by setting the front temperature obtained from the diffusion equation, with the expansion (14), equal to that found from the growth-rate curve via (16).

Laser-Driven Growth[17]

As stated above, the regime in which the time evolution of the crystallization front is slow is given by $\alpha\sim 1/(1-\beta)$, or equivalently $\alpha_{-1}\sim(1-\beta)$, and R small. The natural measure of the rate at which the front evolves is the growth rate (10) of the marginal perturbation at the stability boundary (which is of order $\sqrt{R}$), so we identify Ω as its imaginary part,

$$\Omega^2 = 2R/\beta(2-3\beta). \tag{17}$$

We can also write the laser temperature profile $T_L(x)$ explicitly near the front: small R means that the laser contribution to the front temperature is small, so that for reasonable laser powers it also means that the front is rather far ahead of the laser. Thus, since the steady-state Green's function is exponential, T_L will also be exponential near the front; the power profile P(x) will only determine the prefactor. We may use the definition of R to eliminate this prefactor - R is proportional to the gradient of T_L, and so is proportional to T_L itself because T_L is exponential. Thus the laser contribution to the front temperature will be of order Ω^2. Finally, since a detailed analysis of the stability equation reveals[15] that the one-dimensional stability boundary, for small R and β less than 2/3, differs from the curve $\alpha=1/(1-\beta)$ by an amount of order R, we will write

$$\alpha_{-1} = (1-\beta)(1 - \Omega^2\delta), \tag{18}$$

where δ, which is of order unity, measures how far the steady state under consideration is from the curve $\alpha=1/(1-\beta)$. The steady state should be linearly stable for sufficiently small δ.

We now see that when we match the two expressions for the front temperature, the one coming from the diffusion equation with the expansion (14) and the other from the growth-rate curve (15), the order-Ω terms only contribute at order Ω^3 because α_{-1} is near $1-\beta$, and the laser temperature profile enters at order Ω^2. Thus the leading terms in the evolution equation for ξ come from the Ω^2 terms in the expansions, and the first corrections from the Ω^3 terms, which contain $\dot{\xi}$, $\dddot{\xi}$, $\dot{\xi}\ddot{\xi}$, and $\dot{\xi}^3$ contributions. The corrections can be simplified in two (rather arduous!) steps: first we factor out a time derivative to eliminate the $\dddot{\xi}$ term; then we factor out $1-C\Omega\dot{\xi}$, choosing the constant C to eliminate the $\dot{\xi}\ddot{\xi}$ term. Finally, we rescale ξ by the decay length of the laser temperature profile. This ultimately gives us the evolution equation for the front position ξ,

$$0 = \ddot{\xi} + (a/2)\dot{\xi}^2 + [1 - \exp(-\xi)]$$

$$+\Omega\{-d + f[1 - \exp(-\xi)] + g\dot{\xi}^2\}\dot{\xi} + \ldots \quad (19)$$

The terms proportional to $1-\exp(-\xi)$ are the ones which come from the laser temperature profile. The coefficients a, d, f, and g involve β and the derivatives of the growth-rate curve. Explicitly, we have

$$a = 2[2\alpha_{-2}+3\beta(1-\beta)]/\sqrt{\beta}(2-3\beta)(1+\sqrt{\beta}), \quad (20a)$$

$$d = [2(1-\beta)/\beta(2-3\beta)](\delta - \delta_c), \quad (20b)$$

where $\delta_c=\beta^2(3-5\beta)/2(1-\beta)$ is the value of δ beyond which the steady state $\xi=0$ is linearly unstable, so that d measures how far beyond the onset of instability the steady state under consideration is. The parameter a is proportional to the difference in curvatures of the steady-state and growth-rate curves. The fact that we are taking R to be small and positive means that these two curves do not intersect at (T_s^b,V_s), but miss by a small amount. However, if the growth-rate curve has the larger curvature, so that a is negative, then the two curves (which are approximated as quadratics at this order) will intersect at some higher velocity. This is reflected by the fact that for negative a the leading-order terms of (19) admit a solution which has $\xi\sim\sqrt{2/|a|}$ for long times.

The evolution equation (19) can be put in a tractable form by making the change of variable

$$z = \exp(a\xi/2), \quad (21)$$

so that the leading-order terms in (19) become

$$0 = \ddot{z} + (a/2)z(1-z^{-2/a}) \quad (22)$$

This equation can be interpreted as the equation of motion of a particle in an algebraic potential well. The solutions are oscillatory in general, although for a>1 there are also solutions for which ξ reaches $-\infty$ in finite time, and for a<0 there are solutions which reach $\xi=+\infty$. The oscillations can be calculated as integrals using standard methods of classical mechanics. Samples of the results of such calculations are shown in Fig. 4. Note that the period of the oscillation depends on its amplitude; this is the case for all values of a except a=2, for which (22) becomes the equation of motion for a harmonic oscillator, and a=1/2, for which it is identical to the radial equation for a multidimensional harmonic oscillator. The periods of the oscillations can also be written as explicit integrals.

Note that in a standard bifurcation analysis, one makes the additional assumption that ξ itself is small of order Ω, pushing the nonlinear terms in ξ down into the higher-order corrections. This makes the leading-order terms in (19) a simple harmonic oscillator equation, so that all the unperturbed solutions have the same period. However, this is not necessary here, since we can solve the leading-order equation (22) <u>without</u> assuming that

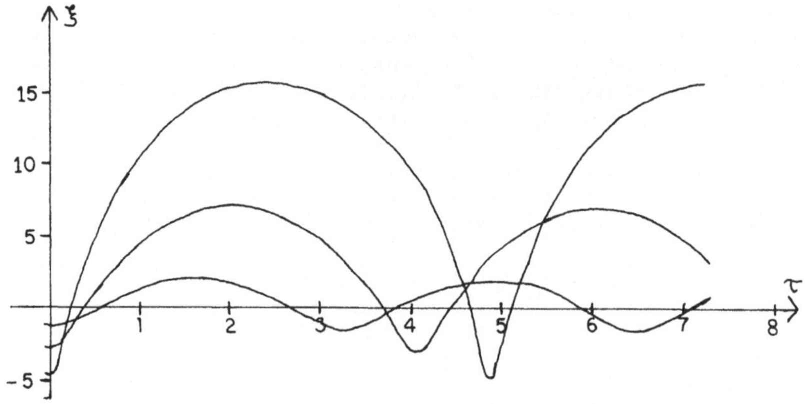

Fig. 4. Oscillatory solutions of the evolution equation (22) with a=0.15.

the excursion of the front from its steady-state position is small. Thus our analysis is richer than the standard analysis, and in particular is able to follow the development of finite-amplitude front oscillations.

The effect of the correction terms in (19) is to destroy all but a discrete set of the periodic solutions of the leading-order equation. This is most easily seen by calculating the corrections to the trajectories in the z-$\dot{z}$ phase plane. Periodic solutions of (22) are represented by closed curves in the phase plane; one can show[17] that the only trajectories which remain closed to first order in Ω are those which satisfy

$$\frac{\int[\dot{z}^3(z)/z^2]dz}{\int\dot{z}(z)dz} = \frac{a}{(4g/a)-(2f/3)}d, \tag{23}$$

where $\dot{z}(z)$ appearing in the integrals is calculated from the leading-order equation (22). This is then an implicit equation for the surviving limit cycles. The left side is positive, so that there is no solution for negative d; this indicates that the steady state $\xi=0$ is globally stable. The left side of (23) can be calculated numerically for arbitrary a and analytically for a=1/2 and a=2 as a function of amplitude, and can be analyzed asymptotically for large and small amplitudes.[17] For a>3/2, it remains finite for infinite amplitudes, so that a finite value of d can drive the amplitude of the surviving oscillation to infinity, forcing the system to follow one of the trajectories that lead to z=0 ($\xi=-\infty$). This may indicate that if the curvature of the steady-state curve is large enough, and we are working deep enough in the unstable region, the oscillation may get too far ahead of the laser and become so cold that the laser subsequently passes it without ever heating enough for it to resume moving at a speed V_s.

Self-Sustained Growth[14]

If we match expansions for T^b with the laser absent, then in accord with the fact that there is now a translational symmetry in the problem, only derivatives of ξ remain. If we choose Ω to be of the order of $1-\beta-\alpha_{-1}$, then the leading-order equation contains only $\dot{\xi}$ and its first derivative $\ddot{\xi}$. This equation has two steady states, representing the two possible steady-state velocities for given growth conditions. The higher-order corrections to these solutions will only modify these two velocities slightly.

In order to obtain oscillations, we must make the Ω^2 term dominate. This requires us to choose Ω so that $1-\beta-\alpha_{-1}$ is of order Ω^2. Furthermore, we must look at the region of the α-β plane near $\beta=2/3$, and in fact we must choose β to be within order Ω^2 of 2/3 in order to be able to see the behavior along lines of finite slope in the α-β plane. Thus we are looking precisely at the point on Fig. 3a at which the solid and dashed curves meet, which is a codimension-2 bifurcation point.[14] We now write

$$\alpha = 3 + \Omega^2\delta\alpha, \tag{24a}$$

$$\beta = 2/3 + \Omega^2\delta\beta. \tag{24b}$$

We must also assume that v itself is of order Ω, and so we set

$$\dot{\xi} = \Omega v. \tag{25}$$

Inserting these expressions into the two expansions of T^b and factoring out a time derivative yields the evolution equation for v,

$$0 = \ddot{v} + [3(9\delta\beta-\delta\alpha)/2]v + [9(3\alpha_{-2}+1)/2]v^2$$
$$+ \Omega[(18\delta\beta-5\delta\alpha)/2 + 9(5\alpha_{-2}+1)v]\dot{v}. \tag{26}$$

We will restrict our attention to the case $3\alpha_{-2}+1>0$, which, like $a>0$ for laser-driven growth, means that the steady-state curve has greater curvature than the growth-rate curve. Note that (26) has two steady states, the one with the lower value of v being unstable; as we saw above, for given growth conditions the self-sustained steady states generally occur in such pairs. We need only consider the case $\delta\beta/\delta\alpha>1/9$, for which $v=0$ is the higher steady-state value. This corresponds to $\alpha<1/(1-\beta)$; the slower steady state corresponds to a point with $\alpha>1/(1-\beta)$.

Once again, in the leading terms of (26) we may interpret v as the position of a particle which moves in a cubic potential well. If the particle has an 'energy' less than the well depth, so that it cannot reach the unstable value of v, then it oscillates about v=0. The period of the oscillation is given by an elliptic integral, which increases to infinity as the 'energy' increases to the well depth, and the form of the oscillation is a Jacobian elliptic function, which spends more time at negative values of v than positive. Again, we may calculate the effect of the corrections in (26) by looking at their effect on the trajectories in the v-$\dot{v}$ phase plane; we find that the surviving limit cycle must satisfy

$$\frac{\int \dot{v}(v)v\,dv}{\int \dot{v}(v)\,dv} = \frac{5\delta\alpha - 18\delta\beta}{18(5\alpha_{-2}+1)}. \tag{27}$$

The left side of this equation is negative. Thus for α_{-2} between $-1/3$ and $-1/5$, there can be no solution for $\delta\beta/\delta\alpha>5/18$. This means that the steady steady state v=0 is stable; $\delta\beta/\delta\alpha=5/18$ is the slope of the curve $\beta=(\alpha^2-1)/4\alpha$, the dashed line of Fig. 3a, at $\alpha=3$. For $\delta\beta/\delta\alpha<5/18$, where oscillatory instabilities exist, (27) is an implicit equation for the amplitude of the surviving limit cycle. It is actually rather easy to calculate the condition for this limit cycle to be the one of infinite period which passes through the unstable value of v, since the integrals in (27) become elementary; the final result is

$$\delta\beta/\delta\alpha = (75\alpha_{-2} + 29)/36(3\alpha_{-2} + 2). \tag{28}$$

This line lies in the region of oscillatory instabilities, and so shows the possibility that the oscillating front may eventually get too cold to continue, even though self-sustained growth is energetically possible.

REFERENCES

1. See, e.g., P. C. Hohenberg, Physica Scr. T9:93 (1985).
2. See, e.g., J. S. Langer, Rev. Mod. Phys. 52:1 (1980).
3. G. Gore, Philos. Mag. 9:73 (1855).
4. J. C. C. Fan and H. J. Zeiger, Appl. Phys. Lett. 27:224 (1975).
5. C. C. Coffin and S. Johnston, Proc. R. Soc., Ser. A 146:564 (1934).
6. C. E. Wickersham, G. Bajor, and J. E. Greene, Solid State Commun. 27:17 (1978); C. E. Wickersham, Ph.D. Thesis, Univ. of Illinois (1978).
7. C. E. Wickersham, G. Bajor, and J. E. Greene, J. Vac. Sci. Technol. A 3:336 (1985).
8. J. C. C. Fan, H. J. Zeiger, R. P. Gale, and R. L. Chapman, Appl. Phys. Lett. 36:158 (1980).
9. H. J. Zeiger, J. C. C. Fan, B. J. Palm, R. L. Chapman, and R. P. Gale, Phys. Rev. B 25:4002 (1982).
10. D. Bensahel and G. Auvert, in: 'Laser-Solid Interactions and Transient Thermal Processing of Materials,' J. Narayan, W. L. Brown, and R. A. Lemons, eds., North-Holland, New York, 1983, p. 165.
11. G. H. Gilmer and H. J. Leamy, in: 'Laser and Electron Beam Processing of Materials,' C. W. White and P. S. Peercy, eds., Academic, New York, 1980, p. 227.
12. V. A. Shklovskii, Dokl. Akad. Nauk SSSR 261:1343 (1981) [Sov. Phys.-Dokl. 26:1155 (1981)]; Zh. Eksp. Teor. Fiz. 82:536 (1982) [Sov. Phys.-JETP 55:311 (1982)].
13. W. van Saarloos and J. D. Weeks, Phys. Rev. Lett. 51:1046 (1983).
14. W. van Saarloos and J. D. Weeks, Physica D 12:279 (1984).
15. D. A. Kurtze, W. van Saarloos, and J. D. Weeks, Phys. Rev. B 30:1398 (1984).
16. M. J. Feigenbaum, J. Stat. Phys. 19:25 (1978).
17. D. A. Kurtze, to be published (Physica D).

FRACTAL BEHAVIOR OF SINGLE-PARTICLE TRAJECTORIES AND ISOSETS IN ISOTROPIC AND ANISOTROPIC FLUIDS*

R. K. Kalia and P. Vashishta

Argonne National Laboratory
Argonne, IL 60439

and

S. W. de Leeuw

Michigan State University
East Lansing, MI 48824

ABSTRACT

Fractal behavior associated with single-particle trajectories and isosets is observed in molecular dynamics simulations of liquids and superionic conductors. Fractal dimensions of trails and isosets are found to be 2 and 0.5, respectively. These values are shown to be universal in that they are independent of the spatial dimensionality, the nature of the interparticle interaction, and the thermodynamic state of the system.

The single-particle motion in liquids is conventionally characterized by the velocity autocorrelation function and diffusion constant. These characteristics vary from system to system and are also functions of the thermodynamic state of the system. However, certain features of the single-particle motion are independent of the nature of the interaction or the thermodynamic state of the system. These universal features are manifested in the fractal behavior of single-particle trajectories and isosets: their fractal dimensions are always 2 and 0.5, respectively.[1]

The fractal behavior of single-particle trajectories is related to their length.[2] When measured in units of a step distance ε, the length of a trajectory, $L(\varepsilon)$, decays algebraically over a certain range of ε:

$$L(\varepsilon) \propto \varepsilon^{-\alpha} . \qquad (1)$$

*Work supported by the U.S. Department of Energy, BES-Materials Sciences, under contract W-31-109-ENG-38.

The exponent α is called Richardson's coefficient, and the fractal dimension D is defined as

$$D = 1 + \alpha . \qquad (2)$$

The fractal behavior of isosets is related to Brown functions, $x_i(t)$, $y_i(t)$....which describe the time variations of coordinates of a particle.[2] The instants of time $\{\tau\}$ when the Brown function is equal to a given value x_o constitute an isoset. If x_o is the origin, the isoset is known as the zeroset. The gaps between successive values of τ are described by a probability function $Pr(G > g)$, for finding a gap of duration G greater than a given value g. It can be shown[2] that

$$Pr(G > g) \propto g^{-\bar{D}} \qquad (3)$$

where $\bar{D}$ is the fractal dimension of the isoset.

Fractal behavior of single-particle trajectories was first observed by Powles and Quirke[3] in molecular dynamics simulations of Lennard-Jones fluid. These authors concluded that the fractal dimension D = 1.65 and that D may be a function of the thermodynamic state of the system. However, Rapaport's[4] simulations for a hard-sphere system revealed that the asymptotic value of D inferred from very long trajectories yields the expected value D = 2.

In this paper we report the results of our molecular dynamics calculations on several liquids whose Hamiltonians may be written as

$$H = \sum_i \frac{p_i^2}{2m} + \frac{1}{2} \sum_{i \neq j} u \left(\frac{\sigma}{r_{ij}}\right)^n . \qquad (4)$$

Here u measures the strength of the interaction, and σ the size of the particles. The exponent n determines the steepness and range of the force law. We have chosen these systems because of the relative simplicity in characterizing their thermodynamic state by a single dimensionless variable

$$\Gamma = u\left(\frac{\sigma}{r_o}\right)^n / k_B T , \qquad (5)$$

where r_o is the mean interparticle separation: $\pi r_o^2 = \rho^{-1}$, ρ being the number density of the system. By varying the exponent n we can study the effects of the range and steepness of the interaction. Special values of n correspond to experimentally studied systems.

(1) n=1. If we write $u\sigma$ in Eq. (4) as e^2 the Hamiltonian describes the interaction between electrons confined to move in a plane. Charge neutrality is maintained by adding a uniform background of positive charge. This Hamiltonian has been used to model electrons on the surface of liquid helium, which has been studied by a variety of experimental techniques and simulations.[5,6] In particular, it has been shown that the system undergoes a fluid-solid transition at $\Gamma_f = 125 \pm 4$.

(2) n=3. Replacing $u\sigma^3$ by μ^2 the energy of interaction in Eq. (4) corresponds to dipoles whose moments, μ, point perpendicular to the plane of motion. To a good approximation, this model describes a system of polystyrene spheres floating on water[7] or immersed in a ferromagnetic fluid sandwiched between glass plates and under the influence of a magnetic field perpendicular to the plates.[8]

Dynamical simulations[9] have shown that the system solidifies at $\Gamma_f = 62 \pm 3$.

(3) n=12. The potential energy in Eq. (4) is the repulsive part of the Lennard-Jones potential. It is used as a model for the interaction of inert-gas atoms at high densities. Computer simulation studies[10] indicate a solid-fluid transition at $\Gamma_f = 0.98\ \pi^6$.

Trajectories of the particles in these systems were obtained by the method of molecular dynamics (MD). As is well known, this technique is used to generate a sequence of dynamic states of the system by numerical integration of the equations of motion. We used a fifth-order predictor-corrector method for this integration.[6] MD calculations were performed for systems containing 256 particles in a rectangular cell whose sides L_x and L_y were in the ratio $L_x/L_y = 2/\sqrt{3}$. Periodic boundary conditions were imposed on the system.

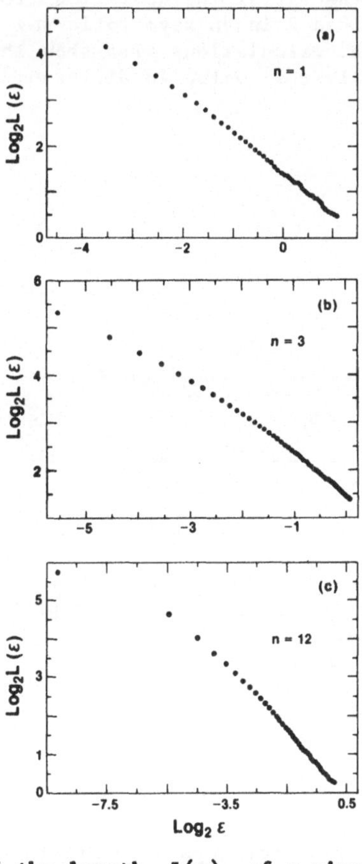

Figure 1: Variation of the length, L(ε), of a single-particle trajectory with the step distance ε for n=1,3 and 12 systems at $T_f/T=0.29$. T_f is the freezing temperature.

Molecular dynamics simulations yield the positions $\vec{r}_i(o)$, $\vec{r}_i(\Delta t)$,-- $\vec{r}_i(n\Delta t)$ of the particles i = 1, 2,--N at regular time intervals Δt where Δt is the time step used in integrating the equations of motion. Since Δt is sufficiently small, the particles may be assumed to travel along straight lines between successive time intervals. The length L(ε) is now calculated for a given value of the step distance ε by counting the number of times a rigid rod of length ε fits into the trajectory. End corrections are taken as fractions. In the limit $\varepsilon \to o$, the length $L(\varepsilon)$ corresponds to the length of the trajectory as if it were stretched out into a straight line. This length is $L_1 = \bar{v}\, t$ where $\bar{v} = (2\, k_B\, T/m)^{1/2}$ is the thermal speed and t the duration of the MD simulation. For large values of ε the length will be given by the separation between the end points of the trajectory and this is close to the root mean square distance of the trajectories.

Figure 1 shows the log-log plot of the length $L(\varepsilon)$ as a function of the step distance ε for n=1, 3, and 12 systems. It is apparent that there is a range of ε (~ 10) over which $L(\varepsilon)$ decreases linearly on the logarithmic scale. The slopes of these linear regions give Richardson's coefficient α which is related to the fractal dimension D by Eq. (2). The data in Fig. 1 show that D is always 2, regardless of the nature of the interparticle interaction in the system. However, this is true only if the trajectories are sufficiently long ($\sim 10^5\ \Delta t$). For shorter trajectories (30,000 Δt) the fractal dimension D is less than 2; the higher the temperature, the larger the deviation from D = 2. Thus, the fractal dimension approaches 2 in an asymptotic way when the trajectories are sufficiently long. MD calculations also show that the fractal dimension has the same universal value in different thermodynamic states.

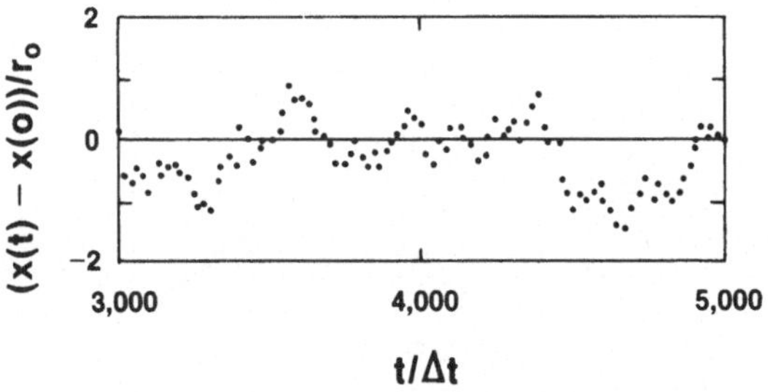

Figure 2: Time variation of a Brown function measured relative to its value at t=0 for n=1 system. The function is plotted at intervals of 20 Δt where $\Delta t = 2.5 \times 10^{-12}$ sec.

The probability function for gaps was calculated from Brown functions $\{x_i(t), y_i(t)\}$ generated by MD simulations. A plot of Brown function is shown in Fig. 2. Members of an isoset are the instants of time when the Brown function equals a given value x_o. These members tend to cluster, but the clusters themselves are distributed sparsely. The gaps G are the durations between successive members of an isoset. The probability function $Pr(G > g)$ is the number of gaps of duration G greater than a given value g, normalized by the total number of gaps. Figure 3 shows a log-log plot of P(g) as a function of g for the three systems at the same reduced temperature $T_f/T = 0.29$. The slopes yield 0.5 for the fractal dimension $\bar{D}$ in the three systems. That $\bar{D} = 0.5$ holds at other values of $\tilde{\Gamma}$ too, implying that $\bar{D}$ is also independent of the thermodynamic state of the system.

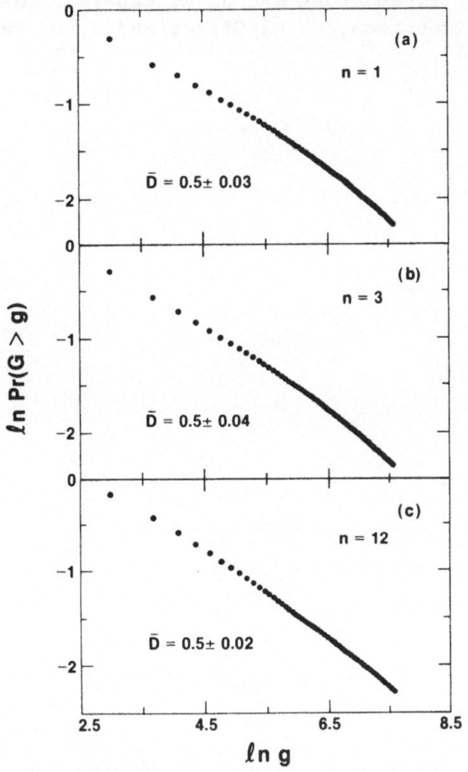

Figure 3: The probability function Pr(G>g) for finding gaps G of duration greater than g at $T_f/T=0.29$. The gaps are measured in units of an MD time step Δt.

In conclusion, MD simulations for a variety of systems in 2 spatial dimensions reveal fractal behavior associated with trajectories and isosets of single particle motion. The fractal dimensions of trajectories and isosets are 2 and 0.5, respectively, irrespective of the nature of the interparticle interaction or thermodynamic state of the system. Recently, we have investigated the fractal behavior of diffusing Ag ions in the superionic phase of Ag_2S. MD calculations have shown that the Ag ions diffuse anisotropically along certain directions in the lattice of S particles. Fractal dimensions D and $\bar{D}$ for Ag ions are again 2 and 0.5, respectively. These results confirm the universal nature of fractal dimensions of trails and isosets.

REFERENCES

1. R. K. Kalia, S. W. de Leeuw, and P. Vashishta, to be published; S. W. de Leeuw, R. K. Kalia, and P. Vashishta, to be published.
2. B. B. Mandelbrot, The Fractal Geometry of Nature (Freeman, San Francisco, 1982).
3. J. G. Powles and N. Quirke, Phys. Rev. Lett. 52, 1571 (1984).
4. D. C. Rapaport, Phys. Rev. Lett. 3, 1965 (1984).
5. C. C. Grimes, Surf. Sci. 73, 379 (1978); R. Mehrotra, B. M. Guenin, and A. J. Dahm, Phys. Rev. Lett. 48, 641 (1982).

6. P. Vashishta and R. K. Kalia, in Melting, Localization and Chaos, ed. R. K. Kalia and P. Vashishta (North-Holland, NY), 1982, p. 43.
7. P. Pieranski, Phys. Rev. Lett. 45, 569 (1980).
8. A. T. Skjeltorp, Phys. Rev. Lett. 51, 2306 (1983).
9. R. K. Kalia and P. Vashishta, J. Phys. C14, L643 (1981).
10. F. van Swol, L. V. Woodcock and J. N. Cape, J. Chem. Phys. 73, 913 (1980); J. Q. Broughton, G. H. Gilmer and J. D. Weeks, Phys. Rev. B25, 4651 (1982).

ENERGY DENSITY FORMALISM,

NUCLEAR MASSES AND HEAVY-ION INTERACTION

Irwin Reichstein, Department of Computer Science, Carleton University, Ottawa, Ontario, K15 5B6, Canada
and
F. Bary Malik*, Department of Physics, Southern Illinois University Carbondale, Illinois 62901, USA and Institut für Theoretische Physik der Universität Tübingen, (D-7400) Tübingen 1, FRG

INTRODUCTION

In the elastic collision of two heavy-ions, we expect a situation depicted schematically in Fig. 1.

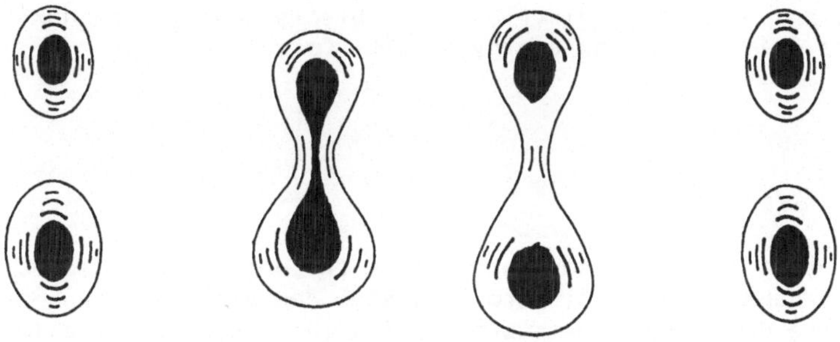

Fig. 1: Schematic description of elastic collision of two nuclei.

An interesting situation is the configuration (b) where two nuclei merge into a single one forming a neck of nuclear matter which is likely to be at a density lower than the saturation density. The calculation of the interaction between two heavy-ions in the London-Heitler approximation is given by the difference in total energies of the configuration like (b) or (c) and (a). Energies of (a) can either be obtained from the observed masses or to a good approximation from a suitable standard mass formula such as the one of Myers and Swiatecki[1]. It is, however, doubtful whether the total energy of a configuration like the one in Fig. 1(b) which has a neck of nucleons at a density other than the saturation one can be obtained with reasonable accuracy within the framework of mass formulae de-

*Partly supported by a grant from the Deutsche Forschungsgemeinschaft

rived from the concept of nuclei being liquid droplets. This is because, in all semi-empirical mass formulae, the mass number dependence is derived primarily under two assumptions. These are: (a) the nuclear density distribution in a nucleus is constant in the first approximation and (b) the nuclear surface is very thin. Let us look at a typical nucleus. It's half density radius is about 1.1 $A^{1/3}$ fm. and the surface is 2.6 fm. For a nucleus with 216 nucleons, the half density radius is 6.6 fm. and the zone of constant density is about 5.3 fm. The density drops from 90% of its value to a 10% value from 5.35 to 7.9 fm. A simple calculation indicates

$$\frac{\text{Volume of Constant Density}}{\text{Volume of Variable Density}} \simeq \frac{1}{2}$$

Assuming that on the average the nuclear density at the surface is about one half of the central density, it is clear that the number of nucleons on the surface is about the same as those in the constant density zone. Clearly, therefore, the fundamental assumptions of the mass A dependence of the mass formula based on the concept of nuclei being liquid droplets should be reexamined. On the other hand, it is true that except for a few light nuclei the semi empirical mass formulae have been very successful in explaining the observed masses. However, it is to be noted that coefficients of A (volume term) and $A^{2/3}$ (surface term) used in these mass formulae are very different from the ones expected from the liquid droplet. Thus, both the A dependence and their coefficients might just be a very good simulation of the actual situation at the saturation density but might not be a sufficient representation of the actual situation at other densities.

In view of these, and particularly, because of the fact that a significant number of nucleons in a nucleus resides on the surface, it is desirable to explore the possibility of calculating the nuclear masses taking into account a proper density distribution. This is done in the next section within the framework of an energy density formalism.

NUCLEAR MASSES

Hohenberg and Kohn[2] have shown that the energy per electron, i.e., the energy density of an electronic system can always be expanded in terms of the density of the system. Following its generalization to any Fermion system[3], Brueckner and his collaborators[4,5,6] extended that to describe nuclear masses. We shall basically follow this approach except for the following modifications: Brueckner et al.[4] attempted to obtain the nuclear density distribution function by solving approximately the equation for the density function that one obtains from the variation of the energy density functional. We shall on the other hand use a trapezoidal density distribution that describe closely the actual observed Fermi distribution function[7,8]. In a number of test cases, we have used the observed Fermi distribution function but the difference in the calculated masses with Fermi and trapezoidal distributions was not significant. The use of a trapezoidal distribution, on the other hand, reduces significantly the computational time and complexity. The advantage of using the observed density function is that it insures correct root mean squared radii which was difficult to get in the Brueckner et al.'s approach[4]. Preliminary results of our calculation were reported in refs. 9 and 10.

The starting point of the energy density formalism is to recognizes that the total energy of a Fermion system can be written as

$$E(\rho) = \int e[\rho(\vec{r})]d^3r \qquad (1)$$

where energy density function e must be calculated from a model based on a many body theory. Following refs. 3, 4, 5 this function is given by

$$e(\rho) = \frac{3}{5}\left(\frac{\hbar^2}{2M}\right)\left(\frac{3\pi}{2}\right)^{2/3}\frac{1}{2}\left[(1-\alpha)^{5/3} + (1+\alpha)^{5/3}\right]\rho^{5/3}$$

$$\rho v(\rho,\alpha) + \frac{1}{2}\Phi_c\,\rho - 0.7386 e^2 \rho_P^{4/3} + \left(\frac{\hbar}{8M}\right)\eta(\nabla\rho)^2 \qquad (2)$$

where α is the neutron excess, M is the nucleonic mass, ρ is the density. The part involving many body theory is in the calculation of the average single particle potential $v(\rho,\alpha)$. In the Brueckner-Hartree-Fock approximation one can calculate this using the t-matrix approach and one of the appropriate two nucleon potentials. The calculation reported here uses Brueckner, Gammel and Thaler types of potential and follows the prescription of ref. 6 to evaluate the energy denominator, and the density dependence. It is then given by

$$v(\rho,\alpha) = b_1(1+a_1\,\alpha^2)\rho + b_2(1+a_2\,\alpha^2)\rho^{4/3}$$

$$b_3(1+a_3\,\alpha^2)\rho^{5/3} \qquad (3)$$

Values of the parameters a_1, a_2, a_3, b_1, b_2, and b_3 are discussed later. The Coulomb potential is given by

$$\Phi_c = e\int \frac{\rho_P(\vec{r}\,')}{|\vec{r}-\vec{r}\,'|}\,d^3r' \qquad (4)$$

The fourth term in (2) is the correction to (4) due to the Pauli-Principle among the protons, and this density dependence is due to Peaslee[11]. Originally Dirac in 1930 showed the existence of such a term arising from the Pauli principle for an electron gas. The last term in (2) can be viewed as a combination of the Weizsäcker inhomogenity term to the kinetic energy and first order correlation energy not included in (3). Actually this term does not have quite the density dependence of the Weizsäcker correction.

In refs. 4 and 6, b_1, b_2 and b_3 were taken from the nuclear matter calculation in the local density approximation and η was taken as a free parameter to reproduce the total mass of a few nuclei. Although the results were reasonable, it was not sufficient for our purpose. Brueckner et al. in ref. 5, then proposed to obtain b_1, b_2 and b_3 from three known properties of nuclear matter namely energy per nucleon E/A, Fermi momentum k_F at the saturation density and compressibility K, and allowed a slight variation of a_1, a_2 and a_3 from those obtained in refs. 4 and 6. Values of E/A, k_F and K used in ref. 5 is different from the ones in ref. 6 and hence b_1, b_2 and b_3 in the two cases differ from each other. Similarly, η in the two cases was different.

We have basically used the E/A, k_F and K used in ref. 5 to determine b_1, b_2 and b_3 and adjusted a_1, a_2, a_3 to obtain a reasonable fit to the observed masses. Our values of a_1, a_2, a_3 and η differ slightly from those of ref. 5. Since we are using a trapezoidal density distribution, which is close to the observed Fermi ones, the root mean squared radii are well reproduced.

The trapezoidal distribution used for the density ρ is given by[15]

$$\rho(r) = \begin{cases} \rho_0 & 0 \leqslant r \leqslant r_0 \\ \rho_0\,(b-r)/(b-r_0) & r_0 \leqslant r \leqslant b \\ \rho_0 & b \leqslant r \leqslant \infty \end{cases} \qquad (5)$$

The parameters b, determining the surface thickness and r_0, the range of the constant density zone are related to the half-density radius C and the 10%-90% surface thickness parameter t by the relations

$$b = C + (5/8)t \quad \text{and} \quad r_0 = C - (5/8)t \tag{6}$$

The usual choices for C_0 defined by $C = C_0A^{1/3}$ and t that are compatible with electron scattering and μ-mesic data on nuclear density are C_0=1.07 fm and t=2.4 fm. In recent years Lombard[12] and Ngô and Ngô[13] have followed similar approaches to calculate total masses of nuclei. Both of them use different values of E/A, k_F and K. Lombard uses a density distribution function similar to the one expected from the Hartree-Fock type of calculations, whereas, Ngô and Ngô use a Fermi density distribution function to evaluate the integration in (1).

In table 1, we note the parameters a_1, a_2, a_3, b_1, b_2, b_3 and η used in ref. 6 (marked as BGT), by us (marked as MR), by Brueckner, Buchler, Clark and Lombard[5] (noted as BBCL), by Lombard[12] (noted as L) and by Ngô and Ngô[13] (marked as NN). In the same table we record E/A, k_F and K used by each of these groups. E/A used in all cases are very similar and the values are well within the uncertainty of our present knowledge of this parameter. k_F used by Lombard and Ngô and Ngô are the same but differ significantly from the ones used by BGT, MR and BBCL. The value of the compressibility used by Ngô and Ngô is very different from that used by others. Our knowledge of this parameter has, however, considerable uncertainty and all the values used are within that uncertainty.

In table 2, we compare calculated nuclear masses by us (noted as MR), Lombard (marked as L), Brueckner, Buchler, Clark and Lombard (noted as BBCL), Ngô and Ngô (marked as NN) with those observed[19] and with the Myers-Swiatecki's mass formula[1] based on the concept of the liquid drop model. Clearly, masses obtained from the energy density functional calculations are at least as good as those obtained from the Myers-Swiatecki's mass formula. In addition MR, L and NN's calculations are compatible with the observed density distribution and root mean square radii. These studies clearly establish that within the framework of a proper density distribution, it is possible to calculate nuclear masses and it is not necessary to use mass formulae derived from the concept of the liquid drop model.

HEAVY-ION HEAVY-ION POTENTIAL

In the London-Heitler approximation the interaction potential at a distance R between two centers V(R) is given by

$$V(R) = E(\rho) - E(\rho_1) - E(\rho_2) \tag{7}$$

where ρ, ρ_1 and ρ_2 are, respectively, density distribution function of the overlapped nucleus such as the ones depicted in Fig. 1(b) and 1(c) and nuclei one and two when they are well separated. E(ρ), E (ρ_1) and $E(\rho_2)$ are the respective energies. ρ_1 and ρ_2 i.e., the density distribution function of two colliding nuclei can be taken to be the appropriate trapezoidal types reproducing the observed density distribution. However, the density distribution function ρ of the composite system is not well known. One can calculate it in the two extreme cases:

(a) The Sudden Approximation: This was initially introduced by Brueckner, Buchler and Kelly[14] to calculate the ^{16}O-^{16}O potential. In this approximation, one generates ρ by simply adding the densities of two colliding nuclei.

Table 1. Parameters used by various groups in evaluating nuclear masses and binding energy per nucleon E/A, in MeV, Fermi wave length K_F in fm^{-1} and compressibility K. BGT, MR, BBCL, L and NN refer, respectively, to refs. 6, this work, 5, 12 and 13.

Para-meters	BGT	MR	BBCL	L	NN
η	8.0	10.3	11.955	15.2	7.23
b_1	-717.6	-741.28	-741.28	-818.25	-588.75
b_2	1142.2	1179.89	1179.89	1371.06	563.56
b_3	-452.6	-467.54	-467.54	-556.55	160.92
a_1	-0.146	-0.1933	-0.2	-0.316	-0.424
a_2	0.23	0.3128	0.316	0.2	-0.0973
a_3	1.2	1.715	1.646	-1.646	-2.25
E/A	-15.23	-16.59	-16.59	-16.0	-15.6
k_F	1.433	1.447	1.447	1.36	1.36
K	172.6	184.7	184.7	180.00	250

$$\rho = \rho_1 + \rho_2 \tag{8}$$

(b) Adiabatic Approximation: As noted in refs. 9, 10, 15, 16, the other extreme case is that the densities, as they interpenetrate, have plenty of time to readjust and hence the parameters of the density distribution can be obtained by minimizing the energy with respect to them at every point of separation. Instead of this tedious point by point variation, one can make a particular ansatz for the dependence of parameter C and t on the separation distance with the stipulation that no where in the composite system the value of the density exceeds its value at saturation. The ansatz used (called Spherical Adiabatic Approximation, SAA) is the following:

$$\begin{aligned} C_i(R) &= C_c \exp\left[\ln(C_i/C_c)\cdot R^2/R^2_{cut}\right] \quad \text{for } R \leq R_{cut} \\ &= C_i \quad \text{for } R > R_{cut} \end{aligned} \tag{9}$$

Here i=1 or 2, and C_c refers to the C parameter of the composite system. Exactly the same ansatz is made for the parameter t. R_{cut} has been chosen to be equal to 0.75 (Half-density radii of two ions).

The parameter ρ_0 of the density distribution of the composite system is then determined from the conservation of mass A

$$A = \int \rho(r)d^3r \tag{10}$$

Table 2. Calculated masses by various groups in the energy density approximation are compared with the experimental ones, marked EXPT. and with those obtained from the standard mass fromula of Myers and Swiatecki, noted as MS (ref. 1). MR, L, BBCL, and NN refer, respectivley, to the calculations reported here, and in refs. 12, 5 and 13.

NUCLEI	EXPT	MR	L	BBCL	NN	MS
$_{8}O^{16}$	127.6	123.3	128.8	127.6	121.2	123
$_{16}S^{32}$	271.8	270.1				
$_{20}Ca^{40}$	342.1	342	342	340	342.1	340
$_{20}Ca^{48}$	416.0	416.2	422	422	348.9	415
$_{26}Fe^{56}$	492.3	496.7				
$_{28}Ni^{60}$	526.9	532.1	524	524	530.8	524
$_{40}Zr^{90}$	783.9	793.4	780	780	792.3	782
$_{46}Pd^{118}$		998.4				
$_{48}Cd^{122}$		1033.3				
$_{58}Ce^{140}$	1172.7	1181.8	1169	1173	1185.1	1171
$_{56}Ba^{142}$	1180.3	1185.7				
$_{67}Ho^{165}$	1344.8	1356.1				
$_{82}Pb^{208}$	1636.5	1628.8	1627	1630	1639	1627
$_{92}U^{234}$	1776	1778				
$_{92}U^{238}$	1801.7	1798.5	1797	1812	1814	1805
$_{94}Pu^{240}$	1813.4	1811.0				
$_{98}Cf^{240}$		1804.8				

In ref. 15 it was already shown that the interaction potential between two O^{16} ions differs considerably in these two limiting cases. Nevertheless, the potential near the surface region in both cases are very similar. In the interior region the sudden approximation generates a short range repulsion and in the SAA, the potential for the O^{16}-O^{16} case is attractive but shallow, about 20 MeV deep.

(c) Deformation: The shape of the density function of the composite configurations can be generated either by two overlapping spheres or by two overlapping spheroids. As already noted earlier[16] the effect of generating the density function by two overlapping spheroids is to influence the shape of the external barrier. This point is important because the elastic scattering data of heavy-ions are sensitive to the shape of the external barrier at energies slightly higher than the barrier height[17].

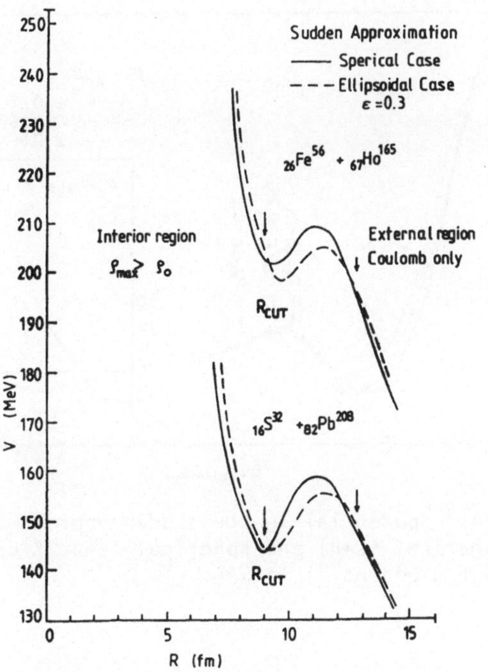

Fig. 2: The interaction potential between Fe^{56} and Ho^{165} and S^{32} and Pb^{208} in the sudden approximation. ε refers to the eccentricity of each nucleus.

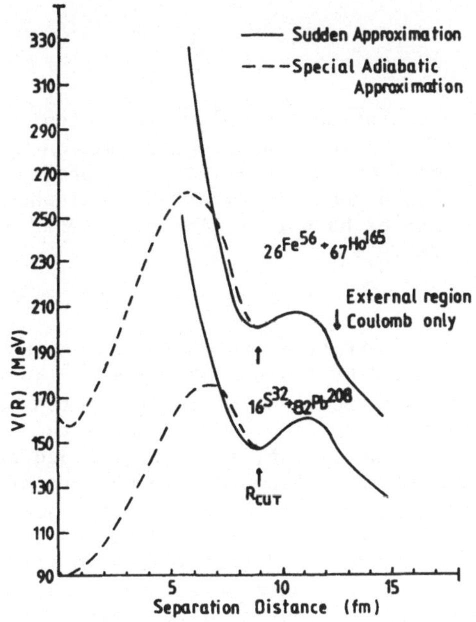

Fig. 3: The interaction potential between Fe^{56} and Ho^{165} and S^{32} and Pb^{208} in the sudden and SAA.

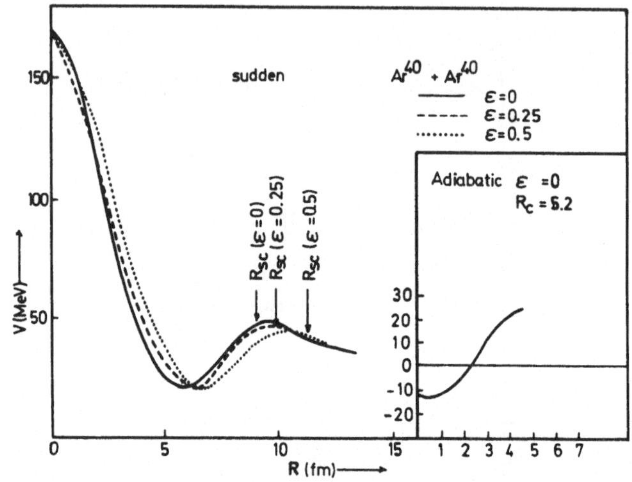

Fig. 4: The Ar^{40}-Ar^{40} potential in the sudden approximation for two overlapped spherical (ε=0) and spherical (ε=0.25 and 0.5) nuclei. The insert is the potential in SAA.

In Fig. 2 we have plotted the calculated potential between Fe^{56} and Ho^{165} and S^{32} and Pb^{208} in the sudden approximation where the composite system is generated once by two super imposed spheres (solid line) and then by two super imposed spheroids (dashed line) each having an eccentricity ε=0.3. The barrier height is clearly lowered in the case of two superimposed spheroids.

In Fig. 3 we examine the effect of adiabaticity on the heavy-ion heavy-ion potential for the S^{32} and Pb^{208} and Fe^{56} and Ho^{165} systems. The solid lines and the dashed lines represent calculated potential in the sudden and the SAA approximation, respectively. Clearly the shape of the potential in the interior region is influenced by the degree of adiabaticity. The potential, inside in the SAA could have pockets of 20 to 80 MeV. Although no calculation is presented here, from the general structure of the potential it is clear that the Pb-Pb, Pb-U and U-U potential in the SAA are likely to have pockets and then likely to stick together. Experimentally, there are some evidence of such a situation in sub-Coulomb collision[21,22]. This then provides an alternative (or additional) explanation to the ones in refs. 23 and 24 for the cause of the formation of a pocket in the potential between two heavy-ions. In fact, the pocket is expected to be much deeper than those calculated in refs. 23 and 24. Despite the formation of the pocket in the SAA, the overall strength of the potential even in the interior region remains positive, i.e., repulsive.

Recently, Trefz, Faessler and Dickhoff[25] have used a similar energy density formalism to calculate the interaction between A^{40} and Pb^{208} for various incident energies. However, they do not have the third (Coulomb potential), fourth (the exchange correction to the Coulomb interaction among protons), and the fifth (the gradiant) terms and the neutron excess term α in the expression (2) for the energy density function $e(\rho)$. They calculate the single particle potential for two superimposed (i.e., sudden approximation) nuclear densities[26] in the local density approximation using Reid's soft core two nucleon potential[27]. They allow the momenta of two colliding nuclei to enhance the Fermi momentum of individual nucleons and obtain a slightly velocity dependent A^{40} and Pb^{208} potential. By calculating the actual probabilities of two nucleons to scatter outside the Fermi sea, they have calculated the imaginary part of the A^{40} and Pb^{208} potential. Alexander and Malik[28] calculated imaginary part of the O^{16}-O^{16} potential in a similar fashion but using the effective range theory instead of the potential between two nucleons.

The question is now which of the two approximations, the sudden or the special adiabatic approximation, is close to the actual physical situation. The actual situation depends very much on the ratio of the global collision time between two heavy-ions and the relaxation time of individual nucleons involved in the collision. If the collision time is short compared to the time taken by nucleons to relax after the impact, the sudden approximation is a good one. On the other hand if the nucleons involved in the collision can relax before the collision is complete, we expect the adiabatic situation. In case these two times are comparable, in the low energy collision we have an adiabatic and in the high energy collision a sudden situation. Calculations of the relaxation time is of interest in this context.

On the other hand one can look at the empirical potentials derived phenomenologically by fitting elastic scattering data. The low energy data i.e., data up to energies slightly higher than the Coulomb energy can be fitted usually by potentials similar to the one derived in both approximations provided the tail of the potential i.e., the shape of the potential near the Coulomb barrier remains the same[17]. Only in two cases namely O^{16} and O^{16} and C^{12} and C^{12} high energy elastic scattering data at difference energies are available. They have been fitted[18,19] using a potential similar to the one expected from the sudden approximation and cannot be fitted with energy independent real potential which does not have a repulsive core.

Authors would like to express their appreciation to Professor Manuel de Llano, Dr. Q. Haider and Professor A. Faessler for discussion. One of the authors (FBM) would like to thank the kind hospitality of the University of Tübingen and the financial support of the Deutsche Forschungsgemeinschaft.

References

1. W. D. Myers and W. J. Swiatecki, Nucl. Phys. 81, 1 (;966).
2. P. Hohenberg and W. Kohn, Phys. Rev. 136, 1384 (1964).
3. H. A. Bethe, Phys. Rev. 167, 879 (1968).
4. K. A. Brueckner, J. R. Buchler, S. Jorna, and R. J. Lombard, Phys. Rev. 171, 1188 (1958).
5. K. A. Brueckner, J. R. Buchler, R. C. Clark and R. J. Lombard, Phys. Rev. 181, 1543 (1969).

6. K. A. Brueckner, S. Coon, and J. Dabrowski, Phys. Rev. 168, 1184 (1968).
7. L. R. B. Elton, Nuclear Sizes (Oxford University Press 1961).
8. Y. N. Kim, Mesic Atoms and Nuclear Structure (North Holland Publishing Co., 1971).
9. I. Reichstein and F. B. Malik, Ann. Phys. (N.Y.) 98, 322 (1976).
10. I. Reichstein and F. B. Malik, Superheavy Elements, ed. M. A. K. Lodhi (Pergamon Press 1978).
11. D. C. Peaslee, Phys. Rev. 95, 717 (1959).
12. R. J. Lombard, Ann. Phys. (N.Y.) 77, 380 (1973).
13. H. Ngô and Ch. Ngô, Nucl. Phys. A348, 140 (1980).
14. K. A. Brueckner, J. R. Buchler, and M. M. Kelly, Phys. Rev. 173, 944 (1958).
15. I. Reichstein and F. B. Malik, Phys. Lett. 37B, 344 (1971).
16. I. Reichstein and F. B. Malik, Workshop on High Resolution Heavy Ion Physics, eds. M. Martinot and C. Volant (CEN/Saclay Publication).
17. L. Rickertsen, B. Block, J. W. Clark and F. B. Malik, Phys. Rev. Lett. 22, 951 (1969).
18. Q. Haider and F. B. Malik, Proc. Int'l. Conf. on Resonant Behavior of Heavy-ion ed. G. Vourvopulos (Demokritos; Athens 1981).
19. Q. Haider and F. B. Malik, J. Phys. G7, 1661 (1981).
20. J. H. E. Mattauch, W. Thiele, and A. H. Wapstra, Nucl. Phys. 67, 1 (1965).
21. M. Clemente, E. Bedermann, P. Kienle, H. Isertos, W. Wagner, C. Koghuharov, F. Bosch and W. Koenig, Phys. Lett. 137B, 41 (1984).
22. J. Schweppe, Phys. Rev. Lett. 51, 2261 (1983).
23 M. Seiwert, W. Greiner and W.T. Pinkston, J. of Phys. G: Nucl. Phys. 11, L21 (1985).
24. M. Ismail, M. Rashdan, A. Faessler, M. Trefz and H. M. M. Mansour, to be published (1985).
25. M. Trefz, A. Faessler, and W. H. Dickhoff, Nucl. Phys. A443, 499 (1985).
26. M. Ismail, A. Faessler, M. Trefz, and W. H. Dickhoff, J. Phys. G: Nucl. Phys. 11, 763 (1985).
27. R. V. Reid, Ann. Phys. (N.Y.) 50, 411 (1968).
28. D. R. Alexander and F. B. Malik, Phys. Lett. 42B, 412 (1972).

THREE-BODY FORCES IN NUCLEI

S. A. Moszkowski†

Department of Physics
University of California
Los Angeles, California 90024

ABSTRACT

This is a review of some aspects of three body forces in nuclei. In particular, I will try to explain why three body forces may play a more important role in nuclear structure than is generally thought to be the case. Among other things I will discuss some exciting new developments in several areas which bear on this problem, namely the relativistic theory of nuclear matter, and the tensor interaction at short distances.

I. INTERACTIONS BETWEEN ATOMS AND BETWEEN NUCLEONS

Let us first very briefly survey the role of two body and three body forces between atoms. For neutral atoms, the Coulomb interaction vanishes, but there are two kinds of two body interactions resulting from the internal structure. These are a.) the short range interactions due to overlap of the electron clouds (short range repulsion and covalent bonding) and b.) the long range interactions due to polarization (Van der Waals interactions). There are also three body interactions, in particular, the long ranged Axilrod-Teller interaction due to three photon exchange. This seems to be required to explain the equation of state of rare gas solids. (However, for liquid Helium 4, the two body Aziz potential by itself seems to fit things very well.)

We can proceed by analogy for the interactions between nucleons. Each nucleon is constituted of three colored quarks which interact via exchange of gluons. Due to color neutrality the first order interaction between nucleons vanishes, just as in the case of neutral atoms. Again, there are short range nucleon-nucleon interactions due to overlap of the quark bags. In particular, we can get non-additive interactions when more than two bags overlap. This could be one of the origins of three body interactions between nucleons. On the other hand, the quark model also leads to long range Van der Waals interactions between nucleons.[1] There is no evidence for such interactions (which have an inverse power law dependence on distance) between nucleons.

---

†Work supported in part by the National Science Foundation Grant 84-20619

Presumably, if and when the quark model is developed further so as to explain the mechanism of confinement, which screens off the long range part of the interaction[2] and also the structure of the pion (as some kind of correlated quark-antiquark pair), then there may be some real hope of deriving the nuclear forces directly from QCD. In the meantime, it seems better to use a different model for the the nucleon-nucleon interaction, especially at large distances, namely one based on meson exchange.

II. SHELL MODELING OF THREE BODY INTERACTION BY EFFECTIVE TWO BODY INTERACTION

I Would like to quote from D. Wilkinson[3] (Lectures at 1977 Summer School on Heavy Ions and Mesons in Nuclear Physics, Les Houches, France):

> "Of course, we have tried very hard to pretend that many-body forces do not exist and have attempted to sweep them under the carpet by, for example, committing ourselves in our shell-modeling of states, to the idea of two body NN-forces only and then adjusting the effective residual NN force to give the best overall fit to data thereby perhaps absorbing into that pragmatic force elements that rightly belong into NNN or higher forces. Who knows?"

This point can be illustrated by a simple example from a classic paper by Skyrme[4]:

Consider particles in a pure j orbit outside closed shells. The energy for n particles in this orbit, relative to that of the closed shells, can be written as follows:

$$E(j^n) = n\,A(y) + \tfrac{1}{2}\,n(n-1)\,B(y) + \tfrac{1}{6}\,n(n-1)(n-2)\,C(y) + \ldots$$

Here A, B, and C denote the effective one, two, and three body interaction energies of the valence nucleons, respectively. (Some of these arise from interactions between these nucleons and those in the core.) The quantity y is a scale parameter, such as, for example, the oscillator spacing, which may depend on the number of particles. A change in y can be absorbed by a modification of the interaction strengths. Suppose that for a single particle we have $y = y_0$ and

$$E(j) = A(y_0) = A_0.$$

For several particles in the shell, we may have a different value of y. In particular, if

$$A(y) = A_0 + \tfrac{1}{2}\,A_2\,(y-y_0)^2 + \ldots$$

$$B(y) = B_0 + B_1\,(y-y_0) + \ldots$$

$$C(y) = C_0$$

then for given n, the energy is minimized for:

$$y(n) = y_0 - \frac{(n-1)}{2}\,\frac{A_2}{B_1}\,.$$

In particular, we find that

$$E(j^n) = n\,A_0 + \tfrac{1}{2}\,n(n-1)\,V_{2(Eff)} + \tfrac{1}{6}\,n(n-1)(n-2)\,V_{3(Eff)} + \ldots$$

where

$$V_{2(Eff)} = B_0 - \frac{1}{4}\frac{B_1^2}{A_2}; \qquad V_{3(Eff)} = C_0 - \frac{3}{4}\frac{B_1^2}{A_2}$$

Note that the effective two and three body interaction parameters may be quite different from the original values B_0 and C_0. In particular, for nuclei in the sd shell with reasonable choice of the parameters, Skyrme found that $V_{3(Eff)}$ is an order of magnitude smaller than C_0. Indeed, in shell model calculations we can do very well with effective two body interactions, without having to put in any three body terms. On the other hand, the $V_{2(Eff)}$ may be quite different not only from the original interaction fitting nucleon-nucleon phase shifts, but also from the calculated G-matrix, since it has absorbed any scale change with the nucleon number.

III. RELATIVISTIC NUCLEAR THEORY

During the last few years, it has become clearly established that a fit of the spin dependent nucleon-nucleus scattering amplitudes (for incident nucleons of energies in the 100 to 500 MeV range) requires a strong non-linear dependence of the single particle potential on densities.[5] Such a density dependence can be easily generated if the nucleons are assumed to be Dirac particles and the single particle potential is assumed to be made up of an attractive Lorentz scalar plus a repulsive Lorentz vector, which happen to nearly cancel. The well-known transformation of the Dirac equation to an equivalent Schroedinger equation gives for the central part of the single particle potential[6]:

$$U_{Eff} = \rho(-S+V) + \rho^2(S^2-V^2)/2Mc^2 + \ldots$$

S and V are due to exchange of scalar and vector mesons.

At normal nuclear density, and low energy, we have $\rho S \approx -350$ MeV and $\rho V \approx 250$ MeV, for a net attractive potential of about 100 MeV. What is notable is that the nonlinear term is repulsive and quite large, of magnitude 30 to 50 MeV, so that the resulting net attractive potential is only about 50 MeV. With increasing energy, the scalar field is quenched, so that the total potential becomes less attractive, and indeed repulsive above about 200 MeV (at normal density).

The strong density dependence implied by the Dirac formalism gives essentially the same results as having (in the nonrelativistic description) a repulsive three body interaction. (This has to qualified somewhat. Density dependence and three body interaction are somewhat different when we consider exchange terms. For example, consider a system of three neutrons interacting via a zero range force. The interaction energy due to a density dependent two body interaction is finite, but that due to a three body interaction vanishes.)

Recently, the Dirac approach has been applied to to calculate the binding energy of nuclear matter as function of density.[7,8] This calculation has been made many times before, using two body interactions which fit nucleon-nucleon phase shifts and the properties of the deuteron. However, in the past, relativistic effects were not taken into account. It is well known that, regardless of details of the two body interaction, the satiration density and binding energy (per particle) lies on a curve, known as the Coester line, which does not pass through the empirical saturation point. ($E/A = -16$ MeV, $k_F = 1.33\ \mathrm{fm}^{-1}$.) For example, for an interaction which yields the correct density, the binding energy is only about 10 to 12 MeV per particle. Many attempts have been made in the past to remedy this problem. Indeed, by allowing some three body interactions in addition to the two body interaction, Carlson, Pandharipande, and Wiringa[9] were able to get some improvement, though they still could not fit the empirical saturation results. However, by using the Dirac formalism, it is possible to fit the saturation properties. Shakin et al.[7] used essentially a mean field approximation, and more recently, Brockmann and Machleidt[8] did a relativistic Brueckner-Hartree-Fock calculation and got excellent agreement.

It is instructive to see just why a relativistic calculation does better than a nonrelativistic one. The basic reason is that the former seems to make a more realistic extrapolation off the energy shell. In particular, we obtain a strong density dependence when we make a nonrelativistic reduction, as was already noted above. If we use a nonrelativistic description to begin with, then we also must have considerable density dependence to get any kind of saturation, but in this case it is largely due to higher order effects of the NN tensor interaction.[10] At very high densities, the tensor interaction is quenched by the action of the Pauli principle. This means that the density dependence from a tensor dominated interaction is much weaker than linear. Indeed, calculations using a reaction matrix, the density dependence is often taken to be something like $\rho^{1/3}$.

The essential feature of a relativistic calculation is that provides something like a linear density dependence, which seems to fit experimental results better. However, could it be that this linear density dependence comes not from relativity, but from some other source? Indeed, Negele[11] has argued that possibly the relativistic description, while a convenient way of parametrizing the nuclear potential, might not necessarily be fundamental. Along these lines, I would like to suggest that perhaps some of the effects described by the Dirac approach might have their origin in three body interactions arising from multimeson exchange.

IV. MULTIMESON EXCHANGE THEORY OF 2 AND 3 BODY FORCES

Figure 1 shows what I think are the important components of two and three body interaction between nucleons arising from meson exchange. First consider the well known case of the two body interaction. We have, of course, the one pi+rho exchange denoted by 1PEP. As is well known its spin-isospin average contribution to the nuclear matter energy is relatively small. Indeed, it cancels exactly in the Hartree approximation. The two meson exchange term 2PEP gives a much larger contribution to the nuclear matter energy, in spite of the fact that it is of higher order in the meson nucleon coupling than the 1PEP. In the 2PEP, the spin-isospin averaged contributions add coherently, and also most of the time the intermediate states are Δ's. The coupling $f^2_{\pi N\Delta}$ is about 4 times as large as $f^2_{\pi NN}$, which makes the 2PEP, an intermediate range attraction, the dominant part of the nucleon-nucleon interaction.

Now let us look at the three body interaction. There is a two meson exchange contribution (2PEP),[12] which was, in fact, used by Carlson et al.[9] in their calculations of nuclear matter and also of nuclei with A = 3 and 4. However, this same spin-isospin cancellations which decrease the 1PEP contribution act for the 2PEP as well, so that this part of the three body interaction contributes only 1 or 2 MeV/A to the nuclear matter binding energy. Finally consider the three meson exchange diagram (3PEP).[13] At first sight it seems reasonable that this diagram should be analogous to the 2PEP two body interaction, and should thus be quite large and repulsive. Indeed, I believe that this is indeed the case. However, a more detailed calculation shows that a strong repulsive 3PEP contribution to nuclear matter results only if the tensor interaction at small distances is weak, indeed weaker than implied by currently accepted models. I will come back to this point in the last part of this talk.

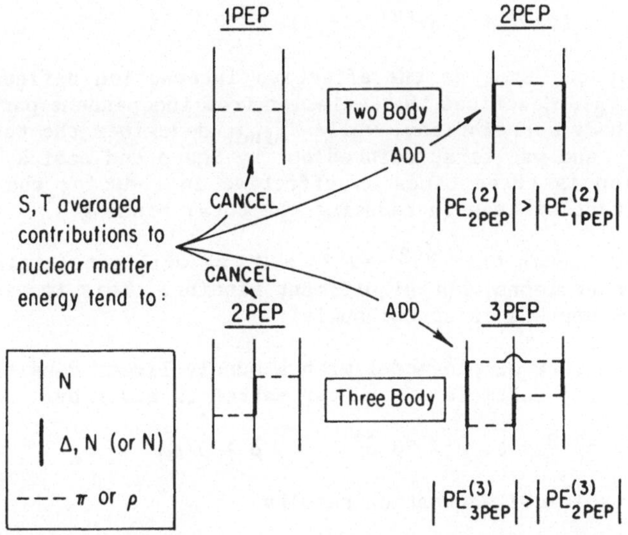

Figure 1. Multimeson Exchange Model of Two and Three Body Interactions Between Nucleons.

V. THREE BODY INTERACTIONS AND DENSITY DEPENDENT 2 BODY INTERACTIONS

For simplicity, let us neglect any finite range effects here. Then we have attractive two body and repulsive three body interactions.

$$V = V^{(2)}\delta(r_{12}) + V^{(3)}\delta(r_{123})$$

Now it is convenient to define an <u>effective</u> two body interaction which reproduces the same results. (We will here not worry about the distinction between neutrons and protons.) This effective interaction can be obtained by integrating over the position of the third particle. This gives:

$$V_{eff} = [V^{(2)}+\rho V^{(3)}]\delta(r_{12})$$

for uniform matter at density ρ. Thus the effective interaction is density dependent.

Next, we consider the total potential energy. This is obtained by integrating over the positions of all the particles. However, we note that a pair 12 is counted twice (as 12 and 21), so we have to divide the two body contribution by two. Similarly, the triplet 123 is counted 3! = 6 times, so we must divide the three body contribution by 6.

We obtain for the potential energy per particle in nuclear matter at constant density ρ:

$$PE/A = \frac{1}{2}\rho V^{(2)} + \frac{1}{6}\rho^2 V^{(3)}$$

This can be reproduced if the effective density dependent two body interaction is given by a term we call the binding interaction.

$$V_{bind.} = [V^{(2)} + \tfrac{1}{3}\rho V^{(3)}]\delta(r_{12})$$

This is not the same as the effective interaction defined above. It turns out that V_{eff} describes the deviation from independent particle motion, i.e., the two body correlations, while $V_{bind.}$ describes the total binding. As we see here, and was first pointed out by Sharp and Zamick,[14] the three body interaction is three times as effective in reducing the correlations due to the attraction than in reducing the total binding.

If it should happen that $V^{(2)}+\rho V^{(3)} \approx 0$, we will get little correlation between valence nucleons, but significant binding. This is, in fact, close to what I think happens in actual nuclei.

Let us take a very simple model with a purely linear density dependence. Then the energy per particle of nuclear matter is given by:

$$W = 22\,\hat{\rho}^{2/3} - 61\,\hat{\rho} + 23\,\hat{\rho}^2 \qquad \hat{\rho} = \rho/\rho_0$$

This fits the empirical saturation results.

For $\rho_0 = 0.16\ \text{fm}^{-3}$, $\hat{\rho} = 1$, $W = -16$ MeV, $dW/d\rho = 0$

Eff. int responsible for binding:

$$V_{bind.} \approx [-122 + 46\,\hat{\rho}]\,\delta(r_{12})$$

Eff. int responsible for correlations:

$$V_{eff.} \approx [-122 + 3\cdot 46\,\hat{\rho}]\,\delta(r_{12}) = [-122 + 138\,\hat{\rho}]\,\delta(r_{12})$$

Thus the effective interaction is attractive at low density, but turns slightly repulsive at normal nuclear density. In actual nuclei, we are somewhere in between. As a rough estimate, the density averaged over the entire nucleus is $3/4\ \rho_0$. Thus the average interaction is:

$$\left[-122 + 138\cdot\tfrac{3}{4}\right]\delta(r_{12}) = -20\,\delta(r_{12}) \ .$$

This corresponds an effective SDI strength of $G = 20/A$ MeV, which is quite close to what is found empirically.[15]

VI. RING DIAGRAM EXPANSION WITH CENTRAL AND TENSOR INTERACTIONS

The 2PEP two body and 3PEP three body interactions are the first two terms in a series of ring diagrams. Assuming closure (i.e., that the energy of the intermediate state is a constant, which we denote by ΔE), we find that the contribution of the nPEP n-body interaction is given by the following, apart from a numerical factor:

$$W_{NM}^{(n)} \approx \left(-\frac{\rho}{\Delta E}\right)^{n-1} \int \left([\tilde{V}_C(q)+2\tilde{V}_T(q)]^n + 2[\tilde{V}_C(q)-\tilde{V}_T(q)]^n\right) d^3q \ .$$

Here $\tilde{V}_C(q)$ and $\tilde{V}_T(q)$ denote the Fourier transforms of the central and tensor component of the transition interaction V_1. Note that the maximum contribution to the integral generally comes from the region for $q \approx 2$ to $4\ \text{fm}^{-1}$, i.e., of quite large momentum transfers. This is for two reasons: i.) the phase space factor $d^3q \approx q^2\,dq$, and ii.) that the tensor interaction also contains a factor q^2, i.e., $V_T(q) \approx q^2 T(q)$.

We see that the second order contribution must be attractive. The same is true for the fourth order contribution. However, the third order contribution can be either attractive or repulsive, depending on the relative importance of the central and tensor interaction. There is also a problem regarding the convergence of the ring diagram expansion which has been discussed by Friman and Nyman.[16)]

We can get a very rough idea of the nature of the second and third order interactions by working in coordinate space. Suppose that the central interaction could be represented by a Gaussian and that the tensor interaction could be neglected. Then if

$$V_C \approx V_1(r) \approx e^{-r^2/a^2}$$

we find that:

$$V^{(2)} \approx V_1{}^2(r_{12}) \approx e^{-2r_{12}{}^2/a^2}$$

$$V^{(3)} \approx V_1(r_{12})V_1(r_{13})V_1(r_{23}) \approx e^{-(r_{12}{}^2+r_{13}{}^2+r_{23}{}^2)/a^2}$$

and the equivalent two body interaction is proportional to $\rho\ e^{-3r_{12}{}^2/2a^2}$. Note that the range of the latter is about the same as the range of $V^{(2)}$, even slightly larger! This appears to be a necessary consequence of assuming that the three body interaction is due to three meson exchange. On the other hand, most interactions of the Skyrme type are taken to have a density dependent part with zero range.

In order to study the role of central and tensor interactions for multi-meson exchange, we will find it convenient to define a quantity x as follows:

$$x = -\ V_T/V_C$$

It is convenient to put the minus sign here, since it is known that V_C is repulsive while V_T is attractive. (The fact that the central even state V_{NN} is attractive is due to the extra factor $\sigma_1 \cdot \sigma_2\ \tau_1 \cdot \tau_2 = -3$.) For exchange of a single pion, it is well known that:

$$x = 1 + 3(\mu r)^{-1} + 3(\mu r)^{-2}.$$

At $r = \mu^{-1} = 1.4$ fm, we find that $x = 7$, a very large value. The n body contribution to the nuclear matter energy is proportional to the integral of the following quantity:

$$V^{(N)} \approx (-)^{n-1} \left[\frac{1}{3}\ (1-2x)^n + \frac{2}{3}\ (1+x)^n\right]$$

where $x \approx -\ \tilde{V}_T(\bar{q})/\tilde{V}_C(\bar{q})$, and $\bar{q}$ is an average momentum transfer. (As has been pointed out, $\bar{q} \approx 2$ to 4 fm^{-1}. In particular, we have:

$$V_2 = -(1 + 2x^2)$$

$$V_3 = \ \ 1 + 6x^2 - 2x^3$$

$$V_4 = -(1 + 12x^2 - 8x^3 + 6x^4)$$

Now define the quantities: $F_n = (-)^{n-1}\ V_n/(V_2)^{n/2}$. Table 1 lists values of F_2, F_3, and F_4 for selected values of x. For the case x = 0, which corresponds to having no tensor interaction, all the F's equal 1. By definition, $F_2 = 1$ for all x, and as we see, F_4 is in the range 1 to 1.5 for all positive values of x. On the other hand, the magnitude and even sign of F_3 depends strongly on the value of x.

Thus, $W^{(2)}$, $W^{(4)} < 0$, but the sign of $W^{(3)}$ depends on x, i.e., on the relative importance of tensor to central interactions.

Table 1. Effect of tensor interaction on 2, 3, and 4 body ring diagrams.

x	F_2	F_3	F_4	
0	1.000	$1+3x^2$	$1+8x^2$	
0.5	1.000	1.225	1.500	
1	1.000	0.962	1.222	$\pi+\rho+A_1$?
1.5	1.000	0.607	1.038	
2	1.000	0.333	1.000	
3	1.000	0.012	1.050	$\pi+\rho$
5	1.000	-0.272	1.173	
7	1.000	-0.397	1.250	π
∞	1.000	-0.707	1.500	

Until about 10 years ago, only pion exchange was considered seriously, and effects of other mesons was only considered crudely. For a pure pion exchange interaction, the tensor component is much larger than the central one, i.e., by a factor 7 even at the pion Compton wavelength. Inspection of Table 1 shows that F_3 is negative for x = 7. Thus, we would obtain a negative value for the third order ring diagram. This was indeed found by Clement,[17] who did a much more detailed calculation. If rho exchange is taken into account, the picture changes somewhat. The contributions of the pion and rho to the tensor interaction tend to cancel.[18] Indeed, calculations made by Friman and Nyman[16] lead to a very small value of the third order contribution. This roughly corresponds to the case that $x \approx 3$. As far as I can tell, this seems to be the "Conventional Wisdom" at this time.

It is my belief, however, that even this picture overestimates the importance of the tensor interaction, and that we should be closer to the region $x \approx 1$, where all the F's are close to 1, and we would then obtain a sizable repulsive three body interaction. I will discuss evidence for this tensor quenching below. From theoretical grounds, we would expect something like this because of chiral symmetry.

It has indeed been shown by other authors [19] that chiral symmetry at large momenta requires a strong coupling of the A_1 (1^+, 1270 MeV) meson to the nucleon. $\rho(1^-)$ and $A_1(1^+)$ are chiral partners. The contributions of π, ρ, and A_1 (the latter two being tensor coupled, as required by the soliton model[20]) are[21]:

$$V_C = V_C(\pi) + 2\,V_C(\rho) - V_C(A_1); \qquad V_T = V_T(\pi) - V_T(\rho) - V_T(A_1) \quad .$$

Thus A_1 exchange further reduces the tensor interaction. It can, of course, be argued that much of the interaction from the A_1 is screened off, on account of its large mass (1270 vs 770 for ρ). On the other hand, if chiral symmetry is approximately valid, then there should be a significant A_1 potential remaining. This point remains to be investigated in more detail.

Finally, I want to point out that calculations of the short range nucleon-nucleon interaction based on the quark model also seem to lead to a very weak tensor term.[22]

VII. SURVEY OF TENSOR FORCES IN NUCLEI

Let me summarize here some of the salient things known about the nucleon-nucleon tensor interaction. Empirical evidence for tensor interaction comes from several sources:

a.) The Deuteron.
We need a non-central interaction AT LARGE r to account for Q, $(\psi_D/\psi_S)_{r\to\infty}$. Indeed, as we have seen, the one pion exchange interaction leads to a large Tensor Interaction at LARGE r. It should be pointed out here that the deuteron is a very unusual nucleus, being very loosely bound. [$\psi \approx e^{-(r/4.3\mathrm{fm})}$]

b.) The detailed behavior of the NN phase shifts clearly requires some tensor interaction.

c.) The cross sections for inelastic scattering to Non-Natural Parity States of Large J, e.g., 6^- in ^{28}Si, also can only be explained with a tensor interaction.[23)] It should be pointed out, however, that all these results only give information concerning the tensor matrix elements for momentum transfers $q \approx 1\ \mathrm{fm}^{-1}$.

On the other hand, there is little evidence regarding the tensor interaction at large q, i.e., at small r. As we have seen, it is the region of 2 to 4 fm^{-1} which is of importance for nuclear matter, so this is still an open problem.

There is evidence of tensor quenching coming from two independent sources:

a.) ISOVECTOR TENSOR MATRIX ELEMENTS DEDUCED FROM ARNDT SCATTERING AMPLITUDES. (Nakayama et al.[24)].) (Similar results are obtained using the Paris potential). For low q, their results are consistent with $\pi+\rho$ exchange. However, there appears to be significant quenching, relative to 1 $\pi+\rho$ exchange, above 2 fm^{-1}. The t-matrix (i.e., for zero density) is quenched more strongly, especially for low energy nucleons. Nakayama[25)] has also shown that there is quenching due to core polarization.

b.) RATIO OF LONGITUDINAL TO TRANSVERSE STRUCTURE FUNCTIONS IN (e,e') SCATTERING. Results for ^{12}C appear to be consistent with NO tensor interaction[26)] at $q = 2\ \mathrm{fm}^{-1}$! This finding would be most interesting, if it stands up under further scrutiny!

As everyone knows, we obtain the ON-ENERGY-SHELL Behavior of V_{NN} from $\delta_{S_{LJ}}(E)$. On the other hand, how do we obtain the OFF-ENERGY-SHELL Behavior of V_{NN}? This brings me to the PRINCIPLE OF MINIMAL TENSOR FORCE: "Use the weakest tensor force allowed by experiment, even though this may require stronger three body interaction."

The alternative of using a $\pi+\rho$ exchange model interaction model appears to overestimate the tensor interaction at large q. As we have seen, this is important for the sign of the three body energy in nuclear matter. (This conclusion is not firm, since the tensor interaction could also be quenched by having a strong ρ coupling. However, this interpretation may be inconsistent with Nucleon-Nucleon Phase shifts.[27)])

I have reason, based on the arguments given above (and guided by the principle of the minimal tensor force) to believe that the three body ring diagram contributes 10 or 15 MeV per particle to the nuclear matter binding energy, i.e., the bulk of the 15 MeV per particle for nuclear for the phenomenological three body interaction, as obtained by Skyrme.[4)]

A Fortune Cookie I got at a Chinese Restaurant a few day ago said: "Your intuition is excellent but another viewpoint could be helpful." In this connection, it is only fair to mention that according to "Conventional Wisdom," it contributes 5 MeV or less per particle. This remains to be worked out in detail.

VIII. CONCLUSIONS AND THINGS TO DO

a. CONCLUSIONS. I have tried to present evidence for:
i.) Effective 3 body interactions between nucleons.
ii.) A weak tensor interaction at high q (i.e., small r).

b.) THINGS TO DO. The further study of some of the following problems may shed some light on the points raised here:
i.) Nuclear 3 body problem. (From ^3H, ^3He, p+d, etc). (Hole in wavefunction?)
ii.) EMC effect (Does the nucleon size increase in nuclear matter?)
iii.) Meson-Nucleon Form factors:
$f_{\rho NN}(q)$ (Vector dominance?)
$f_{\pi N\Delta}(q)$ [Transition potential, relation to $f_{\pi NN}(q)$.]
$f_{A_1 NN}(q)$ (Chiral symmetry?)
iv.) Short range NN repulsion. (Role of short range correlations in NN wavefunction?)
v.) High density nuclear matter. (Equation of state, from heavy ion and neutron star astrophysics results.)

REFERENCES

1. O. W. Greenberg and H. J. Lipkin, Nucl. Phys. A370, 349 (1981).
2. K. F. Liu, Phys. Lett. 131B, 195 (1983).
3. D. Wilkinson, Lectures at 1977 Summer School on Heavy Ions and Mesons in Nuclear Physics, Les Houches, France.
4. T. H. R. Skyrme, Nucl. Phys. 6, 615 (1959).
5. B. C. Clark, S. Hanna, R. L. Mercer, L. Ray, and B. D. Serot, Phys. Rev. Lett. 50, 1643 (1983); S. Wallace, Comments Nucl. Part. Phys. 13, 27 (1984).
6. R. D. Amado, J. Piekarewicz, D. A. Sparrow, and J. A. McNeil, Phys. Rev. C29, 936 (1984).
7. M. Anastasio, L. S. Celenza, W. S. Pong, and C. M. Shakin, Phys. Repts. 100, 327 (1983).
8. R. Brockmann and R. Machleidt, Phys. Lett 149B, 283 (1984).
9. J. Carlson, V. R. Pandharipande, and R. B. Wiringa, Nucl. Phys. A401, 59 (1983); R. B. Wiringa, Nucl. Phys. A401, 86 (1983).
10. H. A. Bethe, Ann. Revs. Nucl. Sci. 21, 93 (1971).
11. J. W. Negele, Talk presented at LAMPF Workshop on Dirac Approaches to Nuclear Physics, Los Alamos, Jan. 1985, LA-10438-C, May 1985.
12. S. A. Coon, et al., Nucl. Phys. A317, 242 (1979); S. A. Coon and W. Glockle, Phys. Rev. C23, 1790 (1981); J. L. Friar, B.F.Gibson, and G. L. Payne, Ann. Rev. Nucl. Part. Sci. 34, 403 (1984).
13. J. Fujita, M. Kawai, and M. Tanifuji, Nucl. Phys. 29, 252 (1962).
14. R. Sharp and L. Zamick, Nucl. Phys. A208, 130 (1973).
15. A. Plastino, R. Arvieu, and S. A. Moszkowski, Phys. Rev. 145, 837 (1966).
16. B. L. Friman and E. M. Nyman, Nucl. Phys. A302, 365 (1978).
17. D. Clement, Nucl. Phys. A205, 398 (1973).
18. J. W. Durso, M. Saarela, G. E. Brown, and A. D. Jackson, Nucl. Phys. A278, 445 (1977).
19. S. Weinberg, Phys. Rev. Lett. 18, 507 (1967); J. Sakurai, "Currents and Mesons," p. 144, University of Chicago Press (1969); J. W. Durso, G. E. Brown, and M. Saarela, Nucl. Phys. A430, 653 (1984). M.C. Birse and M.K. Banerjee, Phys. Rev. D31,118 (1985);
20. A. Jackson, A. D. Jackson, and V. Pasquier, Nucl. Phys. A432, 567 (1985).
21. N. Hoshizaki, I. Lin, and S. Machida, Prog. of Theo. Phys. 26, 680 (1961); S. Ogawa, S. Sawada, T. Ueda, W. Watari, and M. Yonezawa, Suppl. to Prog. Theor. Phys. 39, 140 (1967).
22. K. Isgur, Talk presented at LAMPF Workshop on Dirac Approaches to Nuclear Physics, Los Alamos, Jan. 1985, LA-10438-C, May 1985.

23. R. A. Lindgren and F. Petrovich, in "Spin Excitations in Nuclei," ed. by F. Petrovich, et al., (Plenum, N.Y., 1984).
24. K. Nakayama, S. Krewald, J. Speth, and W. G. Love, Phys. Rev. C31, 2307 (1985).
25. K. Nakayama, Preprint (Jülich, Aug. 1985).
26. S. Fantoni, Talk presented at Conference following this workshop.
27. W. H. Dickhoff, A. Faessler, and H. Müther, Phys. Rev. Lett. 49, 1902 (1982).

PAIRING IN LOW-DENSITY NEUTRON MATTER

J. W. Clark and J. M. C. Chen

McDonnell Center for the Space Sciences
and Department of Physics
Washington University, St. Louis, Missouri 63130

E. Krotscheck and R. A. Smith

Department of Physics, Texas A & M University
College Station, Texas 77843

INTRODUCTION

At least three distinct forms of nucleonic superfluidity are predicted to occur inside neutron stars.[1,2] In some portion of the inner crust, where the mass density increases from 4.3×10^{11} g cm^{-3} to about 2.4×10^{14} g cm^{-3} and a neutron gas coexists with a coulomb lattice of neutron-rich nuclei and a sea of relativistic electrons, isotropic pairing in the $^1S_0$ state will take place.[3,4] Proceeding deeper into the star, the nuclear clusters dissolve and one enters the quantum fluid interior with its interpenetrating seas of neutrons, protons, electrons, and muons. At densities comparable to that in the centers of ordinary nuclei, $\rho_0 = 2.8 \times 10^{14}$ g cm^{-3}, the relatively dilute proton component, with a partial density similar to that of the pair-condensed neutron gas in the inner crust, will form a similar, but superconducting, singlet-S superfluid.[6,7] Finally, at slightly higher densities within the quantum fluid interior, it is predicted that (anisotropic) $^3P_2$-$^3F_2$ pairing of the neutron component will prevail.[3,8,9]

These three manifestations of ordering in momentum space may be traced to special features of the two-nucleon interaction. At low densities, where S-wave collisions overwhelmingly predominate and the interacting gas-phase neutrons of the inner crust (or proton pairs of the quantum fluid interior) see mainly the longer-range attraction and very little of the short-range repulsion, the pairing matrix element at the Fermi surface is negative, and singlet-S pairing is favored. The energetic advantage of neutron pairing in the $^3P_2$-$^3F_2$ state at higher densities may be ascribed to the tensor force (and spin-orbit force) in this channel.

Superfluidity of the nucleonic components of a neutron star will have important consequences for the evolution and for the internal dynamics of the star, and some predictions of models based on nucleon superfluidity are subject to observational tests. Among these are predictions regarding (a) cooling rates and internal temperatures and (b) pulsar timing following sudden spin-ups, or "glitches" (Ref. 1,10-14). In the case of (a) the cooling rates hinge in part on the specific heat of the medium, which may be strongly influenced by superfluidity, while in (b) the timing of pulses after a glitch carries information on the relaxation processes associated with the coupling between the crust and the massive fluid interior.

Here we shall report the results of a microscopic investigation of $^1S_0$ neutron superfluidity in the inner crust. This study is prompted by (i) current activity within the vortex-pinning theory of the post-glitch dynamics of neutron stars, which appears to be successful in correlating observed features of the extant collection of glitches[1,14] and by (ii) recent advances in many-body calculational methods,[15] which allow a much more precise treatment of this problem than has been given to date.

The new techniques that are especially well suited to the problem of pairing in strongly-interacting Fermi systems are correlated-basis perturbation theory[16-18] and Fermi-hypernetted-chain (FHNC) evaluation[17,19] of the matrix elements of that theory. The best earlier estimates[4,7] of the $^1S_0$ energy gap and condensation energy in low-density neutron-star matter were obtained from a variational approach based on a Jastrow-BCS trial function and semirealistic bare two-nucleon interactions, the relevant expectation values being calculated within a two-body cluster approximation. In the present study we shall proceed to second order in a Jastrow-BCS correlated-basis perturbation expansion for the effective pairing matrix elements. The leading term of this expansion is equivalent to the usual variational description, while the second-order correction brings in explicit effects of virtual particle-hole pairs excited by the energy-dependent effective interaction of the theory. The two-body cluster approximation employed in earlier work is upgraded to FHNC summation of cluster diagrams, extending to all orders in the number of closely correlated bodies. While this improvement entails (mainly) a quantitative difference, the incorporation of the second-order perturbation correction in the Jastrow-BCS basis leads to a substantial quenching in the effective pairing matrix element at the Fermi surface and accordingly produces an emphatic depression of the energy gap and condensation energy, relative to the variational results. A third improvement over previous efforts (which causes only modest changes relative to earlier predictions at the variational level) is the adoption of realistic semiphenomenological bare two-nucleon interactions: The Reid v_6 and Bethe-Johnson v_6 potentials.[20]

MICROSCOPIC THEORY OF PAIRING

A straightforward microscopic treatment of pairing phenomena in neutron matter along the lines of the original BCS theory is impeded by the strong short-range repulsion appearing in the bare two-neutron interaction. To deal with this problem of the repulsive core, we may appeal to the method of correlated basis functions,[16,24,17,18,21,22] which serves to transform the original problem of a system of bare particles interacting through strong two-body forces into a more amenable problem of dressed fermions interacting via weak effective two-body, three-body,... forces. We outline the theory of singlet-S pairing in the form presented by Krotscheck, Smith, and Jackson;[22] antecedents may be found in Refs. 24,4, 7,25,21,26,27. Considerations begin with a basis of normalized (but generally nonorthogonal) correlated N-particle states

$$|\Psi_m^{(N)}\rangle = \left(I_{mm}^{(N)}\right)^{-1/2} F_N|\Phi_m^{(N)}\rangle \quad , \tag{1}$$

where the $|\Phi_m^{(N)}\rangle$ are eigenkets of the noninteracting system, F_N is a suitable correlation operator incorporating, at a minimum, the strong short-range dynamical correlations induced by the repulsive core of the bare force, and the $I_{mm}^{(N)}$ are normalization constants whose definition is obvious. A correlated analog of the BCS ansatz $|\Phi_o^s\rangle = \Pi_k(u_k + v_k a^\dagger_{k\uparrow} a^\dagger_{-k\downarrow})|0\rangle$ is constructed as follows:

$$|\Psi_o^s\rangle = \sum_{m,N} |\Psi_m^{(N)}\rangle\langle\Phi_m^{(N)}|\Phi_o^s\rangle \quad . \tag{2}$$

The variational approach to pairing in the presence of strong interactions centers on the expectation value, in the above trial superstate, of the operator $\hat{H} - \mu\hat{N}$, where $\hat{H}$ is the second-quantized Hamiltonian, $\hat{N}$ is the number operator, and μ is a chemical potential, i.e., the usual Lagrange parameter determined so as to ensure a specified average particle number A. It is convenient to expand this expectation value in terms of the deviations of the Bogoliubov amplitudes $u_k = \cos\phi_k$, $v_k = \sin\phi_k$ from their normal-state step-function values [viz. $v_k^o \equiv n(k) = \Theta(k_F - k)$ and $u_k^o = 1 - n(k)$]. To first order in $v_k^2 - n(k)$ and second order in $u_k v_k$, the result is

$$\langle\hat{H} - \mu\hat{N}\rangle_s = H_{oo}^{(A)} - \mu A + 2\sum_k\{[1-n(k)]v_k^2 - n(k)u_k^2\}[e(k) - \mu] + \sum_{k,\ell} u_k v_k u_\ell v_\ell P_{k\ell} \quad , \tag{3}$$

where $H_{oo}^{(A)}$ is the energy expectation value in the normal correlated ground state [m = 0 in (1)], e(k) is the normal-phase variational single-particle energy (or correlated Hartree-Fock single-particle energy) defined in Refs. 21,19,18, and the pairing matrix elements $P_{k\ell}$ are given by

$$P_{k\ell} = \langle k\uparrow - k\downarrow|W(12)|\ell\uparrow - \ell\downarrow\rangle_a + [|e(k) - \mu| + |e(\ell) - \mu|]\langle k\uparrow - k\downarrow|N(12)|\ell\uparrow - \ell\downarrow\rangle_a \quad . \tag{4}$$

The quantities W(12) and N(12) appearing in $P_{k\ell}$ are the two-body interaction operator and the two-body nonorthogonality operator constructed in Refs. 21,19,18 from the bare two-body interaction v(12) and the correlation operator F_N. For a Jastrow choice of F_N, these ingredients, as well as the CBF single-particle energies e(k), may be accurately calculated using Fermi-hypernetted-chain techniques.[19,28,18] Physically, the approximation involved in truncating the aforementioned expansion at the leading terms displayed in (3) boils down to treating the interaction of only one Cooper pair at a time, while considering the background medium as normal. There is ample evidence[29,30,4] that this "decoupling approximation" is an excellent one for the present problem of ^1S pairing in neutron matter. While the errors are nominally of order $m^*\Delta/\hbar^2 k_F^2$, where m* is the effective mass and Δ the superfluid energy gap, they turn out to be numerically inconsequential.

Stability of the normal correlated trial ground state hinges on positive-definiteness of the quadratic form

$$U_{k\ell} \equiv \frac{\delta^2}{\delta\phi_k\delta\phi_\ell}\langle\hat{H} - \mu\hat{N}\rangle_s\Big|_o = [1 - 2n(k)][1 - 2n(\ell)][2|e(k) - \mu|\delta_{k\ell} + P_{k\ell}], \tag{5}$$

where the o subscript implies that after variation the amplitudes u_k, v_k are to be set to their normal-state values. (This statement, with e(k), $P_{k\ell}$ as prescribed above, is *exact* independent of the decoupling approximation.[21,22]) If an instability of the normal state is indicated by a failure of this stability criterion, an *optimal* variational superstate may be determined by finding the minimum of the expectation value $\langle\hat{H} - \mu\hat{N}\rangle_s$ with respect to the phase angle ϕ_k. Working in the decoupling approximation, we define a gap function $\Delta_k = (1/2)\Sigma_\ell(\sin 2\phi_\ell)\, P_{k\ell}$ and, upon functional variation of (3) with respect to ϕ_k, arrive at the gap equation

$$\Delta_k = -\frac{1}{2}\sum_\ell P_{k\ell}\frac{\Delta_\ell}{\{[e(\ell)-\mu]^2+\Delta_\ell^2\}^{1/2}} \tag{6}$$

as the condition for determining the optimal phase angle

$$\phi_k = \frac{1}{2}\arctan\left[\frac{-\Delta_k}{e(k)-\mu}\right] \tag{7}$$

and therefore the optimal superstate of type (2). Formally this equation is precisely the same as the gap equation of standard, weak-interaction BCS theory; however, the key inputs e(k) and $P_{k\ell}$ include the effects of strong dynamical correlations.

These key inputs are easy to calculate only for Jastrow correlations $F = \Pi_{i<j}\, f(r_{ij})$. Such a choice provides a rather good description of the *gross* geometrical correlations in neutron matter and can for example yield a reasonable value of the normal ground-state energy, especially at the low densities of interest for singlet-S pairing in neutron stars. On the other hand, this choice -- which in field-theoretic terms, amounts to an average-propagator approximation -- is obviously deficient when one is concerned with spin- and momentum-dependent correlations and with subtle energy- and momentum-dependent effects taking place near the Fermi surface. In view of the various estimates[31-33] pointing to the importance of polarization processes for pairing in neutron matter (corresponding to exchange of density and spin-density fluctuations), it is clear that a quantitative description of $^1S_0$ superfluidity in this system demands systematic improvement of the above variational approach to pairing.

Such improvement may be sought within correlated-basis perturbation theory.[16-18,34,28,35,36,22] Omitting the details,[22] an appropriate basis of correlated superfluid states $|\Psi_m^s\rangle$ is built from the trial state $|\Psi_o^s\rangle$ above, considered as a "correlated quasiparticle vacuum," by application of an arbitrary number of "correlated quasiparticle creation operators" formed in analogy with the BCS or Bogoliubov theory of superfluid excitations. A straightforward extension of the coupled-cluster formalism[37,34] provides an efficient vehicle for generating successive terms in a perturbation expansion of the exact superstate energy. The exact superfluid ground state may be expressed as $|\Psi^s\rangle = |\exp(S_c)\,\Psi_o^s\rangle$ where the operator S_c, being a linear combination of any number of quasiparticle creation operators with arbitrary labels, has the same structure as the S operator of the ordinary coupled-cluster method. However, the aforementioned correlated Bogoliubov quasiparticle creation operators enter in place of the usual particle-hole creation operators. Forming the expectation value $E_s - \mu A$ of $\hat{H} - \mu\hat{N}$ in the state $|\Psi^s\rangle$ and expanding in powers of S_c and $S_c^\dagger$, one is led to an exact representation of the superfluid ground-state energy E_s in terms of a CBF perturbation expansion. The first term in the expansion, independent of S_c and $S_c^\dagger$, is simply the variational result $\langle\hat{H} - \mu\hat{N}\rangle_s$ obtained above. The second-order CBF perturbative correction is derived by retaining only terms linear in S_c or $S_c^\dagger$, plus the part of the bilinear term involving only *diagonal* CBF matrix elements of $\hat{H} - \mu\hat{N}$. At successive stages of truncation, S_c is to be determined by setting the first variation of the truncated expression with respect to $S_c^\dagger$ equal to zero. Unlinked terms will arise, but may be shown to cancel in groups. Care must be taken to avoid splitting the members of such a group. In a calculation to second order this problem is obviated by dropping all contributions involving four- or higher-body matrix elements; in practice we shall also drop any three-body matrix elements. Accordingly, one arrives at a second-order correction to the variational superstate energy which can be expressed in terms of the CBF ingredients e(k), W(12), and N(12) alone, the three- and more-body cluster-diagram sums W(123), N(123), $W_c(1234)$, $N_c(1234)$, etc.[19,34,36,18] being needed only at more refined levels of approximation.[35]

It is clear that at any stage in this process the approximant for the superfluid ground-state energy, or rather $E_s - \mu A$, may be regarded as a functional of the BCS amplitudes u_k, v_k or of the phase angle ϕ_k. More pointedly, it can be seen that in the decoupling approximation any such functional may be put in the same form (3) as we obtained in the variational theory, but involving improved, or *corrected* versions of the single-particle energies e(k) and pairing matrix elements $P_{k\ell}$ -- these quantities still being independent of the u_k, v_k. Explicit results for the second-order corrections to e(k) and $P_{k\ell}$ are given in Ref. 22. To establish formal contact with well-known weak-interaction notions, it is convenient to introduce Goldstone-like diagrams in which the propagators are defined in terms of the variational single-particle energies and the wavy line symbolizes the energy-dependent CBF two-body effective interaction V(12). (The latter is formed out of W(12), N(12), and single-particle energies as explicated in Refs. 34-36,38,22,18) The pairing interaction at the Fermi surface (known alternatively as the BCS limit of the general two-particle vertex function, i.e., the limit of the four-leg vertex part in the particle-particle (pp) channel as the particle momenta approach $\hbar k_F$ with zero total momentum) consists in second-order CBF perturbation theory of the diagrams shown in Fig. 1. As usual, the pairing instability is driven by the pp- and hh-irreducible diagrams: here the first-order (or "variational") diagram shown in (a) and the second-order ring and particle-hole-ladder diagrams and corresponding exchange diagrams shown as the first four entries in (b). The remaining two diagrams in Fig. 1 (the pp and hh ladders) are seen to be reducible, in the sense that they can be split into two separate pieces merely by cutting a single pair of particle lines or a single pair of hole lines; they are to be ignored at the present stage of calculation.[22]

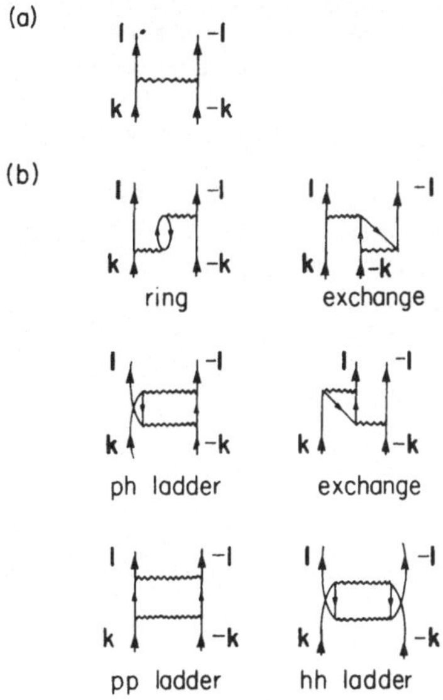

Figure 1. Diagrammatic contributions to the CBF perturbation expansion: (a) first-order (i.e., variational) contribution and (b) second-order CBF perturbation terms, where it is understood that antisymmetrizations are to be performed on the pairs (k,-k) and (l,-l).

With the improved pairing interaction and single-particle energies, Eqs. (6), (7), and (3) may be used to determine an improved gap function Δ_k and to calculate the condensation energy E_c. The latter is given by $(E_s - \mu A)[u_k, v_k]$ with the amplitudes u_k and v_k set to their normal-state versions, *minus* the value of this functional at the optimal u_k, v_k implied by the solution of the gap equation.

CALCULATION AND RESULTS

A second-order CBF treatment of singlet-S pairing has been carried out as outlined above, subject to further assumptions and simplifications. Most important is the restriction to a Jastrow correlation operator. The specific form chosen for the two-body correlation function is that studied by Lam and Clark,[17]

$$f(r) = \exp[-(1/2)(b/r)^m \exp(-(r/b)^n)] \quad , \tag{8}$$

with positive integers m and n and parameter b determined by minimization of the Jastrow ground-state energy expectation value for the given Hamiltonian. In conjunction with the decoupling approximation, we have adopted an effective-mass approximation for the single-particle energies $e(k)$ appearing in the gap equation and condensation energy (but *not* within the pairing matrix elements through expression (4) or through the second-order perturbative diagrams). Earlier work[4,7] gives assurance that this use of an effective mass approximation is quantitatively reliable for superfluid pure neutron matter at the low densities of interest to us. Values of the effective mass m* as a function of Fermi wave number k_F are derived (a) from the normal-state variational single-particle energies calculated as in Jackson et al.[38] for the same choice of Jastrow correlations and (b) from the normal-state self-energy evaluated to second CBF perturbative order, again as in Ref. 38.

The required matrix elements of the two-body CBF operators W(12) and N(12) are also calculated as in Ref. 38. These ingredients are computed within the FHNC/FHNC' approximation at the level of factorable diagrams as described in Ref. 28. (The single-particle energies $e(k)$ of normal-state CBF theory are likewise evaluated via FHNC/FHNC' techniques.)

The integrals over intermediate-state momenta appearing in the second-order correction $(\Delta P_{k\ell})^{(2)}$ to the variational result $(P_{k\ell})_V$ have been performed using the Monte Carlo algorithm outlined in Ref. 28. For the sake of economy, this correction is only evaluated on the Fermi surface, i.e., for $k = \ell = k_F$. The ratio of $(\Delta P_{k_F k_F})^{(2)}$ to $(P_{k_F k_F})_V$, which we denote by $-\lambda$, is then used to simulate the dependence of the correction on k and ℓ away from k_F by simple scaling of the variational result. It can be argued that this procedure is numerically adequate because of the strong concentration of the relevant integrands at the Fermi surface.

The procedure set forth in Ref. 30 is used to solve the gap equation.

Calculations are performed for two of the semiphenomenological models of the bare two-nucleon interaction considered in Ref. 38, namely the Reid and Bethe-Johnson (BJ) v_6 potentials.[20] Both may be regarded as quite realistic at the low densities (and hence low relative energies of nucleon pairs) involved.

Figure 2 shows our results for the pairing matrix element at the Fermi surface, $P_{k_F k_F}$, obtained from the Jastrow-BCS variational scheme and from second-order CBF perturbation theory in a Jastrow-BCS basis. (Some variational results for the OMY4 potential, calculated within the FHNC framework by Krotscheck and Clark,[21] are included for comparison with the Reid v_6 and Bethe-Johnson v_6 data. The OMY4 potential is more schematic than those used here; fitted to the low-energy scattering data, it is purely central with a Serber exponential attraction outside a state-independent hard core of radius 0.4 fm.) The correction factor $\alpha = (P_{k_F k_F})_{CBF} / (P_{k_F k_F})_V = 1 - \lambda$ due to the inclusion of second-order effects is superimposed on the same figure. The two versions (a) and (b) of the effective mass ratio m*/m are given in Table 1 for various k_F values. Working within the decoupling approximation, obviously (a) is to be used together with the variational pairing matrix elements in a variational evaluation of the gap and condensation energy, while a consistent CBF perturbative evaluation of these quantities requires the use of (b) in tandem with our second-order CBF estimate of the pairing matrix elements. However, it is also of interest to carry through a third set of calculations in which the variational effective masses are used with the CBF pairing matrix elements; for this enables us to decide the importance of modifications of the gap and condensation energy due to (i) second-order modifications of the pairing matrix elements separately from (ii) second order modifications of the effective mass.

Table 1. Variational and CBF Effective Masses

k_F (fm^{-1})	Reid		Bethe-Johnson	
	$(m^*/m)^{VAR}$	$(m^*/m)^{CBF}$	$(m^*/m)^{VAR}$	$(m^*/m)^{CBF}$
0.3	0.99	1.10	0.99	0.83
0.4	0.98	1.10	0.98	0.91
0.5	0.97	1.09	0.96	0.97
0.6	0.96	1.09	0.94	1.01
0.7	0.95	1.08	0.92	1.03
0.8	0.94	1.08	0.90	1.04
0.9	0.92	1.06	0.88	1.04
1.0	0.90	1.05	0.85	1.03
1.1	0.88	1.04	0.82	1.01
1.2	0.86	1.02	0.78	0.98
1.3	0.84	1.00	0.75	0.94

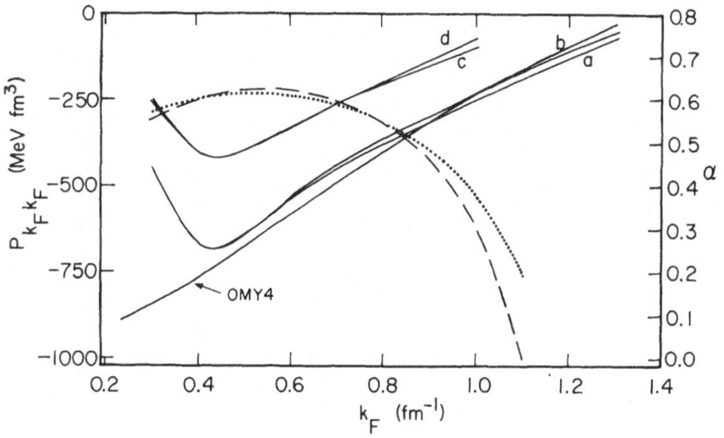

Figure 2. Correction factor α and diagonal pairing matrix elements on the Fermi surface $P_{k_F k_F}$. *Correction factors.* Dashed line: Reid v_6. Dotted line: Bethe-Johnson v_6. *Diagonal pairing matrix elements.* First-order terms for Reid and Bethe-Johnson potentials: (a) and (b), respectively. First- plus second-order terms for Reid and Bethe-Johnson potentials: (c) and (d), respectively.

The converged output from iteration of the gap equation (6), based on variational or CBF input, is summarized in Δ_{k_F}, the gap function evaluated at the Fermi surface. In Fig. 3 we plot the variational and CBF versions of this quantity against k_F for the two "realistic" potentials adopted here. Figure 4 presents corresponding results for the condensation energy (per particle), determined from the k-dependent gap function and k,ℓ-dependent pairing matrix elements via expression (3), the e(k) being simulated by effective mass approximation and the k and ℓ integrations being performed numerically.

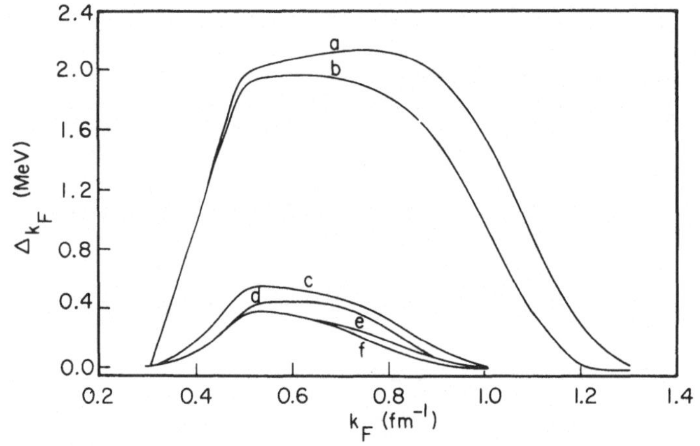

Figure 3. *Energy gaps.* Reid and Bethe-Johnson variational cases: (a) and (b), respectively. Reid and Bethe-Johnson CBF cases with CBF effective masses: (c) and (d), respectively. Reid and Bethe-Johnson CBF cases with variational effective masses: (e) and (f), respectively.

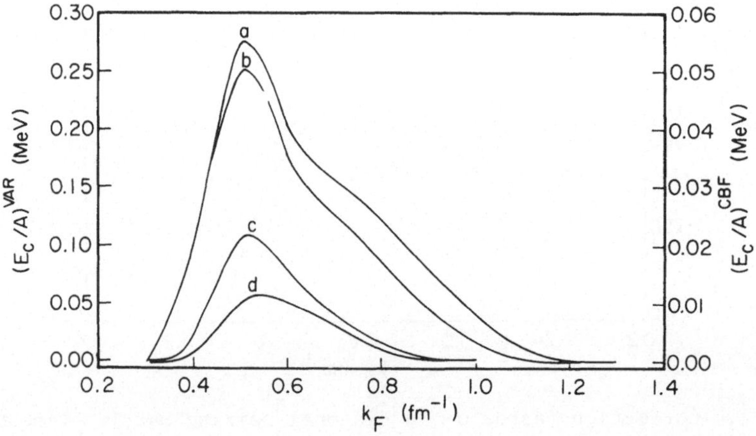

Figure 4. *Condensation energy.* (E_c/A) for Reid and Bethe-Johnson variational cases: (a) and (b), respectively. (E_c/A) for Reid and Bethe-Johnson CBF cases with CBF effective masses: (c) and (d), respectively.

DISCUSSION

The variational results for the gap and condensation energy obtained in the present study are not very different from those of the earlier, more primitive, variational treatments of Yang and Clark[4] and Chao, Clark, and Yang.[7] The path taken by these authors departs from ours in the following respects: (i) the semirealistic OMY4 potential was assumed, (ii) the pairing interaction and single-particle energies were evaluated in two-body cluster approximation (rather than by FHNC cluster-resummation methods), and (iii) the gap was determined by explicit variation of the condensation energy, or else by an algorithm for solving the gap equation which is somewhat different from the one we have adopted. Apparently, these differences have rather minor net consequences, except in deciding the density at which the gap closes. In regard to point (ii), the primary qualitative failing of Refs. 4,7 is that the gap persists out to appreciably higher densities than in the (presumably reliable) FHNC calculation. The peak values of Δ_{k_F} in the old and new calculations are quite similar, some 2.4 MeV being obtained in Refs. 4,7 as opposed to our 2.0 – 2.1 MeV for BJ and Reid v_6 interactions.

Let us now consider the most dramatic result of our work, which emerges from the novel inclusion of CBF-perturbative diagrams describing the virtual excitation of particle-hole pairs. The second-order CBF correction is found to suppress the pairing matrix element $P_{k_F k_F}$ by a factor α^{-1} around 1.7 over the interesting density range, the gap Δ_{k_F} by a factor of 4 or more, and the condensation energy by a factor of roughly 13 or more (see Figs. 2-4). The suppression is a bit more significant for the BJ interaction than for the Reid model. Comparing the perturbed gap results for (a) variational and (b) CBF effective masses (see Fig. 3), we observe that the suppression is appreciably larger (by some 25-50% at $k_F = 0.6\ \mathrm{fm}^{-1}$, depending on potential) when the second-order correction to the effective mass is dropped.

The effect of polarization of the neutron medium on the strength of the singlet-S pairing phenomenon in neutron matter has been debated over a considerable period. At first it was thought that the exchange of density fluctuations (which tends to amplify the pairing interaction) would dominate over the exchange of spin-density fluctuations (which works in the opposite direction) and produce a significant *enhancement* of the energy gap.[31] The induced interaction in the $^1S_0$ state due to such polarization processes, carried to all orders by a random-phase approximation, can be expressed as

$$\mathcal{V}_i(^1S_0) = -\left[\frac{F_0^2}{1+F_0U(q)} - 3\,\frac{G_0^2}{1+G_0U(q)}\right]U(q) \quad , \tag{9}$$

where F_0 and G_0 are the usual dimensionless Landau parameters in density [1] and spin $[\underset{\sim}{\sigma}_1\cdot\underset{\sim}{\sigma}_2]$ channels, q is the wave number associated with the density (spin-density) fluctuation, and U(q) is the normalized Lindhard function. At the time of the Pines-Pethick estimate reported in Ref. 31, the parameter F_0 was considered to be near -1 in the relevant density range, so with a positive G_0 somewhat less than unity, the repulsive second term of (9) was regarded as negligible. However, a subsequent microscopic derivation of Landau parameters for neutron matter by Bäckmann, Källman, and Sjöberg[39] (BKS) within the Babu-Brown theory[40] gave F_0 values safely higher than the singular value -1: around $k_F = 0.6\ \mathrm{fm}^{-1}$ they found $F_0 \sim -0.3$ and $G_0 \sim 0.8$. This was noted by Clark, Källman, Chakkalakal, and Yang[32] (CKCY), who pointed out that the exchange of spin-density fluctuations must then dominate and estimated that the $^1S_0$ pairing matrix element would be *suppressed* by a factor 1.2-1.4, resulting in a reduction of the energy gap by a factor roughly 3 and an order-of-magnitude decrease in the condensation energy. It was established that the suppression factor should carry a substantial density dependence, but this dependence was not explored. The latest development along these lines is due to Niskanen and Sauls[33] (NS), who have calculated the Fermi-liquid interactions in neutron matter in a manner parallel to that followed by Bäckman, Jackson, and Niskanen[41] for symmetric nuclear matter. In this work there are certain detailed refinements upon and departures from the procedure of BKS which we shall not trace here; suffice it to say that (as in Ref. 38) a tensor component (h term) was included in the q-dependent quasiparticle interaction (q being the momentum transfer), a more accurate treatment of the tensor force of the bare interaction was implemented, and estimates of the two-phonon contribution were made. The most important departure for our purposes is that a different driving term (though still derived from a G-matrix approximation to the particle-hole-irreducible ("direct") interaction) was used in the RPA construction of the induced interaction. The relevant driving term is less attractive than that of BKS, and consequently the induced-interaction contribution to F_0 is less repulsive. At $k_F = 0.6$ fm^{-1}, NS find $F_0 = -0.36$ (not much lower than the BKS result), but this parameter drops to a minimum of about -0.7 around $k_F = 1.3$ fm^{-1} and then rises again. The G_0 results of the two calculations agree reasonably well until high densities are reached, and such refinements as the introduction of an h term and inclusion of the two-phonon contribution have essentially no impact on the problem at hand. With their Landau parameters and pursuing an analysis similar to that of CKCY, Niskanen and Sauls reexamined the modification of the pairing interaction due to polarization of the medium. For k_F below about 0.85 fm^{-1} they predict a suppression of the pairing interaction, in agreement with CKCY; however, at higher densities they predict an enhancement of $P_{k_F k_F}$ and hence an increase of the energy gap, over the values corresponding to the direct quasiparticle interaction alone. The amplification of the energy gap, ascribed to the drop of F_0 toward -1, can reach something like a factor 2, near $k_F = 1.0$ fm^{-1}. Thus, the NS analysis indicates that in the competition between density and spin-density fluctuations, the latter dominate at low densities and the former at high densities, within the regime ($k_F =$ 0.35-1.20 fm^{-1}) where singlet-S pairing is expected to be important. It should be noted that all of the microscopic studies under consideration here, and in particular CKCY and NS, are based on the Reid-soft-core interaction.

At any rate, in the interesting region $k_F = 0.6$-0.75 fm^{-1} where the "unperturbed" gap and condensation energies assume their maximum values, there is essential concurrence of the CKCY and NS treatments, and of the CKCY analysis repeated using the Landau parameters of Ref. 38, with regard to the quantitative as well as the qualitative effect of polarization on pairing: one gets a suppression of the pairing interaction by roughly a factor 1.2-1.4. It is reassuring that our (much more elaborate) CBF treatment of virtual particle-hole excitations leads to a similar conclusion, although with a somewhat greater suppression ($\alpha^{-1} \sim 1.7$). However, it remains to be understood why NS predict an enhancement at higher k_F where we still find a suppression. While there are some elements of the NS procedure which may be questioned (e.g., reliance on the weak-coupling formula for the gap with = 1 -- see below), it seems to us most likely that a resolution of this discrepancy is to be found in the fact that there are other important effects in the second-order CBF addend of $P_{k\ell}$ besides those of spin-density fluctuations and (the propagator correction to) density fluctuations. Another possibility is that the CBF expansion has not adequately converged at second order, and that our result substantially overestimates the effect of spin-density fluctuations. However, we consider this alternative as unlikely for basically the same reasons as stated in Ref. 38 with regard to the CBF expansion of the Landau quasiparticle interaction. This issue will be addressed in more detail elsewhere.

The formulas derived in the weak-coupling (WC) approximation by BCS, widely used to estimate superfluid properties, may be tested against our numerically accurate results for the energy gap and condensation energy. Given the pairing matrix element $P_{k_F k_F}$ at the Fermi surface, the formula for the energy gap (at k_F) is

$$\Delta_{WC} = 2\eta\varepsilon_0 \exp\left[1/N(0)P_{k_F k_F}\right] , \tag{10}$$

where $N(0) = m^* k_F/2\pi^2\hbar^2$ is the density of states of one spin at the Fermi surface and ε_0 is the characteristic energy scale, ordinarily taken as the Fermi energy $\varepsilon_F = \hbar^2 k_F^2/2m^*$ in the case of neutron matter. We insert an additional fitting parameter η which will be chosen so that the peak of Δ_{WC} as a function of k_F matches that of Δ_{k_F} determined by solving the gap equation taking account of the k,ℓ dependence of $P_{k\ell}$. Given an energy gap Δ at the Fermi surface, the weak-coupling approximation for the condensation energy (per particle) is

$$[E_c/A]_{WC} = \frac{3}{8}\frac{\Delta^2}{\varepsilon_F} . \tag{11}$$

First we check the accuracy of (11), with our Δ_{k_F} values as input and our values for the condensation energy E_c/A (calculated by doing the k,ℓ integrations in (3) numerically) as the standard of comparison. Remarkably, there is close concurrence in all four cases, namely for variational and CBF gaps based on Reid v_6 or BJ v_6 interactions, the greatest deviation naturally being seen at the maxima of the E_c/A vs k_F curves. Typically, the agreement is better than 1%, and in the CBF cases the weak-coupling approximation to the condensation energy is practically indistinguishable from the precisely calculated quantity.

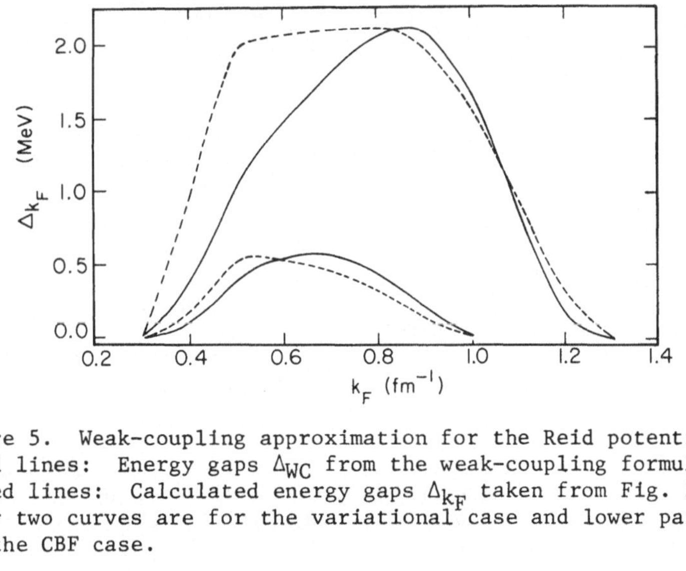

Figure 5. Weak-coupling approximation for the Reid potential. Solid lines: Energy gaps Δ_{WC} from the weak-coupling formula. Dashed lines: Calculated energy gaps Δ_{k_F} taken from Fig. 3. Upper two curves are for the variational case and lower pair for the CBF case.

Turning to the gap formula (10), with our variational and CBF values of $P_{k_F k_F}$ as input, significant deficiencies are seen. Figure 5 compares the weak-coupling results with results from Fig. 3 obtained by numerical solution of the gap equation. The picture is similar for the two potentials studied here. Upon matching the approximate and "exact" peak values by adjustment of η, the weak-coupling approximation tends to underestimate the gap at the lower densities and (usually) overestimate it at higher k_F. The required values of the parameter η are: $\eta = 1.35, 1.49, 1.85$, and 1.68 for Reid variational, BJ variational, Reid CBF, and BJ CBF calculations, respectively.

FUTURE PROSPECTS

The next problem to examine is evidently the effect of the proton contaminant in the inner crustal regime of a neutron star. To some extent, this problem has already been addressed by Takatsuka,[5] but it needs to be studied in the CBF framework. A related project is the extension of the approach developed here to the description of the $^1S_0$ proton superfluid in the quantum fluid interior. We should also mention that there are no apparent technical obstacles to a CBF calculation of the anisotropic $^3P_2$-$^3F_2$ neutron gap in the quantum fluid regime.

It is important to explore the observational ramifications of the modified picture of nucleon pairing which has emerged, or will emerge, from the inclusion of second- and higher-order CBF diagrams in the microscopic description of this class of condensed-matter phenomena. The principal finding of the current work, namely the strong suppression of the isotropic $^1S_0$ neutron gap in the inner crust, will not have any significant impact upon neutron-star cooling, but it will have interesting implications for the modeling of glitch phenomena in terms of the vortex creep theory.[1,14] We intend to discuss these implications in some detail in another place.

ACKNOWLEDGMENTS

This research was supported in part by the National Science Foundation under Grants No. DMR 83-04213 to Washington University and PHY 82-06325 to Texas A & M University. EK acknowledges support from the Deutsche Forschungsgemeinshaft through a Heisenberg fellowship. We thank A. D. Jackson, D. Pines, and J. A. Sauls for informative conversations and communications.

REFERENCES

1. D. Pines, J. de Physique **41**, C2-111-124 (1980); D. Pines, to be published: D. Pines and M. A. Alpar, to be published.
2. G. Baym and C. J. Pethick, Ann. Rev. Nucl. Sci. **25**, 27 (1975); Ann. Rev. Astron. Astrophys. **17**, 415 (1979).
3. M. Hoffberg, A. E. Glassgold, R. W. Richardson, and M. Ruderman, Phys. Rev. Lett. **24**, 75 (1970).
4. C.-H. Yang and J. W. Clark, Nucl. Phys. **A174**, 49 (1971); C.-H. Yang, Ph.D. thesis, Washington University (1971), unpublished.
5. T. Takatsuka, Progr. Theor. Phys. **71**, 1432 (1984).
6. G. Baym, C. J. Pethick, and D. Pines, Nature **224**, 673 (1969).
7. N.-C. Chao, J. W. Clark, and C.-H. Yang, Nucl. Phys. **179**, 320 (1972).
8. T. Takatsuka and R. Tamagaki, Progr. Theor. Phys. **46**, 114 (1971).
9. T. Takatsuka, Progr. Theor. Phys. **48**, 1517 (1972).
10. O. V. Maxwell, Ap. J. **231**, 201 (1979).

11. M. Soyeur and G. E. Brown, Nucl. Phys. **A324**, 464 (1979).
12. K. Nomoto and S. Tsuruta, Ap. J. **250**, L19 (1981).
13. G. Greenstein, Ap. J. **200**, 281 (1975); **208**, 836 (1976); D. Harding, R. A. Guyer, and G. Greenstein, Ap. J. **222**, 991 (1978).
14. P. W. Anderson and N. Itoh, Nature **256**, 25 (1975); M. A. Alpar, Ap. J. **213**, 527 (1977); D. Pines, J. Shaham, M. A. Alpar, and P. W. Anderson, Progr. Theor. Phys. Suppl. **69**, 376 (1980); P. W. Anderson, M. A. Alpar, D. Pines, and J. Shaham, Phil. Mag. **A45**, 227 (1982); M. A. Alpar, S. Langer, and J. A. Sauls, Ap. J. **282**, 533 (1984); M. A. Alpar, P. W. Anderson, D. Pines, and J. Shaham, Ap. J. **276**, 325 (1984); M. A. Alpar, P. W. Anderson, D. Pines, and J. Shaham, Ap. J. 278, 791 (1984); M. A. Alpar, R. Nandkumar, and D. Pines, Ap. J. **288**, 191 (1985); D. Pines and M. A. Alpar, in *Proceedings of the NRAO Workshop No. 8*, ed. S. P. Reynolds and D. R. Stinebring (1984), p. 161.
15. *Recent Progress in Many-Body Theories*, ed. J. G. Zabolitzky, M. de Llano, M. Fortes, and J. W. Clark (Springer-Verlag, Berlin, 1982); *Recent Progress in Many-Body Theories*, ed. H. Kümmel and M. L. Ristig (Springer-Verlag, Berlin, 1984).
16. E. Feenberg, *Theory of Quantum Fluids* (Academic Press, New York, 1969).
17. J. W. Clark, Prog. Part. Nucl. Phys. **2**, 89 (1979).
18. J. W. Clark, in *Proceedings of the Third International Conference on Nuclear Reaction Mechanisms, Varenna, 14-19 June 1982*, ed. E. Gadioli, Ricerca Scientifica ed Educazione Permanente, Universita degli Studi di Milano, Supplemento **n.28**, p. 464 (1982).
19. E. Krotscheck and J. W. Clark, Nucl. Phys. **A328**, 73 (1979).
20. V. R. Pandharipande and R. B. Wiringa, Rev. Mod. Phys. **51**, 821 (1979).
21. E. Krotscheck and J. W. Clark, Nucl. Phys. **A333**, 77 (1980).
22. E. Krotscheck, R. A. Smith, and A. D. Jackson, Phys. Rev. **B 24**, 6404 (1981).
23. J. W. Clark and P. Westhaus, Phys. Rev. **141**, 833 (1966).
24. J. W. Clark and C.-H. Yang, Nuovo Cim. Lett. **3**, 272 (1970); **2**, 379 (1971).
25. T. C. Paulick and C. E. Campbell, Phys. Rev. **B 16**, 2000 (1977).
26. S. Fantoni, Nucl. Phys. **A363**, 381 (1981).
27. K. Nakamura, Progr. Theor. Phys. **21**, 713 (1959); **24**, 1195 (1960).
28. E. Krotscheck, R. A. Smith, J. W. Clark, and R. M. Panoff, Phys. Rev. **24**, 6383 (1981).
29. R. C. Kennedy, Nucl. Phys. **A118**, 189 (1968).
30. E. Krotscheck, Z. Phys. **251**, 135 (1972).
31. D. Pines, in *Proc. XIIth International Conference on Low Temperature Physics*, ed. E. Kandu (Kligatu Publ. Co., Tokyo, 1971), p. 10.
32. J. W. Clark, C.-G. Källman, C.-H. Yang, and D. A. Chakkalakal, Phys. Lett. **61B**, 331 (1976).
33. J. A. Niskanen and J. A. Sauls, private communication and preprint.
34. E. Krotscheck, H. Kummel, and J. G. Zabolitzky, Phys. Rev. **A 22**, 1243 (1980).
35. E. Krotscheck and R. A. Smith, Phys. Rev. **B 27**, 4222 (1983).
36. E. Krotscheck, Phys. Rev. **A 26**, 3536 (1982).
37. H. Kümmel, K. H. Lührmann, and J. G. Zabolitzky, Phys. Rept. **C36**, 1 (1978); Y. Gerstenmaier and D. Schütte, Z. Naturforsch. **A35**, 796 (1980); K. Emrich and J. G. Zabolitzky, in *Recent Progress in Many-Body Theories*, ed. H.Kümmel and M. L. Ristig (Springer-Verlag, Berlin, 1984), p. 271.
38. A. D. Jackson, E. Krotscheck, D. E. Meltzer, and R. A. Smith, Nucl. Phys. **A386**, 125 (1982).
39. S.-O. Bäckman, C.-B. Källman, and O. Sjöberg, Phys. Lett. **43B**, 263 (1973).
40. S. Babu and G. E. Brown, Ann. Phys. **78**, 1 (1973).
41. S.-O. Bäckman, A. D. Jackson, and J. A. Niskanen, preprint.
42. J. Bardeen, L. N. Cooper, and J. R. Schrieffer, Phys. Rev. **108**, 1175 (1957); J.R.Schrieffer, *Theory of Superconductivity* (Benjamin, New York, 1964).

π^0 CONDENSATION IN HOT NUCLEAR MATTER*

Tatsuyuki Takatsuka

College of Humanities
and Social Scineces
Iwate University
Morioka 020, Japan

Ryozo Tamagaki

Department of Physics
Kyoto University
Kyoto 606, Japan

1. INTRODUCTION

The occurrence of pion condensation in nuclear medium, firstly proposed by Migdal[1] and also by Sawyer and Scalapino [2], has been one of the main concern in dense nuclear matter physics[3]. Among possible types of pion condensation, the neutral pion (π^0) condensation is particularly interesting, since it accompanies the drastic change of nucleon system such as the "laminated structure" suggested by Migdal[4]. At the almost same time, such an interesting structure of nucleon system was also considered by Calogero and Palumbo[5], although without relevance to pion condensation.

On the basis of the notion that for the π^0 condensation the field description (abbreviated to FD; explicit use of the pion degrees of freedom) is equivalent to the potential description (PD; conventional potential picture with elimination of the pion degrees of freedom), Tamiya, Tatsumi and the present authors[6] clearly showed that the π^0 condensed phase is nothing but the nucleonic phase which brings about the attractive first order effect of the OPEP tensor component. By introducing the Alternating-Layer-Spin (ALS) model to represent suitably this aspect, they revealed the characteristic features of the phase and clarified the mechanism of the phase transition; the π^0 condensed phase (equivalently the ALS phase) is characterized by the one-dimensional density localization with a specific spin-isospin order and is realizable only when the energy gain due to π^0 condensation overwhelms the kinetic energy increase due to nucleon localization.

At early stage, the transition density (ρ_c) for π^0 condensation was expected to be $\rho_c \sim \rho_0$ ($\equiv 0.17$ fm^{-3}; the nuclear density). Later, as a result of the theoretical efforts [7~12] to take account of the various effects coming from the exchange part of the OPEP, short-range correlation, isobar Δ(1232) mixing and two-nucleon interaction other than the OPEP, considerably higher values of ρ_c have been predicted; at present it is believed that $\rho_c \simeq (2\text{-}3)\rho_0$ which is not inconsistent with the information inferred from the studies on the precritical phenomena in finite nuclei [13].

---

*) Presented by T. Takatsuka

High-density nucleon matter with the density $\rho \gtrsim 2\rho_0$, where π^0 condensation can be expected, exists first of all in neutron stars and so the possible influence of π^0 condensate on neutron star properties such as the superfluidity in the interior core[14] and the extent of proton mixing[15] has been preferentially investigated.

On the other hand, recent active studies on high energy nucleus-nucleus collisions are giving rise to such expectation that high-density matter would be formed in laboratories, although in this case the matter is in "high temperature" instantaneously.[16] Then it becomes important to extend the studies on pion condensation into the case for finite temperature. Up to date, few works have been done on this problem[17~21] as compared with those at zero temperature. Moreover, they are almost restricted to the σ-model. Therefore it would be desirable and complementary to study the problem in different approaches, especially in the framework of the familiar potential description.

In this report, we try to discuss the problem of π^0 condensation in hot nuclear matter from the viewpoint of effective interaction approach. We use PD based on the ALS model and simplify the problem as much as possible so as to provide intuitive picture. Our aims in this article are to show to what extent the onset of the phase is affected by the temperature T and to draw characteristic features of the phase transition. Before going to this subject, we briefly explain the interrelation between the ALS model and a typical π^0 condension.

2. ALS MODEL AND π^0 CONDENSATION [6, 7]

The equivalence between FD and PD has been rigorously shown for the (neutron+proton+π^0) system described by the Hamiltonian $H = H_N + H_\pi + H_{\pi-N}$ where $H_N (H_\pi)$ denotes the free part of nucleon (π^0) and $H_{\pi-N}$ is the π-N P-wave interaction, the driving force for π^0 condensation.
The ground state $|\Phi_0\rangle$ for the system can be shown to be represented by the direct product of the boson (namely π^0) coherent state $|\Phi_B\rangle$ and the model nucleon state $|\Phi_N\rangle$ which is given by the Slater determinant of the single-particle basis set $\{\phi_\alpha\}$:[22]

$$|\Phi_0\rangle = |\Phi_N\rangle \otimes |\Phi_B\rangle \, , \quad |\Phi_N\rangle = \prod_{\alpha}^{(occ)} C_\alpha^+ |0\rangle \qquad (2.1), (2.2)$$

with C_α^+ being the creation operator for the nucleon in α-state. Then the equivalence is expressed in the Hartree (direct part) approximation as*)

$$\langle \Phi_0 | H_\pi + H_N | \Phi_0 \rangle = -\frac{1}{2}\int d\vec{r} \, \{ (\vec{\nabla}\langle \varphi_0 \rangle)^2 + m_\pi^2 \langle \varphi_0 \rangle^2 \} \qquad (2.3)$$

$$= \frac{1}{2}\sum_{\alpha,\alpha'} \langle \alpha\alpha' | V^{OPE} | \alpha\alpha' \rangle \, , \qquad (2.4)$$

*) In this article, we use the following units, $\hbar = c = \kappa_B$ (Boltzmann constant) $= 1$.

where φ_0 is the field operator for π^0 which is static corresponding to the zero chemical potential of π^0, $\langle\varphi_0\rangle \equiv \langle\Phi_B|\varphi_0|\Phi_B\rangle$ and V^{OPE} denotes the OPEP including the δ-function part. Eqs. (2.3) and (2.4) indicate the physical situation that the nucleon structure to utilize the attractive effect of the OPEP in the first order produces the nonvanishing π^0 field (i.e., π^0 condensate $\langle\varphi_0\rangle$ =0). This clearly shows the interplay between the occurrence of π^0 condensation and the structure change of nucleon system. When we go beyound the Hartree approximation, we have the OPEP-exchange energy corresponding to the effect of non-condensed pions as the lowest order quantum correction.[22] Thus we can treat the problem of π^0 condensation by using the conventional PD.

On the basis of the notion shown above, our task is to find out such a new nucleon system as to utilize the OPEP effect which vanishes in the normal phase (that is, the ordinary Fermi gas (FG) phase). Taking account of the characteristics of the OPEP-tensor part which dominates in the OPEP and depends on the spin, isospin and position of the interacting pair, we derive such a localized system with the particular spin-isospin order as sketched in Fig.1 which we call the ALS structure: In the ℓ-th layer, the spin of like nucleons are alligned in the combination (n↑,p↓) or (n↓,p↑) and the spin directions change alternately layer by layer. Denoting the spin (isospin) in the ℓ-th layer by σ_ℓ (τ_ℓ) and assigning a central (n↑,p↓) layer by ℓ =0, we can express this configuration as

$$\sigma_\ell \tau_\ell = -(-)^\ell; \quad \sigma_\ell(n) = (-)^\ell \quad \text{and} \quad \sigma_\ell(p) = (-)^{\ell+1}. \tag{2.5}$$

In order to avoid the large increase of kinetic energy due to localization, with keeping the interaction energy gain, the one-dimensional localization is favorable because of small zero-point energy. In this

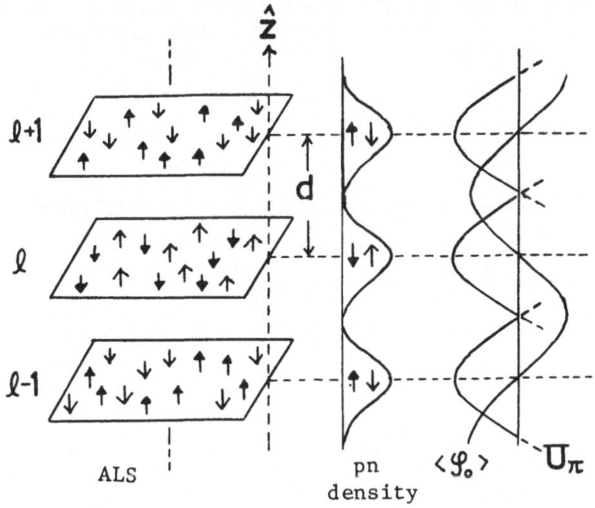

Fig.1. Profile of the ALS structure in symmetric nuclear matter and the resulting aspects of density, single-particle potential U_π and condensed π^0-field $\langle\varphi_0\rangle$.

respect the ALS model is substantially different from the tensor interaction model of Pandharipande and Smith [23] based on three-diemnsional solid structure. Therefore we take $\{\phi_\alpha\}$ in the ALS model as

$$\phi_\alpha(\vec{\xi}) \equiv \phi_{\vec{q}_\perp, \ell}(\vec{r})\chi_{\sigma_\ell \tau_\ell}(\text{spin, isospin}) = \Omega_\perp^{-1/2} e^{i\vec{q}_\perp \cdot \vec{r}_\perp} \phi_{W\ell}(z)\, \chi_{\sigma_\ell \tau_\ell}, \tag{2.6}$$

where $\vec{r}_\perp = \{x, y\}$, $\vec{q}_\perp = \{q_x, q_y\}$, $\alpha \equiv \{\vec{q}_\perp, \ell, \sigma_\ell, \tau_\ell\}$ and $\Omega_\perp$ is the two-dimensional normalization volume. $\phi_{W\ell}(z)$ represents the Wannier-type basis localized around $z=\ell d$ with d being the layer spacing, which is well approximated by the Gaussian form for the well-localized situation:

$$\phi_{W\ell}(z) \simeq \phi_\ell(z) \equiv (a/\pi)^{1/4} \exp\left[-a(z-d\ell)^2/2\right] \tag{2.7}$$

The orthogonality between these basis functions (especially, for the nearest parallel-spin pair) is assured for $\Gamma \equiv ad^2 > 2$, where Γ is an important parameter indicating the degree of localization compared with d. In the ALS model $|\Phi_N\rangle \equiv |\Phi_{ALS}\rangle$ is the Slater determinant of $\{\phi_\alpha\}$ given by Eq. (2.6), where the two-dimensional Fermi gas is the state occupied up to $|\vec{q}_\perp| \leq q_{\perp F}$. In the case of symmetric nuclear matter, the number of neutrons ($N_{\perp n}$) and protons ($N_{\perp p}$) are equal and $q_{\perp F}$ is defined through

$$\rho_\perp = N_\perp / \Omega_\perp = q_{\perp F}^2 / 2\pi = \rho d \tag{2.8}$$

with $N_\perp = N_{\perp n} + N_{\perp p}$.

Let us see how the ALS structure brings about the π^0 condensation or energy gain due to the OPEP. This structure provides the nonvanishing source function $\vec{S}$ to the field equation of $\langle \varphi_0 \rangle$:

$$(\vec{\nabla}^2 - m_\pi^2)\langle \varphi_0 \rangle = \vec{\nabla} \cdot \vec{S}, \tag{2.9}$$

$$\vec{S} = \hat{f} \langle \Phi_{ALS} | \psi^\dagger \tau_3 \vec{\sigma} \psi | \Phi_{ALS} \rangle = \tilde{f} \langle \Phi_{ALS} | \psi_p^\dagger \vec{\sigma} \psi_p - \psi_n^\dagger \vec{\sigma} \psi_n | \Phi_{ALS} \rangle$$

$$\simeq -\hat{z} \hat{f} \rho_\perp \sum_\ell (-)^\ell |\phi_\ell(z)|^2, \tag{2.10}$$

where

$$\psi^\dagger = \sum_\alpha C_\alpha^\dagger e^{i\varepsilon_\alpha t} \phi_\alpha^* \tag{2.11}$$

with ε_α being the single-particle energy for the state α, m_π is the pion mass, $\hat{f} \equiv f/m_\pi$ and we take f =1. In Eq. (2.10) it is remarked that the contributions from protons and neutrons are designed to be additive through the Eq. (2.5). The solution for Eq. (2.9) is obtained as

$$\langle\varphi_0\rangle = \sum_{n\geq 0} A_n \sin k_n z \simeq A_0 \sin k_0 z \,, \tag{2.12}$$

$$A_n = -2\tilde{f}\rho k_n e^{-k_n^2/4a}/\omega_n^2 \,, \quad k_0 \equiv \pi/d \,, \tag{2.13}$$

with $k_n=(2n+1)k_0$ and $\omega_n^2=k_n^2+m_\pi^2$. In most cases $\Gamma\lesssim 10$ and the approximate equality in Eq. (2.12) holds. Thus we can see that $|\Phi_{ALS}\rangle$ generates the condensed π^0-field $\langle\varphi_0\rangle\simeq A_0 \sin k_0 z$ with the momentum $\vec{k}_0=\pm k_0\hat{z}$. This gives rise to the OPEP energy gain for $|\Phi_{ALS}\rangle$ as seen from Eqs. (2.3) and (2.4). On the other hand, the $\langle\varphi_0\rangle$ thus resulted provides a nucleon system with a deep periodic potential $U_\pi=f\hat{\tau}_3\vec{\sigma}\cdot\vec{\nabla}\langle\varphi_0\rangle$ having the spin-isospin dependence. In order to feel U_π efficiently, nucleons arrange to form the ALS structure, which in turn maintains the source function for $\langle\varphi_0\rangle$. That is, the self-consistent relation between the ALS structure of nucleons and $\langle\varphi_0\rangle$ is fulfilled.

So far we have concerned with the zero temperature case. When we go into the finite temperature case, we should take account of $\langle\Phi_N|C_\alpha^+C_\alpha|\Phi_N\rangle=f_\alpha$ with f_α the occupation probability depending on T. Then $\vec{S}$ reads

$$\vec{S} = -\hat{z}\hat{f}\Omega_\perp^{-1}\sum_{q_\perp}\sum_{\ell} f_\alpha (-)^\ell |\phi_\ell(z)|^2 \,. \tag{2.14}$$

As shown later on, f_α for the ALS system depends only on $q_\perp$ ($f_\alpha\equiv f(q_\perp)$) and so Eq. (2.14) concides with Eq. (2.10), because $\sum_{\vec{q}_\perp} f(q_\perp)= N_\perp$. That is, $\langle\varphi_0\rangle$ does not depend on T, consistent with the concept of Bose condensation in the sense that the condensation occurs in a complete fashion. Also by the same reason, the direct contribution from the V^{OPE} is easily shown to be independent of T. This means that in the Hartree approximation the equivalence between PD and FD holds as well even for T>0 case. Here it should be mentioned that the V^{OPE}-direct contribution (or $\langle\varphi_0\rangle$) depends implicitly on T through the parameters (Γ and d) which are to be determined by the free energy minimization and hence have T-dependence.

As for the OPEP exchange part associated with the effect of non-condensed pions, it is, of course, T-dependent primarily through f_α and also through the Bose distribution of non-condensed pions at T>0. The latter point is closely related to the thermal fluctuation, as well as the quantum one, of the pion field other than the classical part $\langle\varphi_0\rangle$. This is an interesting problem to be studied by FD, e.g., such a line as in Ref. 22. But here we do not go into its details. For the present purpose, we are content to neglect the OPEP-exchange part because it is expected to be very small when the δ-part of V^{OPE} is subtracted (hence $\tilde{V}^{OPE}\simeq\tilde{V}_T^{OPE}$ which is defined later on), as inferred from the results at T=0 case.[6,7]

3. π^0 CONDENSATION AT T>0

Here we outline the way to investigate the subject. We consider the hot symmetric nuclear matter in thermal equilibrium. For simplicity, we approach the problem from the following viewpoint:

1) We introduce the effective interaction suitable for the system and assume that it is T-independent.

2) Using the PD we solve the Hartree-Fock (HF) equation at finite temperature both for the FG (normal) and ALS (π^0 condensed) phases and discuss the phase transition by comparing their free energies.

(1) Model Hamiltonian

As has been shown in the previous section, the essential ingredient for the realization of π^0 condensation lies in the strong tensor part V_T^{OPE} in the OPEP. To be realistic, however, the medium corrections to V_T^{OPE}, such as (a) weakening due to the short-range correlations and (b) rho-meson contribution and also (c) enhancement coming from the Δ-mixing, should be included. At the simplest level we introduce the effective OPEP-tensor force $\widetilde{V}_T^{OPE}$ by

$$\widetilde{V}_T^{OPE} = -\frac{1}{3}\eta^2\tilde{f}^2(\vec{\tau}_1\cdot\vec{\tau}_2)\int\frac{d\vec{k}}{(2\pi)^3}\frac{S_{12}(\vec{k})}{k^2+m_\pi^2}e^{i\vec{k}\cdot\vec{r}}\frac{m_\rho^2}{k^2+m_\rho^2}, \quad (3.1)$$

where $S_{12}(\vec{k})=3(\vec{\sigma}_1\cdot\vec{k})(\vec{\sigma}_2\cdot\vec{k})-(\vec{\sigma}_1\cdot\vec{\sigma}_2)\vec{k}^2$ and m_ρ denotes the rho-meson mass. In $\widetilde{V}_T^{OPE}$, the appearance of $m_\rho^2/(k^2+m_\rho^2)$ takes account of the effect (b) and the effect (a) and (c) are simply absorbed into the parameter $\eta^2\equiv(f_{eff}/f)^2$. As a typical case, we adopt $\eta^2=1$ which reproduces fairly well the results obtained by the fundamental approach to include explicitly the Δ degrees of freedom[12].

As for non-tensor part of two-nucleon interaction, we use the effective central one $\widetilde{V}_C$ which is the main ingredient for the interaction energy in usual case, since the spin-orbit part vanishes as a whole in both phases. As a suitable choise for $\widetilde{V}_C$, we use the Go-force of Sprung and Banerjee [24] which is obtained by solving the reaction-matrix equation with the RSC potential[25] in the normal phase, assuming that the short-range correlation is not so different between the two phases:

$$\widetilde{V}_C(r;\rho,\beta)=\sum_{i=1}^{5}W_i(\beta)\exp\left[-(r/\lambda_i)^2\right] \quad (3.2)$$

where $q_F=(3\pi^2\rho/2)^{1/3}$, $W_i(\beta)=a_i(\beta)+\sqrt{q_F}\,b_i(\beta)$ and the parameters $\{a_i, b_i, \lambda_i\}$ are referred to Ref.24. Here β denotes the two-nucleon state; $\beta={}^3O, {}^1E, {}^1O$ and ${}^3E$ according to $(S,t,L)=(1,1,\text{odd})$, $(0,1,\text{even})$, $(0,0,\text{odd})$ and $(1,0,\text{even})$, respectively.

Thus our model Hamiltonian H to represent the essentials relevant to the problem is given by

$$\widetilde{H}=\sum_i T_N(i)+\sum_{i<j}\widetilde{V}(i,j), \quad \widetilde{V}(i,j)=\widetilde{V}_T^{OPE}(i,j)+\widetilde{V}_C(i,j), \quad (3.3), (3.4)$$

with T_N denoting the kinetic energy part. Both $\widetilde{V}_T^{OPE}$ and $\widetilde{V}_C$ are assumed to be independent of T, although, in principle, they are T-dependent through the T-dependence of the nucleon correlations.

(2) HF equation

In the variational theory, the thermal equilibrium state is determined by the minimization of the free energy under the constraint $N=\sum_\alpha f_\alpha$ where N is the total nucleon number. As well known, this leads to the HF equation which determines the optimum values of the single-particle spectra ε_α and the orbitals ϕ_α, together with the chemical potential μ. Since we are concerned with the infinite system, and ϕ_α are given (by the plane wave $\times$ spin-isospin state for the case with FG; by Eq. (2.6) for the case with ALS), HF equation reduces to the one determining ε_α and μ:

$$\varepsilon_\alpha = t_\alpha + \sum_{\alpha'} f_{\alpha'} \langle \alpha\alpha' | \tilde{V} | \alpha\alpha' - \alpha'\alpha \rangle , \tag{3.5}$$

$$f_\alpha = \left[1 + \exp\{(\varepsilon_\alpha - \mu)/T\}\right]^{-1}, \quad N = \sum_\alpha f_\alpha , \qquad (3.6), (3.7)$$

where $t_\alpha \equiv \langle \alpha | T_N | \alpha \rangle$. From the solution ε_α and μ we get directly all the thermodynamic quantities; for instance, the internal energy E, the entropy S, the free energy F, respectively defined as per particle, and the pressure P:

$$E = \sum_\alpha f_\alpha (\varepsilon_\alpha + t_\alpha)/2N , \tag{3.8}$$

$$S = -\sum_\alpha \{ f_\alpha \ln f_\alpha + (1 - f_\alpha) \ln (1 - f_\alpha) \} / N , \tag{3.9}$$

$$F = E - TS , \quad P = \rho^2 \, \partial F / \partial \rho . \qquad (3.10), (3.11)$$

The explicit expressions of Eqs. (3.5) (3.7) are as follows:

<u>FG-Phase</u>

$$\varepsilon_\alpha \equiv \varepsilon(q) = q^2/2m_N + \frac{1}{2\pi} \int_0^\infty q'^2 dq' f(q') \, G_c(q, q') , \tag{3.5'}$$

$$\rho = \frac{2}{\pi^2} \int_0^\infty q^2 dq \, f(q) , \tag{3.6'}$$

$$f_\alpha = f(q) = \left[1 + \exp\{ (\varepsilon(q) - \mu)/T \} \right]^{-1} , \tag{3.7'}$$

where

$$G_c(q, q') = \frac{\sqrt{\pi}}{4} \sum_\beta (2t+1)(2s+1) \sum_{i=1}^{5} W_i(\beta) \times \left[\lambda_i^3 + (-)^L \lambda_i \{ e^{-\lambda_i^2 q_-^2} - e^{-\lambda_i^2 q_+^2} \} / q q' \right] \tag{3.12}$$

with the definition $q_\pm \equiv (q \pm q')/2$.

<u>ALS-Phase</u>

In this case, the two-dimensional nature of the Fermi-gas state manifests itself in the expressions. They read as

$$\varepsilon_\perp(q_\perp) = q_\perp^2/2m_N + \frac{1}{2\pi} \int_0^\infty q_\perp' dq_\perp' f(q_\perp') \, G_c(q_\perp, q_\perp') , \tag{3.5''}$$

$$\rho_\perp \equiv \frac{1}{\pi} \int_0^\infty q_\perp \, dq_\perp \, f(q_\perp) , \tag{3.6''}$$

$$f_\alpha \equiv f(q_\perp) = \left[1 + \exp\{ (\varepsilon_\perp(q_\perp) - \mu_\perp)/T \} \right]^{-1} , \tag{3.7''}$$

where

$$\varepsilon_\perp(q_\perp) \equiv \varepsilon_\alpha - a/4m_N - \overset{\circ}{U}_T , \tag{3.13}$$

$$\mu_\perp \equiv \mu - a/4m_N - \overset{\circ}{U}_T , \tag{3.14}$$

$$G_c(q_\perp, q_\perp') = \Omega_\perp \sum_{\alpha'}' \langle \alpha\alpha' | \tilde{V}_c | \alpha\alpha' - \alpha'\alpha \rangle \tag{3.15}$$

In the above expression, the exchange contribution from $\tilde{V}_T^{OPE}$ is neglected as described in the previous section and $\Sigma'_{\alpha'}$ stands for the summation over ℓ', $\sigma_{\ell'}$' and $\tau_{\ell'}$. The terms $a/4m_N$ and $\overset{\circ}{U}_T$ which are constant (independent of $q_\perp$) are the zero point energy and the direct contribution from $\tilde{V}_T^{OPE}$, respectively. The $\overset{\circ}{U}_T$ is expressed as

$$\begin{aligned} \overset{\circ}{U}_T &\equiv \sum_{\alpha'} f_{\alpha'} \langle \alpha\alpha' | \tilde{V}_T^{OPE} | \alpha\alpha' - \alpha'\alpha \rangle \\ &= -\frac{4}{3}\rho\eta^2\tilde{f}^2 \sum_{n\geq 0} \left(\frac{k_n}{\omega_n}\right)^2 \frac{m_\rho^2}{k_n^2 + m_\rho^2} e^{-k_n^2/2a} , \end{aligned} \tag{3.16}$$

whose T-independence results from the one of $\langle \varphi_0 \rangle$, as discussed in the previous section.

The expression for $G_c(q_\perp, q_\perp')$ is derived straightforwardly by paying attention on the coherence expressed by Eq. (2.5):

$$G_c(q_\perp, q_\perp') = (a\pi^2/32)^{1/2} \sum_\beta \sum_{i=1}^{5} \lambda_i^3 \tilde{\lambda}_i W_i(\beta) \overset{\circ}{G}_c(i,\beta; q_\perp, q_\perp') \tag{3.15'}$$

$$\begin{aligned} \overset{\circ}{G}_c(i,\beta; q_\perp, q_\perp') &= 5R_-(i,\text{even}) + 4R_-(i,\text{odd}) \quad \text{for } \beta = {}^3O, \\ &= 2R_+(i,\text{odd}) + R_+(i,\text{even}) \quad \text{for } \beta = {}^1E, \\ &= R_-(i,\text{odd}) \quad \text{for } \beta = {}^1O, \\ &= R_+(i,\text{even}) + 2R_+(i,\text{odd}) \quad \text{for } \beta = {}^3E, \end{aligned} \tag{3.17}$$

where

$$R_\pm(i,\text{even(odd)}) = \sum_{n=\text{even (odd)}} \Big\{ e^{-n^2\Gamma\tilde{\lambda}/2} \pm e^{-n^2\Gamma/2} \int_0^{2\pi} \frac{d\varphi}{2\pi} e^{-\lambda_i^2/4 (q_\perp^2 + q_\perp'^2 - 2q_\perp q_\perp' \cos\varphi)} \Big\} \tag{3.18}$$

with $\tilde{\lambda}_i = (1 + a\lambda_i^2/2)^{-1}$.

4. NUMERICAL RESULTS AND DISCUSSION

Here we present the preliminary results. In Fig. 2, free energies F for both phases at several temperatures are shown as functions of ρ. The desities ρ_c corresponding to the cross-points of the solid (ALS) and the dashed (FG) curves indicate the conventional transition densities. The ρ_c at T=0 is about 2.7 ρ_0 in agreement with 2.5ρ_0 of Ref. 12. It is remarked that ρ_c goes high as T increases, as already shown in the previous works.[17~21]

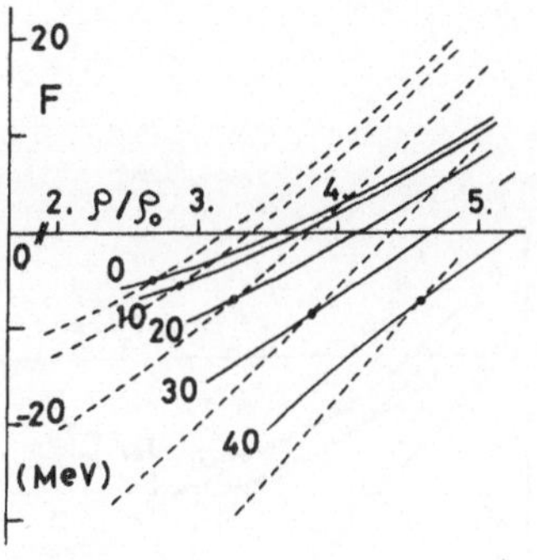

Fig.2. Free energies F versus density in the ALS-phase (solid lines) and in the FG-phase (dashed lines) at the temperature T shown by the attached numbers (MeV).

Here it is worthwhile mentioning that, strictly speaking, these ρ_c are approximate ones in the sense that they are taken without relevance to the character of the phase transition. In contrast to the case with charged pion condensation, the nucleon system under π^0 condensation turns out to be well localized (in z-direction, namely in direction of the condensed π^0 momentum) very soon after the onset of the transition[6], showing a "solid-like" aspect, and the translational invariance in z-direction is broken. Therefore, the nature of the transition should be taken as the first order one, although detailed examinations are still necessary in close vicinity to the transition point. From this viewpoint, we try to determine the transition density ρ_c^* by the Maxwell double tangent construction displayed in Fig. 3. The resulting ρ_c^* are, for instance, as $\rho_c^* \simeq 2.5\,\rho_0$ ($4.2\rho_0$) at T=0 Mev (40 Mev) as compared with $\rho_c \simeq 2.7\,\rho_0$ ($4.6\,\rho_0$). From the results at T=0 ($\rho_c - \rho_c^* \simeq 0.2\,\rho_0$), we can say that for the T=0 case, the current use of ρ_c is allowable as an alternative use of ρ_c^*.

In Fig. 4, we show the phase diagram for the problem. The hatched area indicates the π^0-condensed phase and the dotted narrow region corresponds to the coexistence of the π^0-condensed and normal phases. It is observed that the realization of the new phase is pushed remarkably up to the higher density side as T goes high, clearly indicating the significant effect of T.

In order to understand the reason, the contributions from respective parts of $\widetilde{H}$ are illustrated in Fig. 5, by putting emphasis on their differences between two phases, where several quantities are defined as $\Delta F \equiv F_{ALS} - F_{FG}$, $\Delta\langle\widetilde{V}\rangle \equiv \langle\widetilde{V}\rangle_{ALS} - \langle\widetilde{V}\rangle_{FG}$ and so on. The dashed lines are for T=0 Mev case and the solid ones for T=30 Mev case. The following points are notable: (i)Even for T>0 case, the aspect for the realization mechanism of the ALS phase ($\Delta F<0$) is not essentially changed; it results mainly from the large energy gain of the OPEP tensor part ($\Delta\langle\widetilde{V}\rangle \simeq \Delta\langle V_T^{OPE}\rangle < 0$ being much larger in magnitude than $\Delta\langle\widetilde{V}_c\rangle = (6\to 8)$ MeV at T=0 Mev and ($1\to 6$) Mev at T=30 MeV as $\rho \simeq (3\to 5)\rho_0$) which overwhelms the kinetic energy increase ($\Delta\langle T_N\rangle > 0$). (ii) An important difference from the T=0 case is the presence of the contribution from the entropy part (Δ(-TS)>0) which increases ΔF; by this effect the threshold density for $\Delta F<0$ is drived to a higher value as indicated by the crosses.

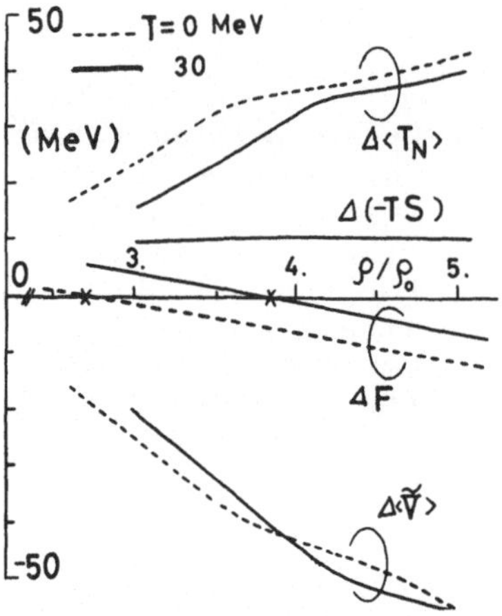

Fig.3. Transition densities ρ_c^* at three temperatures determined by the double tangent method.

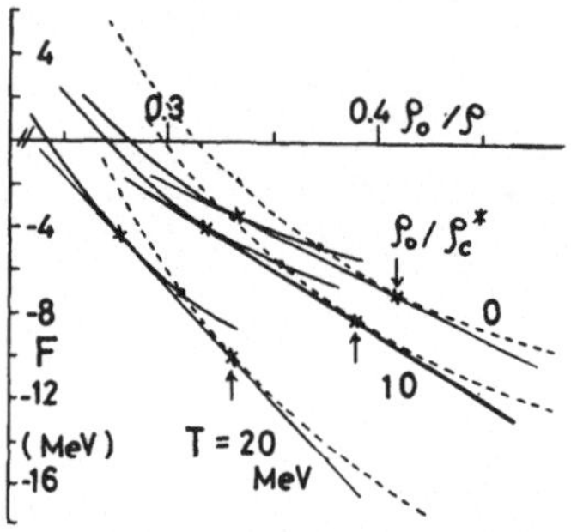

Fig.4. Phase diagram for two phases. Dotted region is the coexistent one.

The point (ii) above is a natural consequence since the ALS-phase is an ordered phase in contrast with the uniform FG-one and so S_{ALS} becomes remarkably smaller than S_{FG}. This aspect is shown in Fig. 6, where the behavior of S in coexistent regions are not traced and illustrated simply by the dashed lines with arrows. We see that the sudden decrease of S occurs according to the phase transition. This means that the latent heat $T\Delta S$ is released at the transition. Closely connected with this, the sudden decrease of E takes place with increasing T as shown in Fig. 7. These features are characteristics of the phase transition between the normal and the π^0-condensed phases at finite temperature.

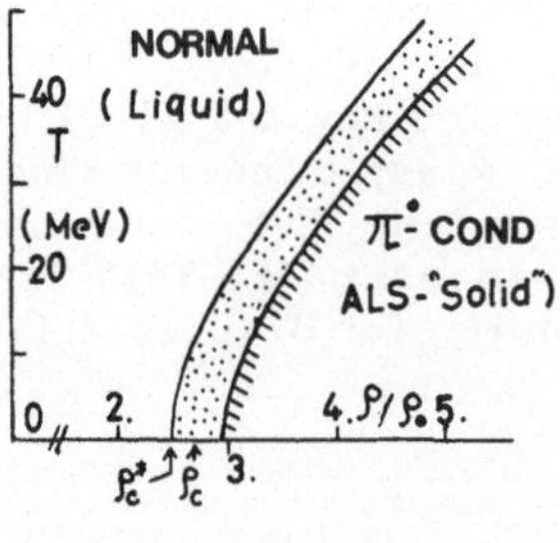

Fig.5. Difference of contribution of each term in the ALS phase from the FG one; $\Delta F = \Delta\langle T_N\rangle + \Delta\langle\tilde{V}\rangle + \Delta(-TS)$.

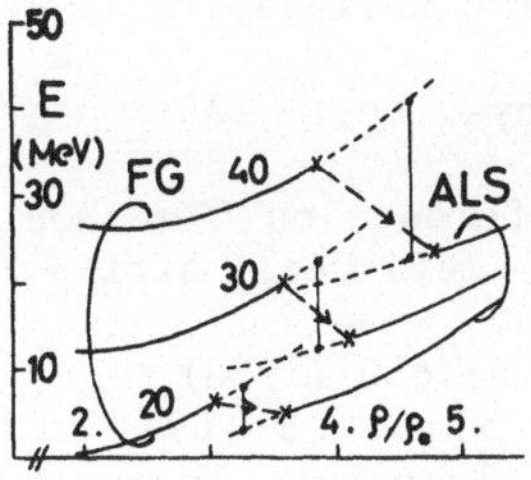

Fig.6. Entropies S versus density in both phases. ✕- - ➔ - - -✕ indicates the coexistent region.

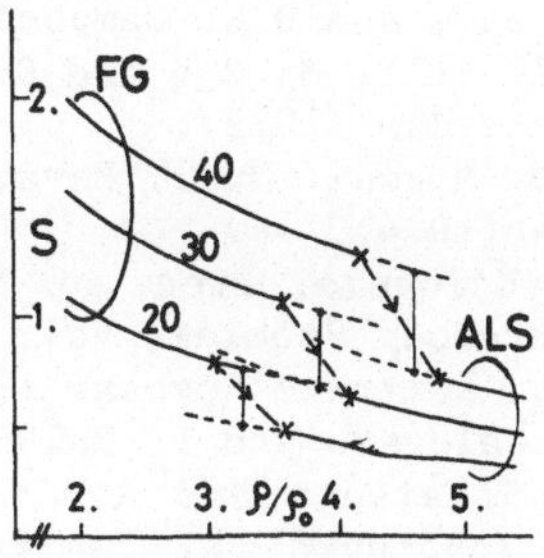

Fig.7. Internal energies E versus density in both phases.

5. CONCLUDING REMARKS

The effect of finite temperature on π^0 condensation is remarkable to the extent that $\rho_c^* \simeq (2.5 \to 4.2)\rho_0$ as $T=(0 \to 40)$ MeV. The increase of ρ_c^* suggests that the realization of the phase becomes unlikely in energetic nucleus-nucleus collisions unless dense nuclear matter would be formed at considerably low temperature; for example, with $\rho \simeq (3\text{-}4)\,\rho_0$ but at $T<40$ MeV. Final conclusion, however, should be postponed untill more quantitative study will be done by taking account of the Δ-effects realistically and by improving the ALS-model wave function. Also, how the effects due to thermally excited pions discussed in Ref.20 act in the case of π^0 condensation should be studied.

We have shown the characteristic aspects of the phase transition; sudden decreases of the entropy and the internal energy and also release of the latent heat. It is of interest to investigate their possible consequences on the phenomena of nucleus-nucleus collisions and the problem of neutron star formation. Especially the sudden decrease of E at the phase transition may be one of the signals for the onset of π^0 condensation[26].

REFERENCE

1 A.B. Migdal, Zh ETF (USSR) 61: 2210 (1971) [JETP (Sov. Phys.) 34: 1184 (1972)]; Zh ETF (USSR) 63: 1933 (1972) [JETP (Sov. Phys.) 36: 1052 (1973)].

2 R.F. Sawyer, Phys. Rev. Lett. 29: 382 (1972).
D.J. Scalapino, Phys. Rev. Lett. 29: 386 (1972).
R.F. Sawyer and D.J. Scalapino, Phys. Rev. D7: 953 (1972).

3 As review articles, e. g.,
G.E. Brown and W. Weise, Phys. Rep. 22C: 292 (1975).
A.B. Migdal, Rev. Mod. Phys. 50: 107 (1978).

G. Baym and D.K. Campbell, Chapter 27 in "Mesons in Nuclei, Vol. III", M. Rho and D.H. Wilkinson ed., North-Holland, Amsterdam (1979).
4 A.B. Migdal. Nucl. Phys. A210: 421 (1973), Phys. Lett. 52B: 264 (1974).
5 F. Calogero, Proc. of the International Conference on "The Nuclear Many-Body Problem, Vol. II", F. Calogero and Ciofi Degli Atti ed., Editrice Compositi, Bologna (1973), p535.
F. Calogero and F. Palumbo, Nuovo Cim. Lett. 6: 663 (1973).
6 T. Takatsuka and R. Tamagaki, Prog. Theor. Phys. 58: 694 (1977).
T. Takatsuka, K. Tamiya, T. Tatsumi and R. Tamagaki, Prog. Theor. Phys. 59: 1933 (1978).
7 As review articles,
R. Tamagaki, Nucl. Phys. A328: 352 (1979); Proc. of the International Summer School, Changchun, July 1983, on "Nucleon-Nucleon Interaction and Nuclear-Many Body Problems", S.S. Wu and T.T.S. Kuo ed., World Scientific, Singapore (1984), p. 97.
8 T. Kunihiro and T. Tatsumi, Prog. Theor. Phys. 65: 613 (1981).
9 K. Tamiya and R. Tamagaki, Prog. Theor. Phys. 66: 948, 1361 (1981).
10 T. Takatsuka, Y. Saito and J. Hiura, Prog. Theor. Phys. 67: 254 (1982).
11 O. Benhar, Phys. Lett. 106B: 375 (1983); Nucl. Phys. A437: 590 (1985).
12 T. Kunihiro, T. Takatsuka and R. Tamagaki, Prog. Theor. Phys. 73: 683 (1985).
13 a)F. Osterfeld. T. Suzuki and J. Speth, Phys. Lett. 100B: 75 (1981).
T. Suzuki, F. Osterfeld and J. Speth, Phys. Lett. 100B: 443 (1981).
b)M. Yabe, K.-I. Kubo and H. Toki, Preprint (Tokyo Metropolitan Univ., 1984).
14 T. Takatsuka and R. Tamagaki, Prog. Theor. Phys. 62: 1655 (1979); 65: 1333 (1981).
R. Tamagaki, T. Takatsuka and H. Furkawa, Prog. Theor. Phys. 64: 2107 (1980).
T. Takatsuka, Lecture Note in Physics 142: 453 (1981).
15 T. Takatsuka, Prog. Theor. Phys. 72: 252 (1984).
16 As review article, e. g.,
S. Nagamiya, Nucl. Phys. A418: 239C (1984).
17 V. Ruck, M. Gyulassy and W. Greiner, Z. Physik A277: 391 (1976).
18 H. Toki, Y. Futami and W. Weise, Phys. Lett. 78B: 547 (1978).
19 P. Hecking, Nucl. Phys. A348: 493 (1980).
20 G. Baym, Nucl. Phys. A352: 355 (1981); K. Kolehmainen and G. Baym, Nucl. Phys. A382: 528 (1982).
21 A.L. Goodman, R.K. Tripathi and A. Faessler, Phys. Lett. 107B: 341 (1981).
22 T. Takatsuka and J. Hiura, Prog. Theor. Phys. 60: 1234 (1978).
23 V. R. Pandharipande and R.A. Smith, Nucl. Phys. A237: 507 (1975).
24 D.W.L. Sprung and P.K. Banerjee, Nucl. Phys. A168: 273 (1971).
25 R.V. Reid, Ann. of Phys. 50: 411 (1968).
26 T. Takatsuka, Prog. Theor. Phys. 73: 1043 (1985).

INDEX